Nonlinear mappings of monotone type

Nonlinear mappings
of monotone type

Dan Pascali

Institute of Mathematics, University of Bucharest

Silviu Sburlan

Pedagogical Institute Constantza, Romania

Editura Academiei, Bucureşti, Romania

Sijthoff & Noordhoff International Publishers
Alphen aan den Rijn, The Netherlands

ISBN-13: 978-94-009-9546-8   e-ISBN-13: 978-94-009-9544-4
DOI: 10.1007/978-94-009-9544-4

Contents

Contents

III. *NONLINEAR MAPPINGS OF MONOTONE TYPE*

IV. *HAMMERSTEIN EQUATIONS*

V. *HOMOTOPY ARGUMENTS*

VI. *VARIATIONAL PROBLEMS AND INEQUALITIES*

Preface

The progress in nonlinear functional analysis has allowed the study of many nonlinear problems in mathematical physics. This book provides basic methods and results for the investigation of the special problems in this area.

The connection between nonlinear analysis and convex analysis gave rise to the important field of monotone operators from a Banach space into its dual space. These mappings extend the properties of compact operators to the infinite-dimensional case.

Generalizations of monotone operators are termed mappings of monotone type. Among these, in the last decade, the pseudo-monotone operators and the mappings of type (M) have provided a more proper tool for solving large classes of nonlinear differential and integral equations. The text dwells upon essentially four interrelated topics: Nonlinear mappings of monotone type, Hammerstein equations, Odd operators and Variational problems. To make the approach easier, we have compiled some basic results on the topological degree and on the Sobolev spaces. In the applications we restrict our discussion to the existence of solutions for nonlinear elliptic equations.

The present English edition was written starting from the Romanian book "Operatori neliniari" (Nonlinear Mappings) by the first author and his lectures delivered at the Universities of Bucharest and Rome. The improved final form of this book is the result of the joint work of the authors.

Throughout the book we have extensively used the contributions of Professor Peters Hess from the Zürich University. We are grateful to him for his valuable suggestions. We are also indebted to Professor Viorel Barbu and his followers, Dan Tiba and Gheorghe Moroşanu, for the valuable remarks they made in reading some chapters.

We hope the book will be useful for universitary courses. Mathematicians and other specialists interested in the present theory of nonlinear equations will find here a presentation of recent and important topics of nonlinear functional analysis.

Organisational notes

The book is divided into six chapters and each of these is further subdivided into sections and subsections. Since each subsection contains at most one lemma, one proposition, one theorem and one corollary, the numbered results identify both the chapter and the subsection. Thus Proposition III.5.1 refers to the proposition in Section 5.1 of Chapter III, while Theorem V.1.1 refers to the theorem in Section 1.1, Chapter V. Exceptions appear whenever we refer to results established in the chapter under reading. For instance, Lemma 2.3 refers to the Lemma in Section 2.3 of the respective chapter.

In each subsection the equations are numbered consecutively, or by a single number starting from (1).

Each chapter is concluded with a section which contains topics and exercises related to the results established in that chapter.

For convenience, bibliographic details and references are given at the end of each chapter, except for some important references that are given in the text where they occur. A list of directions for further study is given at the end of the book.

The symbols for the notions introduced in the text are generally given where they occur. The exceptions concern the sets of natural numbers, integers and real numbers denoted by $\mathbb{N}$, $\mathbb{Z}$ and $\mathbb{R}$. We also denote by $\mathbb{R}_+$ the set of all positive numbers and by $\mathbb{R}^N$, $N \geqslant 1$, the N-dimensional Euclidean space. For $x, y \in \mathbb{R}^N$, with $x = (x_1, \ldots, x_N)$ and $y = (y_1, \ldots, y_N)$ we use the notation $x \cdot y = x_1 y_1 + \ldots + x_N y_N$ and $|x| = (x_1^2 + \ldots + x_N^2)^{1/2}$; these are the scalar product and the induced norm in $\mathbb{R}^N$.

Chapter I

Background in functional analysis

It is the purpose of this chapter to provide the terminology and basic concepts from topology, measure theory and convex analysis used throughout the book. In general, the results presented are stated without proofs. For their complete treatment we refer the reader to universitary textbooks and monographs on linear functional analysis.

1. Topology and measure

We list here some well known theorems for the convenience of later reference and we list the important features of the geometry of Banach spaces. Among these, stress is laid on the equivalent renormings of reflexive Banach spaces. We also present some results concerning the convergence in L^p-spaces which will be useful in applications to nonlinear integral and partial differential equations.

Banach spaces

1.1. Let X be a real Banach space with the norm $\|.\|$. Strong neighbourhoods of a point $x \in X$ are the open balls $B(x, r) = \{y \in X | \| y - x \| < r\}$ for any real number $r > 0$. The boundary and the closure of $B(x,r)$ are $S(x,r) = \partial B(x,r) = \{y \in X | \| y - x \| = r\}$ and $\overline{B}(x,r) = B(x,r) \cup S(x,r)$ respectively.

The space X^* of all linear continuous functionals on X is called the *dual* space of X. Usually, we shall denote the elements of X by small letters at the end part of the alphabet and the elements of X^* by small letters in the middle part of the alphabet. The value of $f \in X^*$ at $x \in X$ will be denoted by (f, x). The space X^* is also Banach with respect to the norm

$$\| f \|_* = \sup \{(f, x)| \ \| x \| \leq 1\}.$$

If there is no danger of confusion, we denote both norms in X and X^* by the same symbol $\|.\|$.

Beside the strong topology defined by the norm we shall also consider the weak topology in X. This topology is given by a base of neighbourhoods of the origin in X:

$$V(0; f_1, \cdots, f_n; \varepsilon) = \{x \in X \mid |(f_j, x)| < \varepsilon, \ 1 \leqslant j \leqslant n\},$$

for any finite subset $\{f_1, \cdots, f_n\} \subset X^*$ and any $\varepsilon > 0$. This locally convex topology is coarser than the strong topology and we denote it by $\sigma(X, X^*)$.

The dual space of X^* is X^{**}, which is also called the *bidual* space of X. Since, in general, $X \subseteq X^{**}$, we shall distinguish two main weak topologies in X^* — the usual weak topology $\sigma(X^*, X^{**})$, defined by the elements of the dual space X^{**}, and a coarser one, $\sigma(X^*, X)$, defined by the elements of X and called the weak* topology on X^*.

A general locally convex separated topological space is metrizable if and only if the topology is defined by a countable base of neighbourhoods of the origin. In a metrizable space the topological properties can be expressed sequentially.

THEOREM *(Uniform-boundedness principle) Any weakly convergent sequence $\{x_n\}$ in a Banach space X is bounded. Further, if $x_n \rightharpoonup x$, then*

$$\|x\| \leqslant \liminf \|x_n\|.$$

(see Dunford and Schwartz [1, p. 68]).

1.2. A *covering* of a set M in a topology of X is a family of open sets whose union contains M. The set M is said to be *compact* if every covering of M contains a finite subfamily which is also a covering.

A family of sets has *finite intersection property* if every finite subfamily has a non-empty intersection. A topological space is compact if and only if every family of closed sets with the finite intersection property has a non-empty intersection. Compact sets are closed and bounded, but closed and bounded sets need not be compact unless X is a finite-dimensional space.

A topological characterization of a compact set of the dual X^* of a Banach space X is given by

THEOREM *(Alaoglu) A subset of X^* is weakly* compact if and only if it is weakly* closed and bounded* (see Dunford and Schwartz [1, p. 424]).

1.3. When $X = X^{**}$, the space X is said to be *reflexive*. A subset M of X is called *weakly sequentially compact* if every sequence in M has a subsequence which converges weakly to a point of X. The reflexivity of a Banach space can be characterized in terms of this property:

THEOREM *(Eberleim-Shmulyan) A Banach space is reflexive if and only if its bounded closed subsets are weakly sequentially compact* (see Yosida [1, p. 141]).

1.4. A set is said to be *dense* in a topological space X if its closure is all of X. A space is *separable* if it contains a countable dense subset. We have the

THEOREM *(Dunford and Schwartz [1, p. 434]) Let X be a Banach space. Then the topology $\sigma(X^*, X)$ on a bounded closed subset of X^* is metrizable if and only if X is separable.*

This theorem allows to establish the following

PROPOSITION *Let M be a bounded set of reflexive Banach space X and let x_0 be a point of the weak closure of M. Then there exists a sequence $\{x_n\} \subset M$ which is weakly convergent to x_0 in X.*

Proof: We show first the existence of a countable subset M_0 of M, whose weak closure contains x_0. Let m be a natural number and let $\bar{B}^m$ be the product of m copies of the closed unit ball in X^*. Since x_0 lies in the weak closure of M, for any element $\{f_1,.., f_m\} \in \bar{B}^m$ and any integer $n > 0$, there exists an element $x \in M$ with the property

$$|(f_j, x - x_0)| < \frac{1}{n}, \quad 1 \leqslant j \leqslant m.$$

For each choice of x, these inequalities define a weak neighbourhood of the element $\{f_1,..., f_m\}$ in $\bar{B}^m$. Theorm 1.2 implies the weak compactness of $\bar{B}^m$ in the product of the weak topologies. Let us fix m and n. We can find a finite subset $F_{m,n}$ of M such that for each element of $\bar{B}^m$ at least one point $x \in F_{m,n}$ serves to verify the above inequalities. Set

$$M_0 = \bigcup_{m,n} F_{m,n},$$

the union being taken over all pairs of natural numbers (m, n). Then x_0 lies in the weak closure of the countable subset M_0.

Let X_0 be the separable closed subspace of X spanned by the elements of M_0. Then x_0 lies in the weak closure of $M \cap X_0$. To show that x_0 is a weak limit of a sequence from M_0 in X, it is sufficient to prove this for X_0. Since a normed linear space is separable if and only if its dual is, (Dunford and Schwartz [1, p. 65]), X_0^* is also separable. By virtue of the previous theorem, the weak topology of X_0 on M_0 is metrizable and each point of the closure of M_0 in the weak topology on X_0 is a weak limit of a sequence from M_0.

In other words, the weak closure of a bounded subset in a reflexive Banach space coincides with its sequentially weak closure.

Elements of the geometry of Banach spaces

1.5. A linear normed space X is called *strictly convex* if its unit sphere contains no line segments, i.e.,

$$\|(1 - t) x + ty\| < 1 \qquad (\forall) \, t \in (0,1)$$

for all $x, y \in S(0,1)$ with $x \neq y$. In other words, X is strictly convex if $x, y \in X$ with $\|x\| = \|y\| = 1$ and $\|(1 - t) x + ty\| = 1$ for a $t \in (0,1)$ holds iff $x = y$.

A simple example of a normed space which is not strictly convex is the Euclidian space $\mathbb{R}^N$ with respect to the norm

$$\|x\| = \sup \sum_{i=1}^{N} |x_i|.$$

PROPOSITION *In a linear normed space X, the following three statements are equivalent:*

(i) *X is a strictly convex space;*

(ii) *Any $f \in X^*$ assumes its minimum value at most in one point on the unit sphere of X;*

(iii) *If $x, y \in X$, $x \neq y$, with $\|x\| = \|y\| = 1$ then* $\left\| \dfrac{x + y}{2} \right\| < 1$.

Proof: (i) $\Rightarrow$ (ii): If there exist $f \in X^*$ and $x, y \in S(0,1)$ such that $(f, x) = (f, y) = \|f\|$, then for all $t \in (0,1)$ we have

$$(f, (1 - t) x + ty) = (1 - t)(f, x) + t(f, y) = \|f\|$$

and $1 \leqslant \|(1 - t) x + ty\| \leqslant (1 - t) \|x\| + t \|y\| \leqslant 1$, which contradicts (i).

(ii) $\Rightarrow$ (iii): Let $x, y \in X$ be such that $x \neq y$, $\|x\| = \|y\| = 1$ and $\left\| \dfrac{x + y}{2} \right\| = 1$. By the Hahn-Banach theorem, there exists an $f \in X^*$ such that $\|f\| = 1$ and $\left(f, \dfrac{x + y}{2} \right) = \left\| \dfrac{x + y}{2} \right\|$, whence $(f, x) + (f, y) = 2$. As $(f, x) \leqslant 1$ and $(f, y) \leqslant 1$, both inequalities are equalities and thus, $(f, x) = (f, y) = \|f\|$, which contradicts (ii).

The implication (iii) $\Rightarrow$ (i) follows since $t \mapsto \|(1 - t)x + ty\|$ is convex. We also mention a basic result due to Asplund [2]:

THEOREM *Let X be a reflexive Banach space. Then the dual spaces X, X^* can be equivalently renormed as strictly convex spaces such that the duality is preserved.*

In virtue of this theorem, the assumption on strict convexity is not an essential restriction for reflexive Banach spaces.

1.6. A linear normed space X is said to be *locally uniformly convex* if for any $\varepsilon > 0$ and $x \in X$ with $\|x\| = 1$ there exists $\delta = \delta(x, \varepsilon) > 0$ such that $\|x - y\| \geqslant \varepsilon$ implies that $\left\| \dfrac{x + y}{2} \right\| \leqslant 1 - \delta$ for all $y \in S(0,1)$. The statement (iii) in proposition 2.1 ensures the strict convexity of a locally uniformly convex space.

We can easily see that a Banach space X is locally uniformly convex if and only if the module of local convexity

$$\delta(x, \varepsilon) = \frac{1}{2} \inf \{2 - \|x + y\| \mid \|y\| = 1, \quad \|x - y\| \geqslant \varepsilon\}, \quad 0 < \varepsilon \leqslant 2,$$

is positive for each $x \in X$ with $\|x\| = 1$.

LEMMA *A Banach space X is locally uniformly convex if and only if for any $x \in X$ with $\|x\| = 1$ and every sequence $\{x_n\} \subset S(0,1)$ the condition $\|x_n + x\| \to 2$ implies that $\|x_n - x\| \to 0$.*

Proof: When X is locally uniformly convex and $(2 - \|x_n + x\|) \to 0$, we have $\|x_n - x\| \to 0$ because $\delta(x, \varepsilon) > 0$.

Now, if $\|x_n + x\| \to 2$ implies that $\|x_n - x\| \to 0$, then $\|x_n - x\| \geqslant \varepsilon > 0$ implies that $\|x_n + x\| < 2$. Hence, by the definition of the module of local convexity, $\delta(x, \varepsilon) > 0$ and so X is locally uniformly convex.

PROPOSITION *Let X be a locally uniformly convex Banach space. If $\{x_n\} \subset X$ is such that $x_n \rightharpoonup x$ in X and $\|x_n\| \to \|x\|$, then $x_n \to x$ in X.*

Proof: For $\|x\| = 0$ the assertion is obvious. Suppose that $\|x\| > 0$ and denote $y = \dfrac{x}{\|x\|}$ and $y_n = \dfrac{x_n}{\|x_n\|}$ for n large enough. By construction, $\|y_n\| = \|y\| = 1$, $y_n \rightharpoonup y$ and thus $y_n + y \rightharpoonup 2y$. Then,

$$2 = 2\|y\| \leqslant \liminf \|y_n + y\| \leqslant \limsup \|y_n + y\| \leqslant \|y\| + \lim \|y_n\| = 2,$$

that is, $\|y_n + y\| \to 2$. By the lemma, $\|y_n - y\| \to 0$. Hence, $\dfrac{x_n}{\|x_n\|} \to \dfrac{x}{\|x\|}$

and $x_n \to x$ because $\|x_n\| \to \|x\|$.

A stronger variant of Theorem 1.5 is the

THEOREM *(Trojanski [1]) Let X be a reflexive Banach space. Then there exist equivalent norms on X and X^* such that both spaces, which are still dual to each other, are locally uniformly convex.*

This result allows to use the previous proposition for reflexive Banach spaces.

Finally, a linear normed space X is said to be *uniformly convex* if for any $\varepsilon > 0$ there exists $\delta(\varepsilon) > 0$ such that $\|x\| = \|y\| = 1$ and $\|x - y\| \geqslant \varepsilon$ imply $\left\| \dfrac{x + y}{2} \right\| \leqslant 1 - \delta(\varepsilon)$.

It is obvious that a uniformly convex space is locally uniformly convex. The Hilbert spaces are examples of uniformly convex spaces. Indeed, let x anp y be two elements in a Hilbert space such that $\|x\|=\|y\|=1$ and $\|x-y\|\geqslant\varepsilon>0$. Since

$$\left\|\frac{x+y}{2}\right\|^2 + \left\|\frac{x-y}{2}\right\|^2 = 1,$$

it follows that

$$\left\|\frac{x+y}{2}\right\| \leqslant 1 - \delta\,(\varepsilon) \text{ with } \delta\,(\varepsilon) = 1 - \left[1 - \left(\frac{\varepsilon}{2}\right)^2\right]^{\frac{1}{2}}.$$

Continuous mappings

1.7. For any set M we denote by 2^M the set of parts in M. Let X and Y be two sets and let $T:X \mapsto 2^Y$ be a multivalued mapping. We set

$D\,(T) = \{x \in X \mid Tx \neq \emptyset\}$, i.e. the effective domain of T,

$R\,(T) = \{y \in Tx \mid x \in D\,(T)\}$, i.e. the range of T,

and

$G\,(T) = \{[x, y] \in X \times Y \mid x \in D\,(T), \quad y \in Tx\}$, i.e. the graph of T.

The mapping S is said to be an extension of T if $G\,(T) \subseteq G\,(S)$. For any mapping $T: X \to 2^Y$ there exists an inverse mapping $T^{-1}: Y \mapsto 2^X$ defined by

$$T^{-1}y = \{x \in X \mid y \in Tx\}.$$

The existence of inverses is the main reason for considering multivalued mappings.

Suppose now that X and Y are two topological vector spaces. We call V a neighbourhood of the set M of a topological space if $M \subset \text{Int } V$.

The mapping $T: X \mapsto 2^Y$ is *continuous* in $x \in D\,(T)$, if to any neighbourhood V of a set $Tx \subset Y$ there corresponds a neighbourhood U of x such that $T\,(U) \subseteq V$. The map T is *sequentially continuous* in $x \in D(T)$ if for any sequence $\{x_n\} \subset D\,(T)$ with $x_n \to x$ in X and for any neighbourhood V of the set Tx in Y, there exists an $N>0$ such that $Tx_n \subset V$ for all $n \geqslant N$. In the case of metrizable topologies these definitions are equivalent to the following: T is continuous if for any $\varepsilon > 0$ there exists $\delta = \delta\,(x, \varepsilon) > 0$ such that $T\,(B\,(x, \delta)) \subseteq B^*\,(Tx, \varepsilon)$. For the multivalued mappings the term *upper-semicontinuity* is frequently used instead of continuity.

The mapping T from X into Y is said to be bounded if it carries bounded subsets of $D\,(T)$ into bounded subsets of Y.

If for every $x \in D(T)$, the set Tx is a singleton, then T is called an *operator* or a single-valued map and denoted by $T: X \mapsto Y$. When $Y = \mathbb{R}$, T is said to be a *function*.

A function T is one-to-one if $Tx = Ty$ implies that $x = y$, T is *surjective* (onto) provided that $R(T) = Y$ and *bijective* if it is both one-to-one and surjective.

The continuity of operators $T: X \mapsto X^*$ can be expressed more precisely in terms of the topologies in the reflexive Banach spaces X and X*. We shall use the symbols "$\to$" and "$\rightharpoonup$" to indicate strong and weak convergence respectively. An operator $T: X \mapsto X^*$ is said to be *continuous* if for any sequence $\{x_n\} \subset D(T)$ such that $x_n \to x$ in X it follows that $Tx_n \to Tx$ in X^* and T is *weakly continuous* if $x_n \rightharpoonup x$ implies that $Tx_n \rightharpoonup Tx$. Similarly, T is *demicontinuous* if $x_n \to x$ implies that $Tx_n \rightharpoonup Tx$ and T is *completely continuous* if $x_n \rightharpoonup x$ implies that $Tx_n \to Tx$. When X is non-reflexive, the weak convergence in X^* is replaced by the weak* convergence denoted by "$\rightharpoonup$".

PROPOSITION *Any linear demicontinuous operator* $L: X \to X^*$ *is continuous.*

Proof: Suppose that, on the contrary, there is a sequence $\{x_n\} \subset D(L)$ with $x_n \to 0$ and an $\varepsilon > 0$ such that $\|Lx_n\| > \varepsilon$ for all $n \in \mathbb{N}$. Denote $t_n = \|x_n\|^{-\frac{1}{2}}$ and $y_n = t_n x_n$. We have $y_n \to 0$ and

$$\|Ly_n\| = t_n \|Lx_n\| \geqslant \varepsilon t_n \to \infty,$$

which contradicts the demicontinuity of L. Thus $Ly_n \rightharpoonup 0$.

A map T is said to be *compact* if T is continuous and carries bounded sets of X into relatively compact sets of Y.

In general, the classes of compact maps and completely continuous operators are not comparable. However, we have the following

THEOREM *Let X and Y be Banach spaces and let C be a closed convex subset of X. If X is reflexive, then every completely continuous operator* $T: C \mapsto Y$ *is also compact.*

Proof: Clearly, T is continuous. Now, let B be a bounded subset in X. We shall prove that every sequence $\{Tx_n\}$ with $x_n \in B \cap C$ contains a strongly convergent subsequence. In fact, by the reflexivity of X, we may assume that $x_n \rightharpoonup x$ at least on a subsequence. Since T is completely continuous, $\{Tx_n\}$ is strongly convergent in Y.

COROLLARY *For any linear mapping from a reflexive Banach space X into a general Banach space Y, complete continuity and compactness are equivalent.*

For a real-valued function φ defined in a topological space X, we define the *support of* φ as

$$\text{supp } \varphi = \overline{\{x \in X \mid \varphi(x) \neq 0\}}.$$

We shall frequently use the partition of unity of class C^m, $m \geqslant 0$, (see Lang [1, p. 25]):

Let K be a compact set of a topological vector space X and let $\{U_0, U_1,...,U_n\}$ be a finite covering of K. Then there exist functions $\beta_j : K \mapsto [0,1]$ $0 \leqslant j \leqslant n$, of class $C^m(X)$, such that:

(i) $\beta_j(x) \geqslant 0$ for all $x \in K$;

(ii)   The support of β_j lies in U_j, $1 \leqslant j \leqslant n$;

(iii) $\displaystyle\sum_{j=0}^{n} \beta_j(x) = 1$ for all $x \in K$.

The family $\{\beta_0, \beta_1, \cdots, \beta_n\}$ is said to be a *C^m-partition of unity* on K subordinated to the covering $\{U_0, U_1, \cdots, U_n\}$.

1.8. Let X be a Banach space with norm $\|.\|$. To any map $T: X \mapsto X$ we assign the seminorm

$$|T| = \sup \left\{ \frac{\|Tx - Ty\|}{\|x - y\|} \,\middle|\, x,\, y \in X,\, x \neq y \right\} .$$

A map T for which $|T| < \infty$ is called *Lipschitz operator* on X and the set of all such operators on X is denoted by $\mathrm{Lip}(X)$.

Clearly, $|T| = 0$ if and only if T is constant on X. For two Lipschitz operators $T: X \mapsto X$ and $S: X \mapsto X$ such that $R(S) \subseteq D(T)$, we have

$$|T \circ S| \leqslant |T|\,|S|.$$

LEMMA *Let $\{T_n\}$ be a sequence of operators in $\mathrm{Lip}(X)$ and let k be a positive constant such that $|T_n| \leqslant k$ for all $n \in \mathbb{N}$. If $T_n x \to Tx$ for all $x \in X$, then $T \in \mathrm{Lip}(X)$ and $|T| \leqslant k$.*

Proof: Indeed, we have

$$\|Tx - Ty\| = \lim \|T_n x - T_n y\| \leqslant k\|x - y\| \qquad (\forall)x, y \in X.$$

Hence $T \in \mathrm{Lip}(X)$ and $|T| \leqslant k$.

We have the following

THEOREM *Let $T \in \mathrm{Lip}(X)$ with $D(T) = X$ and $|T| < 1$. Then $(I - T)^{-1} \in \mathrm{Lip}(X)$ and $|(I - T)^{-1}| \leqslant (1 - |T|)^{-1}$.*

Proof: For $x, y \in X$ we can write

$$\|x - y\| \leqslant \|(I - T)x - (I - T)y\| + \|Tx - Ty\| \leqslant$$

$$\leqslant \|(I - T)x - (I - T)y\| + |T|\,\|x - y\|,$$

whence

$$\|x - y\| \leqslant \frac{1}{1 - |T|} \|(I - T)x - (I - T)y\|.$$

We further remark that $(I - T)^{-1}$ is one-to-one.

When T is linear we have that

$$|T| = \sup \left\{ \frac{\|Tx\|}{\|x\|} \,\Big|\, x \in X, \quad x \neq 0 \right\}$$

and $|T|$ coincides with the usual norm on the space of linear operators on X. If z is a point in X, then $\mathrm{Lip}(X)$ is a Banach space with respect to the norm

$$\|| T \||| = |T| + \|Tz\|.$$

For each $T \in \mathrm{Lip}(X)$,

$$\rho(T) = \{\lambda \in \mathbb{R} \,|\, (\lambda I - T)^{-1} \in \mathrm{Lip}(X)\}$$

is said to be the *resolvent set* of T. Clearly, $\lambda \in \rho(T)$ if and only if $\lambda I - T$ is bijective. The complementary set of $\rho(T)$ is denoted by $\sigma(T)$ and called the *spectrum* of T. For each $T \in \mathrm{Lip}(X)$, the mapping $R(.,T) \colon \rho(T) \mapsto \mathrm{Lip}(X)$ defined by

$$R(\lambda, T) = (\lambda I - T)^{-1}$$

is called the *resolvent* of T.

Suppose that $T \in \mathrm{Lip}(X)$ and $\lambda, \mu \in \rho(T)$. Then the *nonlinear resolvent identity*

$$R(\lambda, T) = R(\mu, T) [I + (\mu - \lambda)R(\lambda, T)] \tag{1}$$

holds. In fact, we can write

$$\mu I - T = (\lambda I - T) + (\mu - \lambda)I = [I + (\mu - \lambda)R(\lambda, T)](\lambda I - T) \tag{2}$$

and multiplying this equality, first on the right by $R(\lambda, T)$ and then on the left by $R(\mu, T)$ we get (1).

We can prove now the following

PROPOSITION *Let $T \in \mathrm{Lip}(X)$. Then $\rho(T)$ is an open set and thus $\sigma(T)$ is closed.*

Proof: Let $\lambda \in \rho(T)$ and $\mu \in \mathbb{R}$ be such that $|\lambda - \mu| < |R(\lambda, T)|^{-1}$. Then, by the theorem,

$$|(I - (\lambda - \mu)R(\lambda, T))| \leqslant (1 - |\lambda - \mu| \, |R(\lambda, T)|)^{-1}.$$

Inverting equality (2), we deduce the equality

$$R(\mu, T) = R(\lambda, T) [I - (\lambda - \mu)R(\lambda, T)]^{-1}$$

which implies that

$$\rho(T) \supset \{\mu \in \mathbb{R} \mid |\lambda - \mu| < |R(\lambda, T)|^{-1}\}.$$

The conclusions of the proposition follow easily.

1.9. Let X be a Banach space and let X^* be its dual space. A function $\varphi : X \mapsto \overline{\mathbb{R}}$ is Gâteaux differentiable at $x \in X$ if there exists an $f \in X^*$ such that

$$\lim_{t \to 0} \frac{1}{t} [\varphi(x + ty) - \varphi(x)] = (f, y) \qquad (\forall) y \in X.$$

The element f, also denoted by $\varphi'(x)$, is called the *G-differential* of φ at x.

When the G-differential exists, it can be regarded as the derivative in the direction y since

$$(f, y) = (\varphi'(x), y) = \frac{d}{dt} \varphi(x + ty)|_{t=0}.$$

For instance, if $\varphi : \mathbb{R}^N \mapsto \overline{\mathbb{R}}$ where $\varphi(x_1, \ldots, x_N)$ has continuous partial derivatives, then

$$(f, y) = \frac{d}{dt} \varphi(x + ty)|_{t=0} = \sum_{i=1}^{N} \frac{\partial \varphi(x)}{\partial x_i} y_i$$

for any $y = (y_1, \ldots, y_N)$ in $\mathbb{R}^N$.

An operator $T : X \mapsto X^*$ is said to be *potential* if it is the G-differential on X of a function $\varphi : X \mapsto \overline{\mathbb{R}}$. Such an operator is also called the *gradient* of φ and denoted by *grad* φ.

The function φ is *Fréchet differentiable* at $x \in X$ if there exists a linear continuous function $d\varphi(x,.): X \mapsto \mathbb{R}$, called *F-differential* of φ at x, such that

$$\lim_{\|y\| \to 0} \frac{1}{\|y\|} |\varphi(x + y) - \varphi(x) - d\varphi(x, y)| = 0, \qquad (\forall) \, y \in X.$$

We note that a Fréchet differentiable function is Gâteaux differentiable. In fact, if we take $y = ty_0$ for any fixed $0 \neq y_0 \in X$, we obtain

$$\lim_{t \to 0} \frac{1}{t\|y_0\|} |\varphi(x + ty_0) - \varphi(x) - t \, d\varphi(x, y_0)| = 0$$

and

$$\lim_{t \to 0} \frac{1}{t} [\varphi(x + ty_0) - \varphi(x)] = d\varphi(x, y_0).$$

Hence, $d\varphi(x,.) \in X^*$ and φ is Gâteaux differentiable.

In general, the existence of the G-differential does not imply the existence of the F-differential. However, a continuous G-differential $\varphi'(x)$ is an F-differential. These concepts can be similarly defined for mappings between two Banach spaces (e.g., Vainberg [1]).

Now, let φ and ψ be two F-differentiable functions on X. For any real number r, denote the level surface

$$N_r(\psi) = \psi^{-1}(r) = \{x \in D(\psi) \mid \psi(x) = r\}.$$

We say that $u \in X$ is a *critical point* of φ if $d\varphi(u, y)=0$ for all $y \in X$, and $u \in N_r(\psi)$ is a *constrained critical point* of φ to $N_r(\psi)$ if u is a critical point of the restriction of φ to this surface. For any $v \in N_r(\psi)$ we set $X_v = \{y \in X \mid d\psi(v, y) = 0\}$.

We establish now a variant of the Lagrange multiplier method:

THEOREM *Assume that* $d\psi(x,.) \neq 0$ *for all* $x \in N_r(\psi)$. *Then there exists a real number* λ *such that the constrained critical points of* φ *to* $N_r(\psi)$ *are solutions of the equation*

$$d\varphi(u,.) = \lambda \, d\psi \, (u,.).$$

Proof: For each $x \in X$ we choose $w \in X$ such that $d\psi(x, w) = 1$. This is possible because we have assumed that $d\psi(x,.) \neq 0$ on $N_r(\psi)$. Then any element $z \in X$ can be uniquely represented as

$$z = d\psi(x, z) \, w + y, \tag{1}$$

where $y \in X_x$. Indeed,

$$d\psi(x, y) = d\psi(x, z - d\psi(x, z)w) = d\psi(x, z) - d\psi(x, z) = 0.$$

The uniqueness of this representation follows by applying $d\psi(x,.)$ to both sides of (1).

Let φ_0 be the restriction of φ to $N_r(\psi)$. By our assumptions, φ_0 is Fréchet differentiable and $u \in N_r(\psi)$ is a critical point of φ_0 if $d\varphi_0(u, y) = d\varphi(u, y) = 0$ for all $y \in N_r(\psi)$. For such a point u and for any $z \in X$, by representation (1) with $x = u$, we can see that

$$d\varphi(u, z) = d\varphi(u, d\psi)(u, z)w + y) = d\psi(u, z)d\varphi(u, w)$$

since $y \in X_v$. Hence, for all z in X, we have

$$d\varphi(u, z) = \lambda \, d\psi(u, z) \text{ with } \lambda = d\varphi(u, w).$$

The theorem is proved.

Vitali's theorem

1.10. We collect here the basic concepts from measure theory required by the main convergence theorems in L^p-spaces.

Let S be an arbitrary set and let 2^S be the set of all subsets of S. A family $\Sigma \subseteq 2^S$ is called *σ-algebra* provided that

  (i) $S \in \Sigma$,

  (ii) $A, B \in \Sigma$ implies that $(A - B) \in \Sigma$,

  (iii) $A_j \in \Sigma$, $j \in \mathbb{N}$, implies that $\bigcup_{j \in \mathbb{N}} A_j \in \Sigma$.

A positive function $\mu : \Sigma \mapsto \mathbb{R}$ is called a *measure* if it is countably additive and there exists at least one set $A \in \Sigma$ such that $\mu(A) < \infty$. The triple (S, Σ, μ) is said to be a *measure space*. The value $\mu(A)$ is called the measure μ of the set A. If $\mu(S) < \infty$, then (S, Σ, μ) is called a finite measure space. When S is a union of a countable family of sets in Σ, each of them with a finite measure, then (S, Σ, μ) is called *σ-finite*.

A pointwise property holds almost everywhere (μ — a.e.) if the set A of all points at which this property is not valid belongs to Σ and $\mu(A) = 0$.

Let X be a Banach space with the norm $\|.\|$. Then any function of the form

$$u = \sum_{i=1}^{n} x_i \chi(A_i),$$

where $x_i \in X$, $A_i \in \Sigma$ and $\chi(A_i)$ denotes the characteristic function of A_i, $1 \leqslant i \leqslant n$, is called a *simple function* on S. A function $u : S \mapsto X$ is said to be *strongly measurable* if there exists a sequence of simple functions $\{u_n\}$ such that

$$\lim_{n \to \infty} \|u_n(t) - u(t)\| = 0 \text{ for almost all } t \in S. \tag{1}$$

THEOREM *(Egorov) Let (S, Σ, μ) be a finite-measure space and let u, u_n ($n \in \mathbb{N}$) be μ-measurable functions which are defined and finite on S. Suppose that $u_n \to u$ μ-a.e. on S. Then for each $\varepsilon > 0$ there exists a set $A_\varepsilon \in \Sigma$ such that $\mu(A_\varepsilon) < \varepsilon$ and $u_n \to u$ uniformly on $S - A_\varepsilon$.*

For a simple function $u : S \mapsto X$, we set

$$S_i = \{t \in S \mid u(t) = x_i\}, \; 1 \leqslant i \leqslant n.$$

If $\mu(S_i) < \infty$ for all $x_i \neq 0$, then the function

$$u = \sum_{i=1}^{n} x_i \chi(S_i)$$

is *μ-integrable* on S and

$$\int_S u \, d\mu = \int_S u(t) \, d\mu = \sum_{i=1}^{n} x_i \mu(S_i)$$

is the *μ-integral* of u.

The set of all simple $μ$-integrable functions is a metric space $I(S, \Sigma, X, \mu)$ with the distance

$$d(u, v) = \int \|u - v\| d\mu.$$

The convergence in the topology of this metric is called *convergence in mean*.

Next, let Σ be a σ-algebra of subsets in $\mathbb{R}^N$ and let μ be a positive measure on Σ with the following properties:

(j) Every open set in $\mathbb{R}^N$ belongs to Σ;

(jj) If $A \subset B$, $B \in \Sigma$ and $\mu(B) = 0$ then $A \in \Sigma$ and $\mu(A) = 0$;

(jjj) If $A = \{x \in \mathbb{R}^N \mid a_k \leqslant x_k \leqslant b_k, \quad 1 \leqslant k \leqslant N\}$, then $A \in \Sigma$ and

$$\mu(A) = \prod_{k=1}^{N} (b_k - a_k);$$

(jv) μ is invariant under translations, i.e., if $x \in \mathbb{R}^N$ and $A \in \Sigma$, then $x + A = \{x + y \mid y \in A\} \in \Sigma$ and $\mu(x + A) = \mu(A)$.

The elements of Σ are called *Lebesgue measurable* and μ is called a *Lebesgue measure*. This measure generalizes the geometrical concepts of length, area and volume.

Let S be a Lebesgue measurable set in $\mathbb{R}^N$ and let $u : S \mapsto X$ be a strongly measurable function. Suppose that a sequence of simple functions $\{u_n\}$, with supp $u_n \subseteq S$ for $n \in \mathbb{N}$, which satisfy (1), can be chosen such that

$$\lim_{n \to \infty} \int_S \|u_n(t) - u(t)\| \, d\mu = 0.$$

Then u is called *Bochner integrable* on S and we define

$$I(u) = \int_S u(t) \, d\mu = \lim \int_S u_n(t) \, d\mu.$$

Here $I(u)$ is independent of the choice of the approximating sequence $\{u_n\}$.

The completion of the convergence in mean of $I(S, \Sigma, X, \mu)$ is the Banach space $\mathcal{L}(S, \Sigma, X, \mu)$ of all integrable functions on S. In this space, two functions are equivalent if they coincide μ-a.e. on S.

For a real positive function $u : S \mapsto \overline{\mathbb{R}}_+$ which is μ-measurable but not integrable, we put $I(u) = \infty$.

LEMMA (*Fatou*) *If* $\{u_n\}$ *is a sequence of non-negative real-valued functions which are μ-measurable on S, then*

$$I(\liminf u_n) \leqslant \liminf I(u_n),$$

and the equality holds only for increasing sequences.

1.11. Let $1 \leqslant p < \infty$ and let $\mathscr{L}^p(S, \Sigma, X\mu)$ be the set of all functions $u: S \mapsto X$ with the following properties:

    (i) u is μ-measurable,

    (ii) $\|u\|^p$ is integrable.

    The space $L^\infty(S, \Sigma, X, \mu)$ is the set of all μ-measurable functions, which are μ-essentially bounded, i.e., bounded functions on the complementary sets in S of all μ-null-measure subsets.

    We denote by $L^p(S, X)$ and $L^\infty(S, X)$ the spaces of equivalence classes corresponding to the above spaces. In particular, when $S = (0, t_0)$ we have $L^p(0, t_0; X)$ and $L^\infty(0, t_0; X)$, or in the case of real functions defined on a domain Ω in $\mathbf{R}^N$, the corresponding spaces are $L^p(\Omega)$ and $L^\infty(\Omega)$. By $L^p_{loc}(S, X)$ we denote the space of all functions which are in $L^p(A, X)$ for any compact subset $A \subset S$.

    The positive exponents p and p' are called *conjugate* to each other provided that

$$\frac{1}{p} + \frac{1}{p'} = 1.$$

Clearly, conjugate exponents are equal if and only if $p = p' = 2$.

    In what follows, we shall assume that $p > 1$.

    For any $u \in L^p(S, X)$ and $v \in L^{p'}(S, X)$, the *Hölder inequality*

$$\int_S \|u\|\,\|v\|\,d\mu \leqslant \left(\int_S \|u\|^p\,d\mu\right)^{\frac{1}{p}} \left(\int_S \|v\|^{p'}\,d\mu\right)^{\frac{1}{p'}}$$

holds. If X is a Banach algebra then $uv \in L^1(S, X)$. Moreover, for any $u \in L^p(S, X)$, $v \in L^q(S, X)$ and $w \in L^r(S, X)$ with $p, q, r > 1$ and

$$\frac{1}{p} + \frac{1}{q} + \frac{1}{r} = 1,$$

we have $uvw \in L^1(S, X)$ and the *generalized Hölder inequality*

$$\int_S \|u\,v\,w\|\,d\mu \leqslant \left(\int_S \|u\|^p\,d\mu\right)^{\frac{1}{p}} \left(\int_S \|v\|^q\,d\mu\right)^{\frac{1}{q}} \left(\int_S \|w\|^r\,d\mu\right)^{\frac{1}{r}}.$$

For any two functions $u, v \in L^p(S, X)$, the sum $u + v \in L^p(S, X)$ due to the *Minkowski inequality*

$$\left(\int_S \|u + v\|^p \, d\mu \right)^{\frac{1}{p}} \leqslant \left(\int_S \|u\|^p \, d\mu \right)^{\frac{1}{p}} + \left(\int_S \|v\|^p \, d\mu \right)^{\frac{1}{p}},$$

where equality occurs if and only if $v = cu$ or $u = cv$ with $c \in \mathbb{R}$. Based on the Minkowski inequality is the fact that $L^p(S, X)$ and $L^\infty(S, X)$ are Banach spaces with respect to the norms

$$\|u\|_p = \left(\int_S \|u\|^p \, d\mu \right)^{\frac{1}{p}} \quad \text{for} \quad 1 \leqslant p < \infty$$

and

$$\|u\|_\infty = \operatorname{ess\ sup} \|u\| = \inf \{\sup [\|u(t)\| \mid t \in S - A, \ \mu(A) = 0\}.$$

We remark that for $\mu(S) < \infty$ and $1 < q < p < \infty$ we have $L^p(S, X) \subset \subset L^q(S, X)$ since

$$\|u\|_q \leqslant \|u\|_p \, (\mu(S))^{\frac{1}{q} - \frac{1}{p}} \quad \text{for all} \quad u \in L^p(S, X).$$

For any reflexive Banach space X, $L^p(S, X)$ with $1 < p < \infty$ is uniformly convex (Hewitt and Stromberg [1, p. 232]) and thus reflexive (Yosida [1, p. 127]).

In case when either $S = (0, t_0)$ and X is a reflexive Banach space or S is a domain Ω in $\mathbb{R}^N$ and $X = \mathbb{R}$, the structure of the elements of the dual space in specified by the

THEOREM *(Riesz — Fréchet) To any $f \in [L^p(S, X)]^*$ there corresponds an element $z \in L^{p'}(S, X^*)$ such that*

$$(f, u) = \int_S zu \, d\mu \qquad (\forall) \ u \in L^p(S, X),$$

and $\|f\| = \|z\|_{p'}$ (see Dinculeanu [1, p. 279]).

Therefore, we can identify up to isomorphism the dual space of $L^p(S, X)$ with $L^p(S, X^*)$. In the special case when $p = p' = 2$ and H is a Hilbert space then $L^2(S, H)$ is also a Hilbert space with respect to the inner product

$$(u, v) = \int_S uv \, d\mu \qquad (\forall) \ u, v \in L^2(S, H).$$

1.12. We shall use the previous facts to prove a main convergence result.

LEMMA *Let $u \in L^1(S, X)$. Then for every $\varepsilon > 0$ there exists a $\delta > 0$, which depends on ε and u, such that*

$$\int_A \|u\| \, d\mu < \varepsilon$$

for all $A \in \Sigma$ with $\mu(A) < \delta$.

Proof: For any number $n \in \mathbb{N}$, let

$$\psi_n(t) = \begin{cases} \|u(t)\| & \text{if } \|u(t)\| \leq n, \\ n & \text{otherwise.} \end{cases}$$

Then $\{\psi_n\}$ is a non-decreasing sequence of μ-measurable functions and $\lim \psi_n = \|u\|$. By Fatou's lemma,

$$\lim \int_S \psi_n d\mu = \int_S \lim \psi_n d\mu = \int_S \|u\| \, d\mu.$$

Select $N \in \mathbb{N}$ so that

$$\int_S (\|u\| - \psi_n) \, d\mu < \frac{\varepsilon}{2} \qquad (\forall) \, n \geq N.$$

Set $\delta = \dfrac{\varepsilon}{2N}$ and choose any $A \in \Sigma$ with $\mu(A) < \delta$. We have

$$\int_A \psi_n d\mu \leq \int_A n \, d\mu = n\mu(A) < \frac{\varepsilon}{2} \qquad (\forall)n \geq N.$$

It follows that

$$\left\| \int_A u \, d\mu \right\| \leq \int_A \|u\| \, d\mu = \int_A (\|u\| - \psi_n) \, d\mu + \int_A \psi_n d\mu < \frac{\varepsilon}{2} + \frac{\varepsilon}{2} = \varepsilon$$

for all $A \in \Sigma$ with $\mu(A) < \delta$.

We prove now the

THEOREM *(Vitali) Let $\{u_n\} \subset L^p(S, X)$ be a sequence such that $u_n \to u$ μ-a.e. Then $u \in L^p(S, X)$ and $u_n \to u$ in $L^p(S, X)$ if and only if*

(i) *for each $\varepsilon > 0$, there exists a set $A_\varepsilon \in \Sigma$ such that $\mu(A_\varepsilon) < \infty$ and*

$$\int_{S-A_\varepsilon} \|u_n\|^p \, d\mu < \varepsilon \; ! \; (\forall)n \in \mathbb{N};$$

(ii) $\lim_{\mu(A) \to 0} \displaystyle\int_A \|u_n\|^p \, d\mu = 0$ *uniformly in $n \in \mathbb{N}$,* *i.e., for each $\varepsilon > 0$ there exists a $\delta > 0$ such that*

$$\int_A \|u_n\|^p \, d\mu < \varepsilon \qquad (\forall) n \in \mathbb{N},$$

for all $A \in \Sigma$ and $\mu(A) < \delta$.

Proof: To prove the necessity of (i), let $\varepsilon > 0$ be given and choose $N \in \mathbb{N}$ such that $\|u_n - u\|_p < \dfrac{\varepsilon}{2}$ for all $n \geqslant N$. As the support of an integrable function is σ-finite, we can pick out $B_\varepsilon, C_\varepsilon \in \Sigma$ of finite measure so that

$$\int_{S-B_\varepsilon} \|u_n\|^p \, d\mu < \frac{\varepsilon}{2}, \quad n = 1, 2, \ldots, N \text{ and } \int_{S-C_\varepsilon} \|u\|^p \, d\mu < \frac{\varepsilon}{2}.$$

Set $A_\varepsilon = B_\varepsilon \cup C_\varepsilon$. Condition (i) follows. The necessity of (ii) results similarly, by using the lemma.

Now, suppose that (i) and (ii) hold. Using (i), we get that for any $\varepsilon > 0$ there exists $A_\varepsilon \in \Sigma$ such that $\mu(A_\varepsilon) < \infty$ and

$$\int_S \|u_m - u_n\|^p \, d\mu \leqslant 2\varepsilon + \int_{A_\varepsilon} \|u_m - u_n\|^p \, d\mu \qquad (\forall) m, n \in \mathbb{N}.$$

By the Egorov theorem, there exists a $B \in \Sigma$, $B \subset A_\varepsilon$ with $\mu(B) < \delta$ and such that $u_n \to u$ uniformly on $A_\varepsilon - B$. Using the Minkowski inequality and (ii) we obtain

$$\limsup \left(\int_{A_\varepsilon} \|u_m - u_n\|^p \, d\mu \right)^{\frac{1}{p}} \leqslant \limsup \left(\int_{A_\varepsilon - B} \|u_m - u_n\|^p \, d\mu \right)^{\frac{1}{p}} + 2\varepsilon^{\frac{1}{p}} = 2\varepsilon^{\frac{1}{p}}.$$

Both these inequalities show that $\{u_n\}$ is a fundamental sequence in $L^p(S, X)$. Since $L^p(S, X)$ is complete, there exists a $u_0 \in L^p(S, X)$ such that $\|u_n - u_0\|_p \to 0$. This implies that $u_n(t) \to u_0(t)$ for a.a. $t \in S$. We may conclude that $u = u_0$ a.e. in S and $\|u_n - u\|_p \to 0$ as $n \to \infty$.

In case $\mu(S) < \infty$, the condition (i) of the theorem can be omitted:

Let $\{u_n\}$ be a sequence in $L^p(S, X)$, which converges a.e. in S to a function $u : S \mapsto X$. Let $\varphi : \mathbb{R}_+ \mapsto \mathbb{R}_+$ be a continuous function such that $\varphi(r) \to 0$ as $r \to 0$. If

$$\int_A \|u_n\|^p \, d\mu \leqslant \varphi(\mu(A)) \quad \text{for any } A \in \Sigma,$$

then $u_n \to u$ in $L^p(S, X)$.

We deduce now directly the

COROLLARY *(Lebesgue's dominated convergence theorem) Let $1 \leqslant p < \infty$ and let $\{u_n\}$ be a sequence in $L^p(S, X)$ which converges almost everywhere to a function u. Assume that there exists a $v \in L^p(S, X)$ such that $\|u_n(t)\| \leqslant \|v(t)\|$ a.e. $t \in S$. Then $u \in L^p(S, X)$ and $\|u_n - u\|_p \to 0$.*

2. Convex analysis

Our aim in this section is to survey some notions and basic results of convex analysis, related to some classes of nonlinear mappings.

Continuity of convex functions

2.1. Let X be a linear space. A set $C \subset X$ is *convex* if $x, y \in C$ implies that $\alpha x + \beta y \in C$ for all non-negative numbers α, β with $\alpha + \beta = 1$. Since the intersection of a family of convex sets is again convex, for any set $M \subset X$ we can define the *convex hull* of M, denoted by *conv M*, to be the intersection of all convex sets in X which contain M.

A finite set $\{x_0, x_1, \ldots, x_n\} \subset X$ is *affinely independent* if the set $\{x_1 - x_0, \ldots, x_n - x_0\}$ is linearly independent. In such a case the convex hull $\sigma = \text{conv} \{x_0, x_1, \ldots, x_n\}$ is said to be the *n-dimensional simplex* with vertices $x_0, x_1, \ldots, x_n$.

A *cone C* with vertex 0 is a subset of X invariant under all homothetic maps $x \mapsto \lambda x$ with $\lambda > 0$; if, additionally, C is convex, then C is called a *convex cone* of vertex 0. Thus, a convex cone is a subset of X such that $C + C \subset C$ and $\lambda C \subset C$ for all $\lambda > 0$.

For any real-valued function φ defined on X,

$$\text{eg } \varphi = \{[x, r] \in X \times \mathbb{R} \mid \varphi(x) \leqslant r\}$$

is the *epigraph* of φ and the projection of eg φ on X, that is,

$$\text{dom } \varphi = \{x \in X \mid \varphi(x) < \infty\}$$

is the *effective domain* of φ.

Let C be a convex set of X. A map $\varphi : C \mapsto \overline{\mathbb{R}}$ is said to be *a convex function* provided that

$$\varphi(\alpha x + \beta y) \leqslant \alpha \varphi(x) + \beta \varphi(y)$$

tor all $x, y \in C$ and $\alpha, \beta \in [0,1]$ with $\alpha + \beta = 1$. The function is *concave* if $-\varphi$ is convex, and φ is *affine* if it is both convex and concave. Clearly, the convexity of φ is equivalent to

$$\varphi[(1 - t)x + ty] \leqslant (1 - t)\varphi(x) + t\varphi(y) \qquad (\forall)t \in [0, 1]. \tag{1}$$

If this inequality is strict whenever x, y are different elements of C, the function φ is *strictly convex*.

In what follows, we adopt the convention that

$$\infty + (-\infty) = \infty \text{ and } 0 \cdot \infty = 0.(-\infty) = 0.$$

A function φ is *proper* on C if $\varphi(x) > -\infty$ for all $x \in C$ and $\varphi(x) < \infty$ at least in one point $x \in C$. We can extend a convex function to all of X, if we set $\varphi(x) = \infty$ for all $x \in X - C$.

PROPOSITION *A function $\varphi : X \mapsto \overline{\mathbb{R}}$ is convex if and only if its epigraph is a convex subset of $X \times \mathbb{R}$.*

Proof: Suppose φ is convex and take $[x, r_1], [y, r_2]$ in eg φ. Then

$$\varphi(\alpha x + \beta y) \leqslant \alpha\varphi(x) + \beta\varphi(y) \leqslant \alpha r_1 + \beta r_2.$$

for all $\alpha, \beta \geqslant 0$, with $\alpha + \beta = 1$. This implies that

$$[\alpha x + \beta y, \alpha r_1 + \beta r_2] \in \text{eg } \varphi.$$

Conversely, assume eg φ is convex. Then dom φ is also convex as the projection of eg φ on X. For any $x, y \in \text{dom } \varphi$, we have that $[x, \varphi(x)]$ and $[y, \varphi(y)]$ are in eg φ and thus,

$$[(1 - t)x + ty, (1 - t)\varphi(x) + t\varphi(y)] \in \text{eg } \varphi \text{ for all } t \in [0, 1],$$

which is equivalent to (1). The proof is complete.

By this proposition, the set of all proper convex functions on X corresponds to a class of convex subsets on $X \times \mathbb{R}$, namely the set of all convex epigraphs. Conversely, to any convex set C of X there corresponds the proper convex function

$$I_C(x) = \begin{cases} 0 & \text{if } x \in C, \\ \infty & \text{otherwise} \end{cases}$$

which is called the *indicator function* of C.

Let $\{\varphi_i | i \in I\}$ be an arbitrary set of functions. The function

$$\varphi(x) = \sup \{\varphi_i(x) \mid i \in I\}$$

is called *upper envelope (supremum)* and has as epigraph $\bigcap\limits_{i \in I} \text{eg } \varphi_i$. The supremum of a family of convex functions is also convex.

For any convex function φ, the level sets defined by

$$\varphi^{\leqslant}(r) = \{x \in X | \varphi(x) \leqslant r\} \text{ and } \varphi^{\geqslant}(r) = \{x \in X | \varphi(x) \geqslant r\},$$

for all $r \in \mathbb{R}$, are convex.

Now let X be a linear topological space.

LEMMA *Let $\mathscr{U}$ be a neighbourhood of a point $x_0 \in X$ and let $\varphi : X \mapsto \overline{\mathbb{R}}$ be a convex function which is bounded from above in $\mathscr{U}$ by a finite constant. Then φ is continuous at x_0.*

Proof: Applying the translations $x \mapsto x - x_0$ and $\varphi(x) \mapsto \varphi(x) - \varphi(x_0)$, if necessary, we may consider the case when $x_0 = 0$ and $\varphi(x_0) = 0$. Let $\mathscr{U}$ be a neighbourhood of the origin such that $\varphi(x) \leqslant a < \infty$ for all $x \in \mathscr{U}$. Setting $\mathscr{W} = \mathscr{U} \cap (-\mathscr{U})$ and $\varepsilon \in (0,1)$ we consider the neighbourhood $\varepsilon\mathscr{W}$. Since $-\dfrac{x}{\varepsilon} \in \mathscr{W}$, by the convexity of φ, we have

$$0 = \varphi(0) = \varphi\left[\frac{1}{1+\varepsilon}x + \left(1 - \frac{1}{1+\varepsilon}\right)\left(-\frac{x}{\varepsilon}\right)\right] \leqslant$$

$$\leqslant \frac{1}{1+\varepsilon}\varphi(x) + \left(1 - \frac{1}{1+\varepsilon}\right)\varphi\left(-\frac{x}{\varepsilon}\right),$$

i.e., $\varphi(x) \geqslant -\varepsilon\varphi\left(-\dfrac{x}{\varepsilon}\right) \geqslant -\varepsilon a.$

Since $\dfrac{x}{\varepsilon} \in \mathscr{W}$, inequality (1) implies that $\varphi(x) \leqslant (1-\varepsilon)\varphi(0) + \varepsilon\varphi\left(\dfrac{x}{\varepsilon}\right) \leqslant$

$\leqslant \varepsilon a.$ Hence, $|\varphi(x)| \leqslant \varepsilon a$ for $x \in \varepsilon\mathscr{W}$, i.e., φ is continuous at x_0.

THEOREM *Under the assumptions of the lemma, φ is continuous on* Int(dom φ).

Proof: It is sufficient to show that to any $y \in$ Int(dom φ) there corresponds a neighbourhood $\mathscr{V}$ such that $\varphi(x) \leqslant M < \infty$ for all $x \in \mathscr{V}$. As dom φ is an open set, there exists a $\rho > 1$ such that $\rho y \in$ dom φ. If $\mathscr{W}$ is the neighbourhood used in the theorem, then for any $x \in y + \left(1 - \dfrac{1}{\rho}\right)\mathscr{W}$ we can write

$$x = y + \left(1 - \frac{1}{\rho}\right)z = \frac{1}{\rho}(\rho y) + \left(1 - \frac{1}{\rho}\right)z \qquad (\forall)\, z \in \mathscr{W}.$$

Since $0 < \dfrac{1}{\rho} < 1$, we have $x \in$ dom φ and $y + \left(1 - \dfrac{1}{\rho}\right)\mathscr{W} \subset$ dom φ.

Then the convexity of φ implies that

$$\varphi(x) \leqslant \frac{1}{\rho} \varphi(\rho y) + \left(1 - \frac{1}{\rho}\right) \varphi(z) \leqslant \frac{1}{\rho} \varphi(\rho y) + \left(1 - \frac{1}{\rho}\right) a = M.$$

As a consequence of the theorem we have the following

COROLLARY *Any convex function defined on a finite-dimensional space X is continuous on* Int(dom φ).

Proof: Assume that dom $\varphi \neq \emptyset$. Let σ be an n-dimensional simplex determined by $(n+1)$ affinely independent vertices $a_0, a_1, \ldots, a_n$. Then each point $x \in \sigma$ is of the form $x = \sum_{j=0}^{n} \lambda_j a_j$ with $0 \leqslant \lambda_j \leqslant 1$ and $\sum_{j=0}^{n} \lambda_j = 1$. Since $a_j \in$ dom φ, we have $\varphi(x) \leqslant \sum_{j=0}^{n} \lambda_j \varphi(a_j) < \infty$. Since the function φ is bounded on σ, the theorem implies that it is continuous on Int (dom φ).

Given a set M of a linear topological space, the closure of conv$M$ is called the *convex closure* of M.

Separation of convex sets

2.2. Let X be a vector space. A subset $C \subset X$ is called *absorbent* if for each $x \in X$ there is some $\lambda > 0$ such that $x \in \mu C$ for all μ with $|\mu| \geqslant \lambda$. A function $p : X \mapsto \mathbb{R}$ is said to be *sublinear* if it is both subadditive

$$p(x + y) \leqslant p(x) + p(y) \qquad (\forall)x, y \in X,$$

and positively homogeneous

$$p(\lambda x) = \lambda p(x) \qquad (\forall) x \in X, \lambda \in \mathbb{R}_+.$$

An example of a sublinear function is the *Minkowski function* or the *gauge* of an absorbent convex set C in X,

$$p_C(x) = \inf \left\{ r \in \mathbb{R}_+ \mid \frac{x}{r} \in C \right\}.$$

Within this framework we may state an analytic version of the Hahn-Banach theorem.

THEOREM (*Cristescu* [1, p. 17]) *Let Y be a subspace of a linear space X and let p be a sublinear function on X. If $f : Y \mapsto \mathbb{R}$ is a linear function such that $f(x) \leqslant p(x)$ for all $x \in Y$, then there exists a linear function $\tilde{f}$ extending f to X such that $\tilde{f}(x) \leqslant p(x)$ for all $x \in X$.*

Now let X be a topological vector space and let X^* be its dual space, i.e., the space of all linear continuous real-valued functions on X. For any $0 \neq f \in X^*$ and any $c \in \mathbb{R}$, the set $H = \{x \in X \mid (f, x) = c\}$ is called a *closed hyperplane*. The level sets $f^{\leqslant}(c) = \{x \in X \mid (f, x) \leqslant c\}$ and $f^{\geqslant}(c) = \{x \in X \mid (f, x) \geqslant c\}$ are called the *closed half-spaces* determined by H. The sets $f^{>}(c)$ and $f^{<}(c)$ are the *open half-spaces*. Two non-empty subsets A, B of X are *separated* by the hyperplane H if either $A \subset f^{\leqslant}(c)$ and $B \subset f^{\geqslant}(c)$ or $B \subset f^{\leqslant}(x)$ and $A \subset f^{\geqslant}(c)$. They are *strictly separated* by H if either $A \subset f^{<}(c)$ and $B \subset f^{>}(c)$ or $B \subset \subset f^{<}(c)$ and $A \subset f^{>}(c)$. A hyperplane H is called a *supporting hyperplane* for a convex subset $A \subset X$ provided that $A \cap H \neq \emptyset$ and A lies on one side of H.

The geometric version of the Hahn-Banach theorem asserts that in a topological vector space, every convex open non-empty set A and every linear subspace Y, not intersecting A, can be separated by a closed hyperplane. The refinements of this result are called separation theorems. The well-known ones are the following

FIRST SEPARATION THEOREM *Let A and B be two non-empty convex subsets of a topological vector space, such that* Int $A \neq \emptyset$ *and* (Int $A) \cap B = \emptyset$. *Then there exists a closed hyperplane H which separates A from B. Moreover, H separates strictly* Int A *from B* (Cristescu [1, p. 32]).

SECOND SEPARATION THEOREM *Let A and B be two non-empty, disjoint convex subsets of a locally convex space, such that A is closed and B is compact. Then there exists a closed hyperplane strictly separating A from B* (Schaefer [1, p. 65]).

Lower semi-continuous functions

2.3. Let X be a topological space. A proper function $\varphi : X \mapsto \overline{\mathbb{R}}$ is said to be *lower-semicontinuous* (l.s.c.) on X if

$$\varphi(x_0) \leqslant \lim_{x \to x_0} \inf \varphi(x) \quad \text{for any } x_0 \in X.$$

Similarly, φ is *upper-semicontinuous* on X if

$$\varphi(x_0) \geqslant \lim_{x \to x_0} \sup \varphi(x) \quad \text{for any } x_0 \in X.$$

Since φ is upper-semicontinuous if and only if $-\varphi$ is lower-semicontinuous it is sufficient to consider only l.s.c. functions.

PROPOSITION *The following three statements are mutually equivalent:*

    (i) φ *is l.s.c. on X;*

    (ii)   *the level set* $\varphi^{\leqslant}(r) = \{x \in X \mid \varphi(x) \leqslant r\}$ *is closed in X for every* $r \in \mathbb{R}$;

    (iii)   eg φ *is closed in* $X \times \mathbb{R}$.

Proof: Let $\mathscr{V}(x_0)$ be the set of all neighbourhoods of a point x_0 in X. We recall that

$$\liminf_{x \to x_0} \varphi(x) = \sup \{\inf \{\varphi(x) \mid x \in V\} \mid V \in \mathscr{V}(x_0)\}.$$

(i) $\Leftrightarrow$ (ii). Let $r \in \mathbb{R}$ and let $x_0 \in \varphi^>(r)$. By (i), there exists a $V_0 \in \mathscr{V}(x_0)$ such that $\inf \{\varphi(x) \mid x \in V_0\} > r$. Then $V_0 \subset \varphi^>(r)$. Consequently $\varphi^>(r)$ is open and hence $\varphi^{\leqslant}(r)$ is closed.

Conversely, for any $\varepsilon > 0$, denote $V_0 = \{x \in X \mid \varphi(x) > \varphi(x_0) - \varepsilon\}$. By (ii), $V_0 \in \mathscr{V}(x_0)$. Since $\inf \{\varphi(x) \mid x \in V_0\} \geqslant \varphi(x_0) - \varepsilon$, it follows that $\liminf_{x \to x_0} \varphi(x) \geqslant \varphi(x_0) - \varepsilon$. As ε is arbitrarily chosen, we conclude that (i) holds.

(i) $\Leftrightarrow$ (iii) results from the above equivalence since eg φ can be considered as a level set of the l.s.c. function Φ on $X \times \mathbb{R}$ defined by $\Phi(x, r) = \varphi(x) - r$.

The upper envelope of a family of l.s.c. functions is an l.s.c. function. This follows from

$$\{x \in X \mid \sup_{i \in I} \varphi_i(x) \leqslant r\} = \bigcap_{i \in I} \{x \in X \mid \varphi_i(x) \leqslant r\}$$

and (iii) in the previous proposition.

Now let X be a Banach space. We distinguish on its dual space X^*, a lower semicontinuity and a weak*-lower semicontinuity corresponding to the strong and the weak* topologies on X^*.

THEOREM *Any convex function $\varphi : X^* \mapsto \overline{\mathbb{R}}$ is l.s.c. if and only if it is weakly l.s.c.*

Proof: By (ii) in the proposition, it is sufficient to show that any convex set $C \subset X^*$ is closed if and only if it is weakly* closed. Denote by $\overline{C}$ and $\tilde{C}$ the strong and weak* closures of C. Clearly $\overline{C} \subset \tilde{C}$.

Conversely, let $f \in \tilde{C}$ be such that $f \notin \overline{C}$. By the second separation theorem, there exists an $x \in X$ such that

$$(x, f) > \sup \{(x, k) \mid k \in \overline{C}\}.$$

As $f \in \tilde{C}$, to each $\varepsilon > 0$ there corresponds a weak* neighbourhood U of f so that $(x, h) \geqslant (x, f) - \varepsilon$ for all $h \in U$. In particular, this is so for all $g \in U \cap \overline{C}$, i.e., $(x, g) \geqslant (x, f) - \varepsilon$. If we choose

$$0 < \varepsilon < (x, f) - \sup \{(f, k) \mid k \in \overline{C}\},$$

then we arrive to the contradiction $\varepsilon < (x, f - g) \leqslant \varepsilon$ for all $g \in U \cap \overline{C}$.

2.4. Let X be a Banach space and let K be a subset in X. For any proper function $\varphi : X \mapsto \overline{\mathbb{R}}$ we call $\varphi(x_0)$ the infimum of φ relative to K provided that

$$x_0 \in K \quad \text{and} \quad \varphi(x_0) \leqslant \varphi(x) \text{ for all } x \in K.$$

The existence of the infimum is derived from the following

THEOREM *(Weierstrass) Let K be a bounded weakly closed set in a reflexive Banach space X. Then every weakly l.s.c. function φ on K is bounded from below and attains its infimum there.*

Proof: Let $\{x_n\} \subset K$ be a minimizing sequence of φ, i.e.,

$$\lim \varphi(x_n) = d, \text{ where } d = \inf \{\varphi(y) \mid y \in K\}.$$

Since X is reflexive and K is bounded weakly closed, we may assume, passing eventually to a subsequence, that $x_n \rightharpoonup x_0$ in X and $x_0 \in K$.
On the other hand, φ is a weakly l.s.c. function, whence

$$\varphi(x_0) \leqslant \lim \inf \varphi(x_n) = d.$$

Hence $\varphi(x_0) = d$ and $d > -\infty$.

COROLLARY *Let $\varphi: X \mapsto \mathbb{R}$ be a weakly l.s.c. function such that*

$$\varphi(x) \to \infty \text{ as } \|x\| \to \infty.$$

Then φ is bounded from below and attains its infimum on X.

When φ is Gâteaux differentiable and φ attains its infimum at an interior point x_0 of K, then clearly $\varphi'(x_0) = 0$. Moreover, a strictly convex function on K has a unique minimizing point.

We can remove the assumption of the reflexivity of X. The above results remain valid, with the same proofs carried out on the dual space X^* whose unit ball is weakly* sequentially compact. By Theorem 1.3, this condition is realized, for instance, when X is separable.

2.5. Let X be a Banach space. The structure of an affine function is given by the following

PROPOSITION *An affine function $\gamma: X \mapsto \mathbb{R}$ is continuous if and only if it has the form*

$$\gamma(x) = (h, x) - r, \tag{1}$$

where $h \in X^*$ is the slope of γ and $r \in \mathbb{R}$.

Proof: Clearly, a real-valued function of type (1) is affine.
Conversely, if γ is affine, then the function $\varphi(x) = \gamma(x) - \gamma(0)$ is affine too. Since

$$0 = \varphi(0) = \varphi\left[\frac{1}{2}x + \frac{1}{2}(-x)\right] = \frac{1}{2}[\varphi(x) + \varphi(-x)],$$

we have $\varphi(-x) = -\varphi(x)$. If $\alpha, \beta > 0$, with $\alpha + \beta = 1$, then $\varphi(\alpha x) =$
$= \varphi(\alpha x + \beta 0) = \alpha\varphi(x)$. For $\alpha > 1$ we also have $\varphi(x) = \varphi\left(\dfrac{1}{\alpha}\,\alpha x\right) =$
$= \dfrac{1}{\alpha}\,\varphi(\alpha x)$. Consequently, $\varphi(\lambda x) = \lambda\varphi(x)$ for $\lambda \in \mathbb{R}$.

Since φ is homogeneous and affine, we get that

$$\varphi(\lambda x + \mu y) = (\lambda + \mu)\,\varphi\left(\frac{\lambda}{\lambda + \mu}\,x + \frac{\mu}{\lambda + \mu}\,y\right) = \lambda\,\varphi(x) + \mu\,\varphi(y)$$

for all $x, y \in X$ and $\lambda, \mu \in \mathbb{R}$. Hence, $\varphi(x)$ is a linear continuous function, i.e., $\varphi(x) = (h, x)$ and $r(x) = (h, x) + \gamma(0)$

An affine continuous function $\gamma(x)$ with slope $h \in X^*$ is a *minorant* at $x \in X$ of the function $\varphi : X \mapsto \mathbb{R}$ if $r \geqslant (h, x) - \varphi(x)$.

THEOREM *Any weakly l.s.c. convex function* $\varphi : X \mapsto \overline{\mathbb{R}}$ *is bounded from below by an affine continuous function.*

Proof: For any fixed $y \in X$ choose $t \in \mathbb{R}$ such that $t < \varphi(y)$. Since $[y, t] \notin \text{eg }\varphi$ and eg φ is a closed convex set, there exists a closed hyperplane

$$\{[x, r] \in X \times \mathbb{R} \mid (h, x) + \alpha r = \beta\}$$

which separates $[y, t]$ from eg φ. Hence, $(h, y) + \alpha t < \beta$ and $(h, x) + \alpha r > \beta$ for all $[x, r] \in \text{eg }\varphi$. If $\varphi(y) < \infty$, then taking $[y, \varphi(y)] \in \text{eg }\varphi$ we obtain $\alpha(\varphi(y) - t) > 0$, i.e., $\alpha > 0$. Dividing by α we obtain

$$t < \frac{\beta}{\alpha} - \frac{1}{\alpha}(h, y) < \varphi(y).$$

Hence, $\gamma(x) = \dfrac{1}{\alpha}\,[\beta - (h, x)]$ is an affine continuous function which is a minorant of the function φ.

The subdifferential

2.6. Let X be a Banach space and let $\varphi : X \mapsto \overline{\mathbb{R}}$ be a proper function. We say that the function φ has as an exact minorant at $x \in X$ the affine function $\gamma(.) = (h,.) + r$ provided that $\varphi(x) = \gamma(x)$. If $\varphi(x) < \infty$, then $\gamma(y) = = (h, y - x) + \varphi(x)$.

The function φ is said to be *subdifferentiable* at a point $x \in X$ provided there exists an affine continuous function which is an exact minorant of φ at this point. The slope $h \in X^*$ of such affine function is said to be a *subgradient* of φ at x, and the set of all subgradients at x is denoted by $\partial\varphi(x)$. The mapping $\partial\varphi : X \mapsto 2^{X^*}$ is called the *subdifferential* of φ and φ is subdifferentiable at x provided that $\partial\varphi(x) \neq \emptyset$. Clearly, $f \in \partial\varphi(x)$, if and only if the *subgradient inequality*

$$\varphi(y) - \varphi(x) \geqslant (f, y - x) \qquad (\forall) \, y \in X, \tag{1}$$

holds.

Let φ be a convex function on X. If φ is Gâteaux differentiable at $x \in X$, then it is subdifferentiable at x and $\partial\varphi(x) = \{\varphi'(x)\}$. Indeed, by the convexity of φ we have

$$\varphi(x + t(y - x)) \leqslant \varphi(x) + t \, [\varphi(y) - \varphi(x)] \qquad (\forall) \, x, y \in X, \; 0 \leqslant t \leqslant 1,$$

and

$$\frac{1}{t} \, [\varphi(x + t(y - x)) - \varphi(x)] \leqslant \varphi(y) - \varphi(x),$$

which implies that $\varphi'(x) \in \partial\varphi(x)$. Now, if $f \in \partial\varphi$ then (1) is equivalent to the inequality

$$\frac{1}{t} \, [\varphi(x + ty) - \varphi(x)] \geqslant (f, y) \qquad (\forall) \, y \in X, \; t > 0,$$

and therefore $(\varphi'(x) - f, y) \geqslant 0$ for all $y \in X$, i.e., $f = \varphi'(x)$.

We collect here some elementary properties of the subdifferential:

(P1) *For any* $x \in X$, *the set* $\partial\varphi(x)$ *is convex and weakly* closed in* X^*. Indeed, for fixed $y \in X$ the subgradient inequality can be put in the from $\{f \in X^* \mid (f, x) \leqslant a, \; x \in X\}$, with $a \in \mathbb{R}$, and this set has the required properties.

(P2) $D(\partial\varphi) \subseteq \mathrm{dom}\,\varphi$, or equivalently, $x \notin \mathrm{dom}\,\varphi$ implies that $\varphi(x) = \infty$. Clearly, inequality (1) cannot be true for any $f \in X^*$.

(P3) φ *has a minimum value at* $x \in D(\partial\varphi)$ *if and only if* $0 \in \partial\varphi(x)$. The assertion is an obvious consequence of (1).

(P4) *A subdifferentiable function* φ *is convex l.s.c. on any open convex set* $U \subset \mathrm{dom}\,\varphi$. Since U is convex, $z = (1-t) x + ty$ lies in U for all $x, y \in U$ and $t \in [0,1]$. Then for $f \in \partial\varphi(z)$ we have

$$\varphi(x) \geqslant \varphi(z) + (f, x - z) \qquad \text{and} \qquad \varphi(y) \geqslant \varphi(z) + (f, y - z).$$

We multiply these inequalities by $(1 - t)$ and t, and add the results. Then

$$(1 - t) \, \varphi(x) + t \, \varphi(y) \geqslant \varphi(z) = \varphi((1 - t) x + ty),$$

i.e., we get the convexity of φ. Now, let $x_0 \in U$ and $\{x_n\} \subset U$ be such that $x_n \to x_0$ in X. There exists an $f \in X^*$ such that

$$\varphi(x_n) \geqslant \varphi(x_0) + (f, x_n - x_0).$$

Hence, $\liminf \varphi(x_n) \geqslant \varphi(x_0)$, i.e., φ is also l.s.c. on U.

We remark that property (P3) expresses the equivalence between the solvability of functional equations defined by subdifferentials and the minimization of functions.

In order to establish an assertion converse to (P4) we recall some other topological concepts. Thus, a set C is said to be *balanced* if $\lambda C \subseteq C$ whenever $|\lambda| \leqslant 1$. A set in a locally convex space is called a *barrel* if it is convex, absorbent and closed. Every locally convex space has a base of neighbourhoods consisting of barrels and such a space is called *barrelled* if each barrel is a neighbourhood. Any Banach space is barrelled (see Cristescu [1, p. 73]).

LEMMA *Any proper convex l.s.c. function φ on a Banach space X is continuous on* Int (dom φ).

Proof: Indeed, without loss of generality we may suppose that $0 \in$ Int (dom φ) and that $\varphi(0) < a$ with $a \in \mathbb{R}$. The level set $W = \varphi^{\leqslant}(a)$ is convex closed and so is the set $U = W \cap (-W)$. We shall check whether U is a barrel.

The set U is balanced since $x \in U$ and $|\lambda| \leqslant 1$ implies that $\varphi(\lambda x) = = \varphi[\lambda x + (1 - \lambda)0] \leqslant a$. To any $x \in X$ there corresponds a $\lambda_0 > 0$ such that $|\lambda| \leqslant \lambda_0$ implies that $\lambda x \in$ Int (dom φ). By Corollary 2.1, the restriction of φ to the segment $[-\lambda_0 x, \lambda_0 x] \subset$ Int (dom φ) is continuous. In particular, the restriction of φ is continuous at the origin and there exists a $\lambda_x > 0$ such that

$$0 \leqslant \varphi(\lambda x) - \varphi(0) < a - \varphi(0) \text{ for all } |\lambda| \leqslant \lambda_x.$$

Therefore $\varphi(\lambda x) \leqslant a$, which proves that U is absorbent.

Now Proposition 2.1 implies that φ is continuous on Int (dom φ).

PROPOSITION *Let X be a Banach space and let $\varphi \colon X \mapsto \overline{\mathbb{R}}$ be a proper convex l.s.c. function. Then φ is subdifferentiable on* Int (dom φ).

Proof: By the lemma, φ is continuous on Int (dom φ) and thus Int (eg φ) $\neq \emptyset$. If $x \in$ Int (dom φ), then by Theorem 2.5, the affine continuous function which is an exact minorant at x of φ has the form $\gamma(y) = (f, y - x) + \varphi(x)$, (geometrically $\gamma(y)$ is a supporting hyperplane of eg $\varphi(x)$ at the point $[x, \varphi(x)]$). The condition $\gamma(y) \leqslant \varphi(y)$ implies that $(f, y - x) \leqslant \varphi(y) - \varphi(x)$ for all $y \in X$, i.e., $f \in \partial\varphi(x)$. The proof is complete.

We remark that $\partial(\lambda \varphi) = \lambda \partial\varphi$ for any $\lambda > 0$. Moreover, we have the

THEOREM *If φ_1, φ_2 are two convex functions on X and there exists at least one point $z \in$ dom $\varphi_1 \cap$ dom φ_2 in which one of them, say φ_1, is continuous, then*

$$\partial(\varphi_1 + \varphi_2) = \partial\varphi_1 + \partial\varphi_2.$$

Proof: Inequality (1) implies that

$$\partial\varphi_1(x) + \partial\varphi_2(x) \subseteq \partial(\varphi_1 + \varphi_2)(x) \qquad (\forall)\, x \in X.$$

Conversely, we show that the elements $f \in \partial(\varphi_1 + \varphi_2)(x)$ may be written as $f = f_1 + f_2$ with $f_1 \in \partial\varphi_1(x)$ and $f_2 \in \partial\varphi_2(x)$. Indeed, by inequality (1), the element f satisfies the inequality

$$\varphi_1(y) - \varphi_1(x) - (f, y - x) \geqslant \varphi_2(x) - \varphi_2(y) \qquad (\forall)\, y \in X. \tag{2}$$

27

Now consider the sets

$$A = \{[y, r] \in X \times \mathbb{R} \mid \varphi_1(y) - \varphi_1(x) - (f, y - x) \leqslant r\},$$
$$B = \{[y, r] \in X \times \mathbb{R} \mid \varphi_2(x) - \varphi_2(y) \geqslant r\},$$

which by (2) have only common boundary points. Since A is the epigraph of the convex function

$$\varphi(y) = \varphi_1(y) - \varphi_1(x) - (f, y - x),$$

and φ is continuous in $z \in \operatorname{dom} \varphi_1 \cap \operatorname{dom} \varphi_2$, it follows that $\operatorname{Int} A = \operatorname{Int}(\operatorname{eg} \varphi) \neq \varnothing$. Therefore, by the first separation theorem, Int A and B can be strictly separated by a hyperplane, say H. This hyperplane cannot be vertical since in this case it would separate dom φ_1 from dom φ_2, which contradicts our assumption. As H is the graph of an affine continuous function

$$\gamma(y) = (g, y) + a, \qquad g \in X^*, \ a \in \mathbb{R},$$

it satisfies the inequalities

$$\varphi_2(x) - \varphi_2(y) \leqslant (g, y) + a \leqslant \varphi_1(y) - \varphi_1(x) - (f, y - x) \quad (\forall) \, y \in X.$$

For $y = x$ we get $a = -(g, x)$ and

$$\varphi_1(y) - \varphi_1(x) \geqslant (f + g, y - x), \ \varphi_2(y) - \varphi_2(x) \geqslant (-g, y - x) \quad (\forall) \, y \in X,$$

whence $f + g \in \partial \varphi_1(x)$ and $-g \in \partial \varphi_2(x)$. Therefore, $f = f_1 + f_2$ with $_1 = f + g$ and $f_2 = -g$. The proof is complete.

Smooth norms

2.7. In this section we study the differentiability of the norm in a Banach space X.

Let $\varphi: X \mapsto \overline{\mathbb{R}}$ be a convex function. Then the function $f(t) = \varphi(x + ty)$ is convex on $\mathbb{R}$ for any fixed $x, y \in X$. For any three real numbers $t_1 < t_2 < t_3$ we can write

$$t_2 = \frac{t_3 - t_2}{t_3 - t_1} t_1 + \frac{t_2 - t_1}{t_3 - t_1} t_3.$$

The convexity of f implies that

$$f(t_2) \leqslant \frac{t_3 - t_2}{t_3 - t_1} f(t_1) + \frac{t_2 - t_1}{t_3 - t_1} f(t_3),$$

whence

$$\frac{f(t_2) - f(t_1)}{t_2 - t_1} \leqslant \frac{f(t_3) - f(t_1)}{t_3 - t_1} \leqslant \frac{f(t_3) - f(t_2)}{t_3 - t_2}. \tag{1}$$

LEMMA *Any convex function* $f: \mathbb{R} \mapsto \overline{\mathbb{R}}$ *has one-sided derivatives at any* $t \in \text{Int}(\text{dom} f)$. *Moreover,* $f'_-(t) \leqslant f'_+(t)$ *and* f'_+ *is monotone increasing there.*

Proof: Let $0 < s_1 < s_2$. By the inequalities (1) we obtain

$$\frac{f(t) - f(t - s_2)}{s_2} \leqslant \frac{f(t) - f(t - s_1)}{s_1} \leqslant \frac{f(t + s_1) - f(t)}{s_1} \leqslant \frac{f(t + s_2) - f(t)}{s_2}.$$

These inequalities show that $\dfrac{f(t) - f(t - s)}{s}$ does not decrease as $s \to 0$

and therefore it has a limit $f'_-(t)$. Similarly, $\dfrac{f(t + s) - f(t)}{s}$ does not increase

as $s \to 0$ and it has the limit $f'_+(t)$. Further, we have $f'_-(t) \leqslant f'_+(t)$.

Let now $t_1 < t_2$. Then there exists a number $s > 0$ which is small enough such that $t_1 + s < t_2 - s$. Inequalities (1) applied to the points $t_1 < t_1 + s < t_2 - s < t_2$ yield

$$\frac{f(t_1 + s) - f(t_1)}{s} \leqslant \frac{f(t_2) - f(t_2 - s)}{s}.$$

Letting $s \to 0$, we get $f'_+(t_1) \leqslant f'_-(t_2)$. This together with the inequality $f'_-(t_2) \leqslant f'_+(t_2)$ yields $f'_+(t_1) \leqslant f'_+(t_2)$, as claimed.

Now, the G-differential of φ at x can be obtained as the derivative of f at $t = 0$, because

$$f'(0) = \lim_{t \to 0} \frac{f(t) - f(0)}{t} = \lim_{t \to 0} \frac{1}{t} [\varphi(x + ty) - \varphi(x)] = (\varphi'(x), y).$$

When f is not differentiable at the origin, we can define the one-sided G-differentials of φ by setting

$$(\varphi'_-(x), y) = f'_-(0) \text{ and } (\varphi'_+(x), y) = f'_+(0).$$

Since $f'_-(0) \leqslant f'_+(0)$, by the above lemma we deduce that

$$(\varphi'_-(x), y) \leqslant (\varphi'_+(x), y) \qquad (\forall) \, y \in X.$$

In particular, for the convex function $\varphi(x) = \|x\|$, the one-sided G-differentials

$$(D_+\|x\|, y) = \lim_{t \to 0^+} \frac{1}{t} (\|x + ty\| - \|x\|)$$

and

$$(D_- \|x\|, y) = \lim_{t \to 0^-} \frac{1}{t} (\|x + ty\| - \|x\|) = \lim_{t \to 0^+} \frac{1}{t} (\|x\| - \|x - ty\|)$$

exist and are finite. Moreover, we have the following chain of inequalities

$$- \|y\| \leqslant \frac{1}{t} (\|x\| - \|x - ty\|) \leqslant (D_- \|x\|, y) \leqslant \tag{2}$$

$$\leqslant (D_+ \|x\|, y) \leqslant \frac{1}{t} (\|x + ty\| - \|x\|) \leqslant \|y\| \qquad (\forall) \, t > 0.$$

On the other hand, the subdifferential $\partial \|.\|$ is defined by

$$\partial \|x\| = \{g \in X^* \mid \|x + y\| - \|x\| \geqslant (g, y) \qquad (\forall) \, y \in X\}.$$

This set is non-empty convex and weakly* closed, in virtue of $(P1)$ and Proposition 2.6. Moreover, if $g \in \partial \|x\|$ and $t > 0$, we have

$$\frac{1}{t} (\|x\| - \|x - ty\|) \leqslant (g, y) \leqslant \frac{1}{t} (\|x + ty\| - \|x\|),$$

whence

$$(D_- \|x\|, y) \leqslant (g, y) \leqslant (D_+ \|x\|, y) \qquad (\forall) \, y \in X. \tag{3}$$

Conversely, if $g \in X^*$ verifies (3), then by (2) we have

$$(g, y) \leqslant \frac{1}{t} (\|x + ty\| - \|x\|).$$

For $t = 1$ we obtain $\|x + y\| - \|x\| \geqslant (g, y)$, i.e., $g \in \partial \|x\|$.

Therefore we may conclude that

$$\partial \|x\| = \{g \in X^* \mid (D_- \|x\|, y) \leqslant (g, y) \leqslant (D_+ \|x\|, y) \quad (\forall) \, y \in X\}. \tag{4}$$

PROPOSITION *Let $x \neq 0$ be an element of a Banach space X. Then*

$$\partial \|x\| = \{g \in X^* \mid (g, x) = \|x\|, \, \|g\| = 1\}. \tag{5}$$

Proof: Indeed, when $(g, x) = \|x\|$ and $\|g\| = 1$, we have $(g, y) = (g, x + y) - (g, x) = (g, x + y) - \|x\| \leqslant \|x + y\| - \|x\|$, i.e., $g \in \partial \|x\|$. Conversely, suppose that $g \in \partial \|x\|$. Then, by (4) and (2) we obtain

$$- \|y\| \leqslant \frac{1}{t} (\|x\| - \|x - ty\|) \leqslant (g, y) \leqslant \frac{1}{t} (\|x + ty\| - \|x\|) \leqslant \|y\|$$

which yield for $t < 1$ and $y = x$ that $(g, x) = \|x\|$ and $\|g\| = 1$. The proof is complete.

A Banach space X is said to be *smooth* provided its norm is Gâteaux differentiable on $X - \{0\}$.

THEOREM *A Banach space X is smooth (strictly convex) if and only if the dual space X* is strictly convex (smooth).*

Proof: Indeed, by the above proposition, X is smooth if and only if the set (5) is a singleton. Therefore the space X is smooth if and only if x, as element of X^{**}, assumes its minimum value only at one point on the unit sphere of X^*. By (ii) in Proposition 1.5, this is equivalent to the strict convexity of X^*. The alternating equivalence results similarly.

In virtue of Theorem 1.5, we may consider always the reflexive Banach spaces endowed with Gâteaux differentiable norms.

Next, let H be a Hilbert space with the linear product $(.,.)$ and let $\varphi : H \mapsto \overline{\mathbb{R}}$ be a proper convex l.s.c. function. To each $u \in H$, we associate the strictly convex l.s.c. function

$$\Phi_u(v) = \frac{1}{2}\|v - u\|^2 + \varphi(v) \qquad (\forall)\, v \in H.$$

Since φ is bounded from below by a hyperplane $\gamma(v) = (y, v) + a$, we have that

$$\Phi_u(v) \geqslant \frac{1}{2}\|v - u\|^2 + (y, v) + a = \frac{1}{2}\|v + y - u\|^2 -$$

$$-\frac{1}{2}\|y - u\|^2 + \frac{1}{2}\|u\|^2 + a.$$

Since $\Phi_u(v) \to \infty$ as $\|v\| \to \infty$ there exists, by Corollary 2.4, a unique $x \in H$ which minimizes Φ_u on H. As the norm in H is Gâteaux differentiable, this point is characterized by the inequality

$$(x - u, v - x) + \varphi(v) - \varphi(x) \geqslant 0 \qquad (\forall)\, v \in H. \tag{6}$$

The map $u \mapsto x$ of H into itself is called the *proximity map* with respect to φ and is denoted by $x = P_\varphi u$. Since inequality (6) is equivalent to $u - x \in \partial \varphi(x)$ we have $P_\varphi u = (I + \partial \varphi)^{-1} u$.

3. Related topics and exercises

3.1. (*The Tietze extension theorem*) Let φ be a real bounded continuous function defined on a closed set C of a Banach space X. Then there exists a continuous function ψ defined on X such that $\psi|_C = \varphi$ and $\psi(X) \subseteq \varphi(C)$ (see Dunford and Schwartz [1, p. 15]).

3.2. *(The Ascoli-Arzela theorem)* Let $C(K)$ denote the Banach space of all continuous functions defined on a topological compact space K. A subset M of $C(K)$ is relatively compact if and only if the following conditions are satisfied:

(i) M is bounded in $C(K)$;

(ii) M is equicontinuous on K.

(See Yosida [1, p. 85].)

3.3. Let K be a convex subset of a strictly convex Banach space X. Then there exists at most one $x \in K$ such that $\|x\| = \inf \{\|u\| \mid u \in K\}$. In particular, if X is reflexive and K closed, there exists a unique $x \in X$ such that $\|x\| = \inf \{\|u\| \mid u \in K\}$.

HINT Indeed, if there exist $x, y \in K$, $x \neq y$ such that $\|x\| = \|y\| = \inf \{\|u\| \mid u \in K\}$, then, for $t \in (0, 1)$, the point $z = (1-t)x + ty$ lies in K and $\|z\| < \|x\|$. In the case of a reflexive Banach space, apply Corollary 2.4.

3.4. If a convex l.s.c. function $\varphi : X \mapsto \overline{\mathbb{R}}$ assumes the value $-\infty$, then it takes no finite values.

SOLUTION Indeed, suppose that there exists an $x_0 \in X$ such that $\varphi(x_0) \in \mathbb{R}$. Choose $a_0 \in \mathbb{R}$ such that $a_0 < \varphi(x_0)$. Then there exists a hyperplane strictly separating $[a, \varphi(x_0)]$ and the closed convex set eg φ, that is, there exist $h \in X^*$ and $r \in \mathbb{R}$ such that

$$(h, x_0) + ra_0 < (h, x) + ra \quad (\forall) \ [x, a] \in \text{eg } \varphi. \tag{1}$$

Taking $x = x_0$ and $a = \varphi(x_0)$, we obtain $r(\varphi(x_0) - a_0) > 0$ and thus $r > 0$. Then we can write (1) in the form

$$\frac{1}{r} (h, x_0 - x) + a_0 < \varphi(x), \tag{2}$$

thus arriving at a contradiction.

3.5. Let X be a Banach space and let $\varphi : X \mapsto \overline{\mathbb{R}}$ be a proper function. The greatest l.s.c. function $\overline{\varphi}$ which minorizes φ is said to be the *l.s.c.-regularization* of φ. Prove that

$$\text{eg } \overline{\varphi} = \overline{\text{eg } \varphi} \quad \text{and} \quad \overline{\varphi}(x) = \liminf_{y \to x} \varphi(y).$$

HINT By (iii) in Proposition 2.3, eg $\overline{\varphi}$ is closed in $X \times \mathbb{R}$. Moreover, eg $\overline{\varphi}$ contains eg φ, and thus also $\overline{\text{eg } \varphi}$. Let ψ be a function such that eg $\psi = \overline{\text{eg } \varphi}$. By (i) in Proposition 2.3, ψ is l.s.c. on X and therefore $\psi \leqslant \overline{\varphi}$. Then eg $\psi \supseteq$ eg $\overline{\varphi}$ and hence eg $\overline{\varphi} = \overline{\text{eg } \varphi}$.

3.6. Denote by $\Gamma(X)$ the set of all functions $\psi : X \mapsto \overline{\mathbb{R}}$ which are upper envelopes of a family of affine continuous functions. A function $\psi \in \Gamma(X)$ which minorizes the function $\varphi : X \mapsto \overline{\mathbb{R}}$ is said to be a Γ-*regularization* of φ. Prove that

(i) $\psi \leqslant \overline{\varphi} \leqslant \varphi$;

(ii) if φ is convex and admits an affine continuous minorant $\gamma : X \mapsto \mathbb{R}$, then $\overline{\varphi} \in \Gamma(X)$.

HINT Since conv eg φ is the epigraph of a convex function, we have that eg $\varphi \subseteq$ $\subseteq$ conv eg $\varphi \subseteq$ eg ψ, and (i) follows easily. For (ii) we refer the reader to Problem 3.5.

3.7. The function $\varphi^* : X^* \mapsto \overline{\mathbb{R}}$ defined by

$$\varphi^*(f) = \sup \{(f, x) - \varphi(x) \mid x \in X\}$$

is called the *conjugate* of the convex function $\varphi : X \mapsto \overline{\mathbb{R}}$. Prove that:

(j) φ^* exists if and only if there exists a closed non-vertical hyperplane H in $X \times \mathbb{R}$ such that eg φ lies in one of the closed half-spaces determined by H;

(jj) φ^* is a convex l.s.c. function;

(jjj) $\varphi^*(f) + \varphi(x) \ge (f, x)$ (the Young inequality);

(jv) If $\psi \ge \varphi$, then $\psi^* \le \varphi^*$.

HINT (j) Suppose that φ^* exists and let $[f_0, t_0] \in$ eg φ^*. Then $t_0 \ge \varphi^*(f_0) \ge (f_0, x) - \varphi(x) \ge (f_0, x) - t$ for all $[x, t] \in$ eg φ. Hence $(f_0, x) - t = t_0$ is a closed hyperplane with the properties required by (j). Conversely, let φ and $H = \{(x, t) \mid (f_0, x) - t = t_0\}$ have the assumed properties. Then for any $[x, t] \in$ eg φ, we have $(f_0, x) - t \le t_0$ and $(f_0, x) - \varphi(x) \le t_0$, i.e., $[f_0, t_0] \in$ eg φ^*.

(jj) Let $\{f_n\} \subset \varphi^{*\le}(r)$ be such that $f_n \to f$. If $f \notin \varphi^{*\le}(r)$, then there exist $\varepsilon > 0$ and $x \in X$ such that $\varphi^*(f) > r + \varepsilon$ and $(f, x) - \varphi(x) > r + \varepsilon$. For n large enough we have $|(f - f_n, x)| < \dfrac{\varepsilon}{2}$, whence $\varphi^*(f_n) \ge (f_n, x) - \varphi(x) > r + \dfrac{\varepsilon}{2}$. Therefore $f_n \notin \varphi^{*\le}(r)$, and this contradicts our assumption. Thus, by Proposition 2.3, φ^* is an l.s.c. function.

The remaining statements follow from the definition.

3.8. Let φ^{**} be the conjugate of φ^*. Prove that $\varphi \in \Gamma(X)$ if and only if $\varphi = \varphi^{**}$ (the Moreau-Rockafellar theorem).

SOLUTION The inequality $\varphi^{**} \le \varphi$ is obvious. To prove the converse inequality, let $z \in X$ and $r < \varphi(z)$. Then there exist $f \in X^*$ and $r_0 \in \mathbb{R}$ such that $\varphi(x) \ge (f, x) - r_0$ for all $x \in X$, and $r_0 < (f, z) - r$. Hence, $\varphi^*(f) \le r_0$ and

$$\varphi^{**}(z) \ge (f, z) - \varphi^*(f) \ge (f, z) - r_0 > r.$$

Since r is arbitrary we have that $\varphi^{**} \ge \varphi$.

3.9. Let $\varphi : X \mapsto \overline{\mathbb{R}}$ be a convex function. Prove that $f \in \partial \varphi(x)$ implies $x \in \partial \varphi^*(f)$. If further $\varphi \in \Gamma(X)$, then $f \in \partial \varphi(x)$ if and only if $x \in \partial \varphi^*(f)$, i.e. $\partial \varphi^* = (\partial \varphi)^{-1}$.

SOLUTION If $f \in \partial \varphi(x)$, then $(f, x) - \varphi(x) \ge (f, y) - \varphi(y)$ for all $y \in X$ and $\varphi^*(f) = (f, x) - \varphi(x)$. We have that

$$\varphi^*(g) - \varphi^*(f) \ge (g, x) - \varphi(x) - (f, x) + \varphi(x) = (f - g, x) \quad (\forall) g \in X^*,$$

i.e., $x \in \partial \varphi^*(f)$. When $\varphi \in \Gamma(X)$, then $\varphi = \varphi^{**}$ by 3.8, and $x \in \partial \varphi^*(f)$ implies that $f \in \partial \varphi^{**}(x) = \partial \varphi(x)$.

3.10. Let $\varphi, \psi \in \Gamma(H)$ be two proper convex functions on a Hilbert space H, such that $\varphi^* = \psi$ and let x, y, z be three elements in H. Then the following statements

(i) $z = x + y$, $\qquad \varphi(x) + \psi(y) = (x, y)$;

(ii) $x = P_\varphi x$, $\qquad y = P_\psi z$;

are mutually equivalent (see Moreau [1]).

Bibliographical comments

The results of Section 1 are standard and can be found in most textbooks on functional analysis. We mention here Cristescu [1], Dunford — Schwartz [1], Kato [2], Marinescu [1] and Yosida [1]. The sequential characterization in proposition 1.3 has been pointed out by Browder [10, p. 81] and Browder-Hess [1]. For the Proofs of the basic theorems of Asplund and Trojanski we refer to I. Ciorănescu [1] and Diestel [1]. The presentation of the notions of spectral theory for nonlinear operators follows Da Prato [2] and Martin jr. [1]. The elements of measure theory in Section 1, were taken from Hewitt-Stronberg [1]. The interested reader should also consult the books of Kufner — John-Fučik [1] and Nicolescu [1, vol. III.].

The presentation of convex analysis in Section 2 follows Bourbaki [1], Asplund [3], Ekeland — Temam [1], Holmes [1], Ioffe — Tihomirov [1], and Rockafellar [3]. Some special results in this area can be found in Barbu — Precupanu [1] or Castaing — Valadier [1]. The definition of subdifferentiable norms follows Da Prato [3]. The prox mapping as well as the Γ-regularization of a function are due to Moreau [1]. For a detailed exposition of these see also Moreau [2].

Chapter II

Methods of compactness

The first attempts to solve nonlinear functional equations have involved various aspects of compactness. The fixed point theory has played a decisive role in this area. We proceed from the properties of compact maps in finite-dimensional spaces. As we shall see further on, the operators of monotone type extend some of these properties to infinite-dimensional spaces.

This introductory chapter collects those features of compact operators, Sobolev inbeddings and the fixed point theory which are required in the sequel.

1. Compact operators

1.1. Let X, Y be two Banach spaces and let P be a continuous operator from X into Y. The operator P is said to be *compact* if $P(A)$ is a relatively compact subset of Y whenever A is a bounded subset of X. It is easy to see that P is compact if the image $\{Px_n\}$ of any bounded sequence $\{x_n\}$ in X contains a Cauchy subsequence in Y.

Clearly, the set of all compact operators $P: X \mapsto Y$ is a linear space. Moreover we have the

PROPOSITION *Let* $\{P_n\}$ *be a sequence of compact operators from* X *into* Y *and let* $P: X \mapsto Y$ *be such that* $P_n x \to Px$ *in* Y, *uniformly on bounded subsets of* X. *Then* P *is a compact operator.*

Proof: Let $\{x_k\}$ be a bounded sequence in X. For each $n \in \mathbb{N}$, the sequence $\{P_n x_k\}$ contains a convergent subsequence $\{P_n x_{k_n}\}$. If we denote by $\{y_k\}$ the diagonal sequence $\{x_{k_n}\}$, then $\{P_n y_k\}$ converges as $k \to \infty$ for every n. By hypothesis, $P_n y_k \to P y_k$ in Y and

$$\|Py_j - Py_k\|_Y \leqslant \|Py_j - P_n y_j\|_Y + \|P_n y_j - P_n y_k\|_Y + \|P_n y_k - Py_k\|_Y,$$

which proves that $\{Py_k\}$ is a Cauchy sequence in Y. The proof is complete.

A continuous operator P is said to be of *finite rank* if its range $R(P)$ lies in a finite-dimensional subspace of Y. A bounded operator of finite rank is compact, because it carries a bounded sequence into a bounded sequence in a finite-dimensional space; thus the image of the bounded sequence contains a convergent subsequence.

THEOREM *Let X and Y be two Banach spaces and let A be a closed bounded subset of X. Then a continuous operator $P\colon A \mapsto Y$ is compact if and only if P is a uniform limit of operators of finite rank.*

Proof: Suppose P compact. Then $\overline{P(A)}$ is a compact subset of Y and for any given $\varepsilon > 0$, $\overline{P(A)}$ can be covered by a finite number of open balls $B_i = B(y_i, \varepsilon)$ with centres $y_i \in P(A)$, $1 \leqslant i \leqslant n_\varepsilon$, i.e., $\overline{P(A)} \subset \bigcup_{i=1}^{n_\varepsilon} B_i$. Consider on $\overline{P(A)}$ a partition of unity $\{\beta_i\}$, subordinate to this covering, that is, $\beta_i \in C(B_i)$, $0 \leqslant \beta_i(y) \leqslant 1$ and

$$\sum_{i=1}^{n_\varepsilon} \beta_i(y) = 1, \text{ for } y \in \overline{P(A)}.$$

Define the ε-*approximation* of P by

$$P_\varepsilon x = \sum_{i=1}^{n_\varepsilon} \beta_i(Px)\, y_i \qquad x \in A.$$

If $\beta_i(Px) > 0$ then $Px \in B_i$ and

$$\|P_\varepsilon x - Px\|_Y = \left\| \sum_{i=1}^{n_\varepsilon} \beta_i(Px)\,(y_i - Px) \right\|_Y < \varepsilon \text{ uniformly in } x \in A.$$

The 'only if' part of our assertion follows now since for any $\varepsilon > 0$ the map P_ε is continuous in A and its range lies in the finite-dimensional simplex $\{y_1, ..., y_n\}$.

The converse assertion follows by the above proposition because P_ε are bounded continuous operators,

$$\|P_\varepsilon x\|_Y = \sum_{i=1}^{n_\varepsilon} \|y_i\|_Y,$$

and they have finite ranks. Hence, P_ε are compact and so is P.

Linear integral operators

1.2. Let $\Omega \subset \mathbb{R}^N$ and $\Omega' \subset \mathbb{R}^{N'}$ be two bounded domains with $N \geqslant N'$. Troughout this section $\|\cdot\|$ denotes the norm in $C(\bar{\Omega}')$. For any real number $t > 0$, we denote by t' its conjugate number, so that $\dfrac{1}{t} + \dfrac{1}{t'} = 1$.

If $k : \bar{\Omega}' \times \Omega \mapsto \mathbb{R}$ is a continuous kernel, then the linear map $u = Pv$ defined by

$$u(x) = \int_{\Omega} k(x, y) \, v(y) \, dy \tag{1}$$

is a compact operator from $L^1(\Omega)$ into $C(\bar{\Omega}')$.

In fact, if we denote sup $\{k\,(x, y) \mid [x, y] \in \bar{\Omega}' \times \bar{\Omega}\}$ by K, then

$$\|u\| \leqslant K \|v\|_1,$$

that is, for a bounded subset M of $L^1(\Omega)$, the image $P(M)$ is bounded in $C(\bar{\Omega}')$. Moreover, for every $v \in M$ we have

$$|u(x) - u(x_0)| \leqslant \int_{\Omega} |k(x, y) - k(x_0, y)| \, |v(y)| \, dy \leqslant$$

$$\leqslant \|v\|_1 \int_{\Omega} |k(x, y) - k(x_0, y)| \, dy.$$

Hence, for x sufficiently close to x_0, the right-hand term can be made small and the functions in $P(M)$ are equi-continuous. Now, by the Ascoli-Arzelà theorem (see I.3.2), the set $P(M)$ is relatively compact in $C(\bar{\Omega}')$.

We shall study the linear integral operator (1), which maps $L^p(\Omega)$ into $L^q(\Omega')$, with $p, q \geqslant 1$. To avoid proliferation of constants, we assume that $\mu(\Omega) = 1$.

We give first the following criterion of compactness.

LEMMA *Suppose that P maps $L^p(\Omega)$ into $L^q(\Omega')$ and*

$$\left(\int_{\Omega'} \int_{\Omega} |k(x, y)|^{t'} \, dy \, dx \right)^{\frac{1}{t'}} \leqslant c, \tag{2}$$

for $t = \min(p, q')$. Then P is compact and

$$\|P\| \leqslant Ac,$$

where $A = (\mu(\Omega'))^{\frac{1}{\tau}}$ and $\tau = q \left(\dfrac{t'}{q} \right)'$. $\tag{3}$

Proof: By Hölder's inequality, we have

$$|u(x)| \leqslant \left(\int_{\Omega} |k(x, y)|^{t'} \, dy \right)^{\frac{1}{t'}} \left(\int_{\Omega} |v(y)|^{t} \, dy \right)^{\frac{1}{t}} \leqslant$$

$$\leqslant \left(\int_{\Omega} |k(x, y)|^{t'} \, dy \right)^{\frac{1}{t'}} \left(\int_{\Omega} |v(y)|^{p} \, dy \right)^{\frac{1}{p}} \leqslant \left(\int_{\Omega} |k(x, y)| \, dy \right)^{\frac{1}{t'}} \|v\|_p$$

because $t \leqslant p$. This implies that

$$\|u\|_q \leqslant \|v\|_p \left[\int_{\Omega'} \left(\int_{\Omega} |k(x, y)|^{t'} \, dy \right)^{\frac{q}{t'}} dx \right]^{\frac{1}{q}}.$$

Since $t \leqslant q'$, we may again apply Hölder's inequality with the exponents $\dfrac{t'}{q}$ and $\left(\dfrac{t'}{q} \right)'$, we obtain

$$\|u\|_q \leqslant \|v\|_p \left[\int_{\Omega'} \left(\int_{\Omega} |k(x, y)|^{t'} \, dy \right) dx \right]^{\frac{1}{t'}} \left(\int_{\Omega'} 1^{\left(\frac{t'}{q}\right)'} dx \right)^{\frac{1}{\tau}}.$$

Hence, according to (2) we have

$$\|u\|_q \leqslant Ac \, \|v\|_p,$$

and this establishes (3).

Condition (2) implies that $k\,(x, y)$ belongs to $L^{t'}(\Omega' \times \Omega)$. Hence, $k(x, y)$ can be approximated by a sequence $\{k_n\,(x, y)\}$ of continuous kernels, i.e.,

$$\left(\int_{\Omega'} \int_{\Omega} |k(x, y) - k_n\,(x, y)|^{t'} \, dy dx \right)^{\frac{1}{t'}} \leqslant \varepsilon_n$$

where $\varepsilon_n \to 0$. The integral operators P_n with the kernels $k_n(x, y)$ are linear compact operators and by the previous estimate we have

$$\|P_n - P\| \leqslant A\varepsilon_n.$$

By Proposition 1.1, the operator P is compact.

Now, we shall give conditions for the kernel $k(x, y)$, which make P a mapping from $L^p(\Omega)$ into $L^q(\Omega')$.

THEOREM *Suppose that there exist two numbers s, $t > 1$ with the following properties:*

(i) $\left(\int_{\Omega'} |k\,(x,y)|^s \, dx \right)^{\frac{1}{s}} \leqslant C_1$ a.e. $y \in \Omega$;

(ii) $\left(\int_{\Omega} |k(x, y)|^t \, dy \right)^{\frac{1}{t}} \leqslant C_2$ a.e. $x \in \Omega'$;

(iii) $q \geqslant p, q \geqslant s$ and $\left(1 - \dfrac{s}{q} \right) p' < t.$

Then P is a compact operator from $L^p(\Omega)$ into $L^q(\Omega')$ and

$$\|P\| \leqslant C_1^{\frac{s}{q}}\, C_2^{1-\frac{s}{q}}.$$

Proof: We may write

$$|u(x)| \leqslant \int_\Omega |k(x,y)|\,|v(y)|\,dy =$$

$$= \int_\Omega (|k(x,y)|^s\,|v(y)|^p)^{\frac{1}{q}}\,|v(y)|^{p\left(\frac{1}{p}-\frac{1}{q}\right)}\,|k(x,y)|^{1-\frac{s}{q}}\,dy.$$

Let us apply Hölder's generalized inequality with the exponents

$$\lambda_1 = q, \quad \lambda_2 = \frac{1}{\dfrac{1}{p}-\dfrac{1}{q}}, \quad \lambda_3 = p' = \frac{p}{p-1}, \quad \frac{1}{\lambda_1}+\frac{1}{\lambda_2}+\frac{1}{\lambda_3}=1,$$

thus obtaining

$$|u(x)| \leqslant \left(\int_\Omega |k(x,y)|^s\,|v(y)|^p\,dy\right)^{\frac{1}{q}} \left(\int_\Omega |v(y)|^p\,dy\right)^{\frac{1}{p}-\frac{1}{q}} \times$$

$$\times \left(\int_\Omega |k(x,y)|^{p'\left(1-\frac{s}{q}\right)}\,dy\right)^{\frac{1}{p'}}.$$

Denote $\lambda = p'\left(1-\dfrac{s}{q}\right)$. Since $\lambda < t$ and $\mu(\Omega) = 1$, it follows that

$$\left(\int_\Omega |k(x,y)|^\lambda\,dy\right)^{\frac{1}{p'}} = \left(\int_\Omega |k(x,y)|^\lambda\,dy\right)^{\frac{1}{\lambda}\left(1-\frac{s}{q}\right)} \leqslant$$

$$\leqslant \left(\int_\Omega |k(x,y)|^t\,dy\right)^{\frac{1}{t}\left(1-\frac{s}{q}\right)} \leqslant C_2^{1-\frac{s}{t}},$$

and thus

$$|u(x)| \leqslant \left(\int_\Omega |k(x,y)|^s\,|v(y)|^p\,dy\right)^{\frac{1}{q}}\|v\|_p^{1-\frac{p}{q}}\,C_2^{1-\frac{s}{q}}.$$

The above inequality leads to

$$\|u\|_q = \left(\int_{\Omega'} |u(x)|^q \, dx\right)^{\frac{1}{q}} \leq \left[\int_{\Omega'} \int_{\Omega} |k(x,y)|^s \, |v(y)|^p \, dydx\right]^{\frac{1}{q}} \|v\|_p^{1-\frac{p}{q}} \times C_2^{1-\frac{s}{p}} =$$

$$= C_2^{1-\frac{s}{q}} \left[\int_{\Omega} |v(y)|^p \left(\int_{\Omega'} |k(x,y)|^s \, dx\right) dy\right]^{\frac{1}{q}} \|v\|_2^{1'-\frac{p}{q}} \leq C_1^{\frac{s}{q}} C_2^{1-\frac{s}{q}} \|v\|_p,$$

and this establishes the above estimate.

To prove the compactness of P, choose a positive number ρ such that $\left(1 - \frac{s}{q}\right) p' < \rho < t$ and introduce the *truncated kernels*

$$k_n(x,y) = \begin{cases} -n & \text{if} \quad k(x,y) < -\!\!-\, n \\ k(x,y) & \text{if} \quad -n \leq k(x,y) \leq n \\ n & \text{if} \quad k(x,y) > n. \end{cases}$$

Denote

$$C_n = \text{ess sup} \left\{ \left(\int_{\Omega'} |k(x,y) - k_n(x,y)|^\rho \, dy\right)^{\frac{1}{\rho}} \,\middle|\, x \in \Omega' \right\} \leq$$

$$\leq \text{ess sup} \left\{ \left(\int_{e_n(x)} |k(x,y)|^\rho \, dy\right)^{\frac{1}{q}} \,\middle|\, x \in \Omega' \right\},$$

where $e_n(x) = \{y \in \Omega \mid |k(x,y)| > n\}$. If $y \in e_n(x)$, then

$$|k(x,y)|^\rho \leq -\frac{1}{n^{q-\rho}} |k(x,y)|^q$$

and

$$C_n < \text{ess sup} \left(\int_{e_n(x)} \frac{1}{n^{t-\rho}} |k(x,y)|^t \, dy\right)^{\frac{1}{t} \cdot \frac{t}{\rho}} \leq n^{1-\frac{t}{\rho}} C_2^{\frac{t}{\rho}}.$$

Let P_n be as above the integral operator with kernel $k_n(x,y)$ and apply the above estimate; then

$$\|P - P_n\| \leq C_1^{\frac{s}{q}} (n^{1-\frac{t}{\rho}} C_2^{\frac{t}{\rho}})^{1-\frac{s}{q}}$$

and $\lim \|P - P_n\| = 0$. Since the kernel $k_n(x,y)$ is bounded, it is summable to any power and, by the previous lemma, P_n is compact. Thus P is compact as a uniform limit of compact operators.

The second inequality in (iii) can be omitted. Indeed, it $q < s$, the operator P maps $L^p(\Omega')$ into $L^s(\Omega')$ because $L^s(\Omega') \subset L^q(\Omega')$.

REMARK If the third condition in (iii) is weakened to become

$$\left(1 - \frac{s}{q}\right) p' \le t,$$

then the operator $P: L^p(\Omega) \mapsto L^q(\Omega')$ need not be compact but it is still continuous.

Integral operators of potential type

1.3. In order to prove the Sobolev imbedding theorem we shall establish an integral representation of continuous differentiable functions and discuss some properties of integral operators with a polar kernel.

Let Ω be a bounded convex domain in $\mathbb{R}^N$ of diameter δ and let $u \in C^1(\Omega)$. Then we have the representation

$$u(x) = \frac{1}{\mu(\Omega)} \int_\Omega u(y)\, dy - \int_\Omega \frac{k(x, y)\cdot \nabla u(y)}{|x - y|^{N-1}}\, dy, \qquad (1)$$

where $k: \Omega \times \Omega \mapsto \mathbb{R}^N$ is bounded and continuous for $x, y \in \Omega$, $x \ne y$ and

$$|k(x, y)| \le \frac{\delta^N}{\mu(\Omega)}. \qquad (2)$$

Indeed, for any fixed $x \in \Omega$, we can describe any other point $y \in \Omega$ by specifying the distance $R = |x - y|$ and the unit vector ρ directed from x to y, i.e., $y = x + R\rho$. Obviously,

$$u(x) = u(y) - \int_0^R \frac{\partial u}{\partial \rho}\, d\rho.$$

Multiply this equality by $R^{N-1}\, dR$ and integrate it with respect to R between 0 and $d = d(\rho)$ from point x along the radius vector ρ, where $d(\rho)$ is the length in this direction within Ω. Thus,

$$u(x) \int_0^d R^{N-1}\, dR = \int_0^d u(y)\, R^{N-1}\, dR - \int_0^d R^{N-1} dR \int_0^R \frac{\partial u(y)}{\partial \rho}\, d\rho =$$

$$= \int_0^d u(y)\, R^{N-1}\, dR - \int_0^d \frac{\partial u(y)}{\partial \rho} \left(\int_\rho^d R^{N-1}\, dR\right) d\rho =$$

$$= \int_0^d u(y)\, R^{N-1}\, dR - \int_0^d \frac{d^N - \rho^N}{N} \frac{\partial u(y)}{\partial \rho} \frac{1}{\rho^{N-1}}\, \rho^{N-1}\, d\rho.$$

Let us change the notation in the last integral by putting R instead of ρ and multiply both sides of the equality by the elementary solid angle $d\sigma$, that is the surface element of the unit sphere $S \subset \mathbb{R}^N$. Using the formula

$$\int_\Omega v(y)\, dy = \int_S d\sigma \int_0^d v(y)\, R^{N-1}\, dR,$$

we obtain

$$u(x)\, \mu(\Omega) = \int_\Omega u(y)\, dy - \int_\Omega \frac{d^N - R^N}{N} \frac{\partial u(y)}{\partial \rho} \frac{1}{R^{N-1}}\, dy.$$

Since

$$\frac{u(y)}{\partial \rho} = \rho \cdot \nabla u,$$

we get the representation (1) with

$$k(x, y) = \frac{d^N - R^N}{N\, \mu(\Omega)}\, \rho.$$

Inequality (2) and the fact that $k(x, y)$ is continuous are among the consequences of this equality.

Representation (1) calls for an investigation of the integral operators of potential type

$$u(x) = \int_\Omega \frac{k(x, y)}{|x - y|^\lambda}\, v(y)\, dy \quad \text{where} \quad \lambda < N.$$

This map defines a linear operator $u = P_\lambda v$ in $L^p(\Omega)$, $p \geqslant 1$. Let p' be the conjugate number, of p i.e., $\dfrac{1}{p} + \dfrac{1}{p'} = 1$.

THEOREM *If* $\lambda < \dfrac{N}{p'}$, *then the map* $P_\lambda : L^p(\Omega) \mapsto C(\bar{\Omega})$ *is compact (in particular bounded).*

Proof: By hypothesis, $|k(x, y)| \leqslant k$, where k is a positive constant. Let $B(0, R) \equiv B$ be a ball which contains Ω and let $r = |x - y|$. Hölder's inequality yields

$$|u(x)| \leqslant k\, \|v\|_p \left(\int_B r^{-\lambda p'}\, dy \right)^{\frac{1}{p'}}.$$

In spherical coordinates with pole at x we have $dy = r^{N-1} dr d\sigma$, and thus

$$\int_B r^{-\lambda p'} dy \leqslant \int_0^{2R} r^{N-\lambda p'-1} dr \int_S d\sigma = \frac{|S|}{N - \lambda p'} (2R)^{N-\lambda p'}. \tag{3}$$

Hence,

$$|u(x)| \leqslant k \left(\frac{|S|}{N - \lambda p'} \right)^{\frac{1}{p'}} (2R)^{\frac{N}{p'} - \lambda} \|v\|_p,$$

whereby the boundedness of P is proved.

By the Arzela-Ascoli theorem, it suffices to prove the equicontinuity of all functions $u(x)$. Thus, let $\delta > 0$ and $\Omega_\delta = \{y \in \Omega | \; |y - x| \geqslant \delta\}$.

For $h \in \mathbb{R}^N$ with $|h| < \dfrac{\delta}{2}$ and $x + h \in \Omega$ one has

$$|u(x+h) - u(u)| \leqslant \int_{\Omega_\delta} \left| \frac{|k(x, y)|}{|x - y|^\lambda} - \frac{k(x+h, y)}{|x+h-y|^\lambda} \right| |v(y)| \, dy +$$

$$+ k \int_{\Omega - \Omega_\delta} \frac{|v(y)|}{|x - y|^\lambda} \, dy + k \int_{\Omega - \Omega_\delta} \frac{|v(y)|}{|x+h-y|^\lambda} \, dy.$$

For each $\varepsilon > 0$, by the continuity of $k(x, y)$, there exists a $\delta > 0$ such that the first integral is less than ε provided that $|h| \leqslant \eta$. For the second and the third integral we have $|x-y| \leqslant \delta$ and $|x + h - y| \leqslant \dfrac{3\delta}{2}$. Now apply the esti-

mate (3) with $2R = \delta$ and $2R = \dfrac{3\delta}{2}$. Hence

$$|u(x+h) - u(x)| \leqslant \left\{ \varepsilon [\mu(\Omega)]^{\frac{1}{p'}} + k \left(\frac{|S|}{N - \lambda p'} \right)^{p'} \left[\delta^{\frac{N}{p'} - \lambda} + \left(\frac{3\delta}{2} \right)^{\frac{N}{p'} - \lambda} \right] \right\} \|v\|_p.$$

Since the choice of ε and δ is arbitrary, the images $u = P_\lambda v$ are equicontinuous in $C(\bar{\Omega})$ and thus P_λ is compact.

For $\dfrac{N}{p'} \leqslant \lambda < N$, we apply Theorem 1.2 to integral operators of potential type. The integrability of the kernel of P_λ, of power s with respect to x and of power t with respect to y, requires that

$$\lambda s < N' \text{ and } \lambda t < N.$$

Since we may replace s and t by any numbers close to $\dfrac{N'}{\lambda}$ and $\dfrac{N}{\lambda}$, assumption

(iii) of Theorem 1.2 becomes

$$\left(1 - \frac{N'}{\lambda_q}\right) p' < \frac{N}{\lambda}.$$

Solving the above inequality with respect to q and using $q \leqslant p$, we get

$$q < q^* = \frac{N'p}{N - (N - \lambda)\,p} \quad \text{and } N' > N - (N - \lambda)\,p. \qquad (4)$$

When $N - (N - \lambda)\,p = 0$, the number q may be arbitrary. Hence we have the

PROPOSITION *Let* $\dfrac{N}{p'} \leqslant \lambda < N$. *Suppose that conditions* (4) *are fulfilled. Then the operator* P_λ *is compact from* $L^p(\Omega)$ *into* $L^q(\Omega')$, *where* Ω' *is an* N'-*dimensional plane section of* $\Omega(N' \leqslant N)$.

By Remark 1.2, the operator P_λ from $L^p(\Omega)$ into $L^{q*}(\Omega')$ is at least continuous.

2. Sobolev spaces

In current investigations of partial differential equations many useful concepts are supplied by the theory of distributions. To make the understanding of the subsequent chapters easier, we shall give here an outline of some aspects of this theory relevant for our purposes.

Mollifiers

2.1. Let Ω be a domain in $\mathbb{R}^N$ and let $\partial\Omega$ and $\bar{\Omega}$ be the boundary and the closure of Ω. Recall that the *support* of a function u defined on Ω is the closure of the subset $\{x \in \Omega \mid u(x) \neq 0\}$. In other words, supp u is the smallest relatively closed subset of Ω outside of which u vanishes.

Denote by $C^k(\Omega)$, $k \geqslant 0$, the set of all functions u defined in Ω, whose partial derivatives of order not exceeding k are continuous. We also set $C^0(\Omega) \equiv C(\Omega)$. $C_0^k(\Omega)$ denotes the subset of $C^k(\Omega)$ containing the functions with compact support in Ω. Of particular interest is $C_0^\infty(\Omega)$ which will be called the *space of test functions*.

Let $x = (x_1, ..., x_N)$ be a generic point in $\mathbb{R}^N$, $D_j = \dfrac{\partial}{\partial x_j}$ and $D^\alpha = D_1^{\alpha_1}... D_N^{\alpha_N}$.

Here the exponent is an N-tuple of non-negative integers, $\alpha = (\alpha_1, ... \alpha_N)$ with $|\alpha| = \alpha_1 + ... + \alpha_N$ and we set $\alpha \leqslant \beta$ whenever $\alpha_j \leqslant \beta_j$, $1 \leqslant j \leqslant N$.

Denote by $\mathscr{D}(\Omega)$ the linear space $C_0^\infty(\Omega)$ endowed with the following structure of convergence (pseudotopology): A sequence $\{\varphi_n\}$ of functions belonging to $C_0^\infty(\Omega)$ is said *to converge in the sense of the space* $\mathscr{D}(\Omega)$ to the function $\varphi \in C_0^\infty(\Omega)$ provided the following conditions are satisfied:

(i) there exist a compact set $K \subset \Omega$ such that supp $(\varphi_n - \varphi) \subset K$ for every $n \in \mathbb{N}$, and

(ii) $\lim D^\alpha \varphi_n(x) = D^\alpha \varphi(x)$ uniformly on K for each multi-index α.

The elements of the dual space $\mathscr{D}'(\Omega)$ are called *distributions*. Let $(.,.)$ be the pairing between $\mathscr{D}(\Omega)$ and $\mathscr{D}'(\Omega)$. For any $f \in \mathscr{D}'(\Omega)$, $D^\alpha f$ denotes the distribution defined by

$$(D^\alpha f, \varphi) = (-1)^{|\alpha|} (f, D^\alpha \varphi). \tag{1}$$

The derivation on $\mathscr{D}'(\Omega)$ *is (weakly* sequentially) continuous.* In fact, if $\{f_n\}$ is a sequence of distributions which converges to f, then

$$(D^\alpha f_n, \varphi) = (-1)^{|\alpha|} (f_n, D^\alpha \varphi) \to (-1)^{|\alpha|} (f, D^\alpha \varphi) = (D^\alpha f, \varphi) \text{ for all } \varphi \in \mathscr{D}(\Omega).$$

Now, let $\omega : \mathbb{R}^N \mapsto \mathbb{R}_+$ be a test function which satisfies

$$\text{supp } \omega \subset \{x \in \mathbb{R}^N \mid |x| \leqslant 1\} \quad \text{and} \quad \int_{\mathbb{R}^N} \omega(x)\, dx = 1 \tag{2}$$

The simplest example of such a function is

$$\omega(x) = \begin{cases} \dfrac{1}{C} e^{\frac{1}{|x|^2 - 1}} & \text{for } |x| \leqslant 1, \\[2ex] 0 & \text{otherwise,} \end{cases}$$

where

$$C = \int_{|x| \leqslant 1} e^{\frac{1}{|x|^2 - 1}}\, dx.$$

Given $\varepsilon > 0$, the function

$$\omega_\varepsilon(x) = \frac{1}{\varepsilon^N} \omega\left(\frac{x}{\varepsilon}\right)$$

45

is said to be a *mollifier*. For any real-valued locally integrable function u in Ω the *mollification* or *regularization* of u is given by

$$u_\varepsilon(x) = \mathscr{M}_\varepsilon u(x) = \int_\Omega \omega_\varepsilon(x - y)\, u(y)\, dy.$$

The mollifier operator $\mathscr{M}_\varepsilon$ is obviously linear and $u_\varepsilon = \mathscr{M}_\varepsilon u$ vanishes at all points $x \in \mathbb{R}^N$ for which $\varepsilon < \mathrm{dist}\,(x,\, \mathrm{supp}\, u)$.

The importance of the mollification lies in the fact that u_ε behaves much like u but it is very smooth. Since the differentiation can be carried out under the integral sign any number of times, it results that $u_\varepsilon \in C^\infty(\mathbb{R}^N)$. Therefore, if $\varepsilon < \mathrm{dist}\,(\mathrm{supp}\, u, \partial\Omega)$, then $u_\varepsilon \in C_0^\infty(\Omega)$.

Let us summarize some properties of mollification.

PROPOSITION *If $u \in C(\Omega)$, then $u_\varepsilon(x) \to u(x)$ as $\varepsilon \to 0$ uniformly on any compact subdomain Ω' with $\bar{\Omega}' \subset \Omega$. If $u \in L^p(\Omega)$, then $\|u_\varepsilon\|_p \leqslant \|u\|_p$ and $u_\varepsilon \to u$ in $L^p(\Omega)$ as $\varepsilon \to 0$.*

Proof: For $x, y \in \Omega'$ and $\varepsilon < \mathrm{dist}\,(\Omega', \partial\Omega)$ we have

$$|u_\varepsilon(x) - u(x)| = \left| \int_\Omega \omega_\varepsilon(x - y)\, [u(y) - u(x)]\, dy \right| \leqslant$$

$$\leqslant \sup \{|u(y) - u(x)| \,\big|\, |y - x| < \varepsilon\}$$

with $\varepsilon > 0$ chosen arbitrarily.

Now, for $u \in L^p(\Omega)$, by Hölder's inequality,

$$|u_\varepsilon(x)| \leqslant \left(\int_\Omega \omega_\varepsilon(x - y)\, dy \right)^{\frac{1}{p'}} \left(\int_\Omega \omega_\varepsilon(x - y)\, |u(y)|^p\, dy \right)^{\frac{1}{p}} =$$

$$= \left(\int_\Omega \omega_\varepsilon(x - y)\, |u(y)|^p\, dy \right)^{\frac{1}{p}},$$

and by Funini's theorem,

$$\|u_\varepsilon\|_p = \left(\int_\Omega |u_\varepsilon(x)|^p\, dx \right)^{\frac{1}{p}} \leqslant \int_\Omega \int_\Omega \omega_\varepsilon(x - y)\, |u(y)|^p\, dy\, dx \Big)^{\frac{1}{p}} =$$

$$= \left(\int_\Omega |u(y)|^p\, dy \int_\Omega \omega_\varepsilon(x - y)\, dx \right)^{\frac{1}{p}} = \|u\|_p. \tag{3}$$

By the density of $C_0(\Omega)$ in $L^p(\Omega)$, for each $\eta > 0$ there exists a $v \in C_0(\Omega)$ such that $\|u - v\|_p < \dfrac{\eta}{3}$; by (3) it follows that $\|u_\varepsilon - v_\varepsilon\|_p < \dfrac{\eta}{3}$. Hence, by the first part of the proof we have

$$\|u_\varepsilon - u\|_p \leqslant \|u_\varepsilon - v_\varepsilon\|_p + \|v_\varepsilon - v\|_p + \|u - v\|_p < \frac{\eta}{3} + \frac{\eta}{3} + \frac{\eta}{3} = \eta$$

which proves that $u_\varepsilon \to u$ in L^p-norm.

In particular, this proposition shows that $C_0^\infty(\Omega)$ *is dense in* $L^p(\Omega)$.

Generalized derivatives

2.2. The replacement of the usual derivative by its adjoint operation permits an extension of the definition of the derivative.

For any u in $L^1_{\text{loc}}(\Omega)$ we take the distribution

$$(u, \varphi) = \int_\Omega u(x) \cdot \varphi(x) \, dx \qquad (\forall) \; \varphi \in \mathscr{D}(\Omega)$$

by means of which we can define the derivatives in the distribution sense of such a function. Thus, we say that $v = D^\alpha u$ in the *weak sense* and call v the *α-weak derivative of* $u \in L^1_{\text{loc}}(\Omega)$ provided that $v \in L^1_{\text{loc}}(\Omega)$ and

$$\int_\Omega u(x) D^\alpha \varphi(x) \, dx = (-1)^{|\alpha|} \int_\Omega y(x) \varphi(x) \, dx \qquad (\forall) \; \varphi \in C_0^\infty(\Omega). \tag{1}$$

It is clear that the weak and usual derivatives are the same in the case of functions u from $C^{|\alpha|}(\Omega)$.

The uniqueness of the weak derivative follows from the above definition. In fact, if v_1 and v_2 are the α-th weak derivatives of the same function u, then

$$\int_\Omega [v_1(x) - v_2(x)] \, \varphi(x) \, dx = 0 \qquad (\forall) \; \varphi \in C_0^\infty(\Omega),$$

which implies that $v_1(x) = v_2(x)$ for almost every $x \in \Omega$.

We list the essential properties of weak derivatives:

(i) $D^\alpha u$ *does not depend on the order of derivation in the first term in* (1);

(ii) *If* $D^\alpha u = v$ *and* $D^\beta v = w$, *then* $D^{\alpha+\beta} u = w$;

(iii) *If* $v = D^\alpha u$, *then* $\mathscr{M}_\varepsilon v(x) = D^\alpha \mathscr{M}_\varepsilon u(x)$ *for all* $x \in \Omega$ *with* $\varepsilon < \text{dist}(x, \partial\Omega)$.

The first two properties are obvious. For proving (iii) it is enough to take ω instead of φ in (1). Then

$$D^\alpha \mathscr{M}_\varepsilon u(x) = \int_\Omega u(y) \, D^\alpha_x \omega_\varepsilon(x - y) \, dy = \tag{2}$$

$$= (-1)^{|\alpha|} \int_\Omega u(y) D^\alpha_y \omega_\varepsilon(x - y) dy = \int_\Omega \omega_\varepsilon(x - y) D^\alpha \, u(y) dy = \mathscr{M}_\varepsilon D^\alpha u(x),$$

which results from the properties the mollifiers.

Now, let us pass to another definition of the generalized derivatives. Let $u, v \in L^p_{\text{loc}}(\Omega), p \geqslant 1$. We say that $D^\alpha u = v$ in the *strong L^p sense* provided there exists a sequence of functions $\{\psi_n\} \subset C^{|\alpha|}(\Omega)$ such that

$$\int_A |\psi_n - u|^p \, dx \to 0, \quad \int_A |D^\alpha \psi_p - v|^p \, dx \to 0 \quad \text{as } n \to \infty$$

for any compact subset A in Ω.

The connection of both these definitions is given by the following

THEOREM *Let $u, v \in L^1_{\text{loc}}(\Omega)$. Then $v = D^\alpha u$ in the strong L^p sense if and only if v is the α-th weak derivative of u.*

Proof: Indeed, if $v = D^\alpha u$ in the strong L^p sense, then, passing to the limit in

$$\int_\Omega \psi_p(x) D^\alpha \varphi(x) \, dx = (-1)^{|\alpha|} \int_\Omega \varphi(x) D^\alpha \psi_n(x) \, dx \qquad (\forall) \, \varphi \in C_0^\infty(\Omega),$$

as $n \to \infty$, we obtain $D^\alpha u = v$ in the weak sense.

Conversely, suppose that $u(x) = 0$ outside of Ω. If $D^\alpha u = v$ in weak sense, then by virtue of (iii) and Proposition 2.1, it follows that a sequence $\{\psi_n\}$ with the properties stated in the definition of a strong derivative is for intance the sequence of mollifications $\{u_{\varepsilon_n}(x)\}$, where $\{\varepsilon_n\}$ is a sequence of positive numbers converging to zero.

Next, based on this theorem we use the term *generalized derivative* instead of weak derivative or strong derivative.

The above notion of distribution can be extended to functions from $[0, t_0]$ to a reflexive Banach space X, with norm $\| \cdot \|$, by setting

$$\mathscr{D}'(0, t_0; X) = \mathscr{L}(\mathscr{D}(0, t_0); X).$$

For any $f \in \mathscr{D}'(0, t_0; X)$ the derivative of f is the distribution

$$\frac{df}{dt}(\varphi) = -f\left(\frac{d\varphi}{dt}\right) \qquad (\forall)\varphi \in \mathscr{D}((0, t_0)). \tag{1}$$

If $f \in L^p(0, t_0; X)$, then one can define a distribution denoted again by f such that $f \in \mathscr{D}'(0, t_0; X)$ and

$$f(\varphi) = \int_0^{t_0} f(t)\, \varphi(t)\, dt, \qquad \varphi \in \mathscr{D}((0, t_0)).$$

Moreover, using (1) we can define $\dfrac{df}{dt}$ as an element of $\mathscr{D}'(0, {}_0t; X)$.

A function $u: [0, t_0] \mapsto X$ is said to be *absolutely continuous* if for each $\varepsilon > 0$ there exists $\delta > 0$ such that

$$\sum_{i=1}^{n} \|u(b_i) - u(a_i)\| < \varepsilon$$

whenever (a_i, b_i), $1 \leqslant i \leqslant n$, are non-overlapping subintervals of $[0, t_0]$ with

$$\sum_{i=1}^{n} |b_i - a_i| < \delta.$$

Each absolutely continuous function $u(t)$ may be expressed in the form (the Radom-Nikodym theorem)

$$u(t) = u(0) + \int_0^t \omega(s)\, ds, \qquad 0 < t \leqslant t_0,$$

where $\omega(t) = \dfrac{du(t)}{dt}$ for almost all $t \in [0, t_0]$ and $\omega \in L^1([0, t_0])$.

Let us show that *the generalized derivative of a single variable function $u(t)$ exists if and only if $u(t)$ is absolutely continuous*. In fact, let u be absolutely continuous on $[0, t_0]$. Then, for all $\varphi \in \mathscr{D}((0, t_0))$ we obtain that

$$\int_0^{t_0} u(t)\, \frac{d\varphi(t)}{dt}\, dt = -\int_0^{t_0} \omega(t)\, \varphi(t)\, dt.$$

Thus $\omega = \dfrac{du}{dt}$ is the generalized derivative of u.

Conversely, if u has the generalized derivative ω, then by definition

$$\int_0^{t_0} u(t)\, \frac{d\varphi(t)}{dt}\, dt = -\int_0^{t_0} \omega(t)\, \varphi(t)\, dt$$

for all smooth functions φ with compact support contained in $(\varepsilon, t_0 - \varepsilon)$, where $0 < \varepsilon < \dfrac{t_0}{2}$. Set

$$U(t) = \int_0^t \omega(s)\, ds.$$

Then

$$\int_0^{t_0} \omega(t)\, \varphi(t)\, dt = \int_0^{t_0} U(t)\, \frac{d\varphi}{dt}\, dt$$

and thus,

$$\int_0^{t_0} [u(t) - U(t)] \frac{d\varphi}{dt} \, dt = 0.$$

Since $u(t) - U(t) = $ constant, by the fundamental lemma of variational calculus, u is absolutely continuous.

2.3. For a non-negative integer m, we define the *Sobolev space*

$$W^{m,p}(\Omega) = \{u \mid D^\alpha u \in L^p(\Omega), \cdot | \alpha | \leqslant m\},$$

where $1 < p < \infty$ and the derivatives of u are taken in the generalized sense. In particular, $W^{0,p}(\Omega) \equiv L^p(\Omega)$.

PROPOSITION *The space $W^{m,p}(\Omega)$ is a separable uniformly convex Banach space with the norm*

$$\|u\|_{m,p} = \sum_{|\alpha| \leqslant m} \left(\int_\Omega |D^\alpha u(x)|^p \, dx \right)^{\frac{1}{p}}. \tag{1}$$

Proof: Let $\{u_n\} \subset W^{m,p}(\Omega)$ be a Cauchy sequence with respect to the norm (1). Then for each α with $|\alpha| \leqslant m$, the sequence $\{D^\alpha u_n\}$ is Cauchy with respect to the L^p-norm and thus there exists a $v_\alpha \in L^p(\Omega)$ such that $D^\alpha u_n \to v_\alpha$ in $L^p(\Omega)$. Since $L^p(\Omega) \subset \mathscr{D}'(\Omega)$ and the derivation is continuous in the distributional sense, $u_n \to v$ in $\mathscr{D}'(\Omega)$ and thus $D^\alpha u_n \to D^\alpha v$ in $\mathscr{D}'(\Omega)$. Hence, $D^\alpha v = v_\alpha$, $v \in W^{m,p}(\Omega)$ and $u_n \to v$ in $W^{m,p}(\Omega)$, i.e. $W^{m,p}(\Omega)$ is a Banach space. The uniform convexity and the separability of $W^{m,p}(\Omega)$ follows from the fact that this space can be identified with a closed subset of the cartesian product space $\prod_{|\alpha| \leqslant m} L^p(\Omega)$. We note also that this inclusion implies that $W^{m,p}(\Omega)$ *is a reflexive space*. Moreover $W^{m,2}(\Omega)$ is a Hilbert space with respect to the inner product

$$(u, v)_m = \sum_{|\alpha| \leqslant m} \int_\Omega D^\alpha u(x) \, D^\alpha v(x) \, dx, \qquad (\forall) u, v \in W^{m,2}(\Omega) \tag{2}$$

In general, the set $C_0^\infty(\Omega)$ *is not dense* in $W^{m,p}(\Omega)$. In fact one can imagine a simple example to show that $C_0^\infty(\Omega)$ is not dense in $W^{1,2}(\Omega)$ whenever Ω is bounded. To do this, let us find the orthogonal complement of $C_0^\infty(\Omega)$ in $W^{1,2}(\Omega)$ with respect to the inner product (2). We say that $u \in [C_0^\infty(\Omega)]^\perp$ if and only if $(u, \varphi)_1 = 0$ for all $\varphi \in C_0^\infty(\Omega)$, i.e.,

$$\int_\Omega u(x) \, \varphi(x) + \sum_{|\alpha| \leqslant 1} \int_\Omega D^\alpha u(x) \, D^\alpha \varphi(x) \, dx = 0. \tag{3}$$

By Green's formula applied to the second term,

$$\int_\Omega [-\triangle u(x) + u(x)] \; \varphi(x) \, dx = 0 \qquad (\forall) \; \varphi \in C_0^\infty(\Omega),$$

where $\triangle$ is the Laplace operator. Therefore,

$$u \in [C_0^\infty(\Omega)]^\perp \text{ if and only if } \begin{cases} u \in W^{1,2}(\Omega) \\ - \triangle u + u = 0. \end{cases}$$

We can easily verify that $u(x) = e^{\xi x}$ with $\xi \in \mathbb{R}^N$, $|\xi| = 1$ is a function of this kind.

Nevertheless, *if Ω is all of $\mathbb{R}^N$, then $C_0(\mathbb{R}^N)$ is dense in $W^{m,p}(\mathbb{R}^N)$.* Thus let $\varphi \in C_0^\infty(\Omega)$ be such that $\varphi(x) = 1$ for $|x| \leqslant 1$ and $\varphi(x) = 0$ whenever $|x| \geqslant 2$. Denote by

$$u_r(x) = u(x) \; \varphi \left(\frac{x}{r} \right)$$

the truncation at level r of $u \in W^{m,p}(\mathbb{R}^N)$. It is obvious that $u_r \to u$ in $W^{m,p}(\mathbb{R}^N)$ as $r \to \infty$. By Proposition 2.1 and Property (iii) in Section 2.2, we have $\mathscr{M}_\varepsilon u_r \to u_r$ in $W^{m,p}(\mathbb{R}^N)$ as $\varepsilon \to 0$. Since $\mathscr{M}_\varepsilon u_r \in C_0^\infty(\mathbb{R}^N)$, the assertion is proved.

The closure of $C_0^\infty(\Omega)$ in the norm (1) is denoted by $W_0^{m,p}(\Omega)$, which, in general, is a proper subspace of $W^{m,p}(\Omega)$.

For any integer $m \geqslant 0$ we have the chain of imbeddings

$$W_0^{m,p}(\Omega) \subseteq W^{m,p}(\Omega) \subseteq L^p(\Omega),$$

and by definition $\| \cdot \|_{0,p} \equiv \| \cdot \|_p$.

In order to give equivalent norms on $W^{m,p}(\Omega)$, we shall prove the following

THEOREM *Suppose that $\Omega \subseteq B(0, R)$ and $u \in W^{m,p}(\Omega)$. Then the Poincaré inequality*

$$\int_\Omega \sum_{|\alpha|=k} |D^\alpha u(x)|^p \, dx \leqslant N^{k-m} R^{(m-k)p} \int_\Omega \sum_{|\alpha|=m} |D^\alpha u(x)|^p \, dx, \; 0 \leqslant k \leqslant m, \tag{1}$$

holds.

Proof: We use an induction argument. Consider first $m = 1$ and $k = 0$ and prove that

$$\int_\Omega |u(x)|^p \, dx \leqslant \frac{R^p}{N} \int_\Omega \sum_{|\alpha|=1} \left| D^\alpha u(x) \right|^p dx. \tag{2}$$

By the definition of $W_0^{m,p}(\Omega)$ it is sufficient to prove this inequality for all u in $C_0^\infty(\Omega)$. Introduce the polar coordinates with pole at the origin of $\mathbb{R}^N$ by writing $x = r\zeta$, where ζ is a unit vector.

Since

$$u(R\zeta) - u(r\zeta) = u'(t\zeta)\ (R - r)\ \zeta, \quad r < t < R,$$

we have

$$|u(x)|^p = |u(R\zeta) - u(r\zeta)|^p \leqslant (R - r)^{p-1} \int_r^R |u'(t\zeta)|^p\ dt.$$

Since

$$\int_{B(0,\ R)} |u(x)|^p\ dx = \int_0^R r^{N-1} \int_S |u(r\zeta)|^p\ dr d\zeta,$$

where S is the unit sphere in $\mathbb{R}^N$, it follows that

$$\int_\Omega |u(x)|^p\ dx = \int_{B(0,\ r)} |u(x)|^p\ dx =$$

$$= \int_S \left\{ \int_0^R dr \int_r^R r^{N-1} (R - r)^{p-1} |u'(t\zeta)|^p\ dt \right\} d\zeta.$$

Now, by the Dirichlet formula we obtain

$$\int_{\cdot r}^R dr \int_r^R r^{N-1} (R - r)^{p-1} |u'(t\zeta)|^p\ dt =$$

$$= \int_{\cdot r}^R |u'(t\zeta)|^p\ dt \int_0^t \sum_{k=1}^{p-1} (-1)^k C_{p-1}^k R^{p-1-k} r^{k+N-1} dr =$$

$$= \int_{\cdot 0}^R |u'(t\zeta)|^p\ t^{N-1} \sum_{k=0}^{p-1} (-1)^k \frac{C_{p-1}^k}{k+N} R^{p-1-k} t^{k+1} dt \leqslant$$

$$\leqslant \frac{1}{N} \int_0^R t^N |u'(t\zeta)|^p\ (R - t)^{p-1}\ dt.$$

We estimate the last integral by

$$\int_{\cdot 0}^R t^{N-1} |u'(t\zeta)|^p (R - t)^{p-1} t dt \leqslant R^p \int_0^R t^{N-1} |u'(t\zeta)|^p dt.$$

Then,

$$\int_{\cdot \Omega} |u(x)|^p\ dx \leqslant \frac{R^p}{N} \int_0^R t^{N-1} \int_S |u'(t\zeta)|^p\ d\sigma = \frac{R^p}{N} \int_\Omega \sum_{|\alpha|=1} \left| D^\alpha u(x) \right| dx.$$

Applying m-times successively (2), we get

$$\int_\Omega |u(x)|^p\ dx \leqslant \frac{R^{mp}}{N^m} \int_\Omega \sum_{|\alpha|=m} \left| D^\alpha u(x) \right| dx.$$

This implies easily (1).

COROLLARY *If Ω is bounded, then the norm*

$$\|u\|_{m,\,p}^{0} = \left(\int_{\Omega} \sum_{|\alpha|=m} \left| D^{\alpha}u(x) \right|^{p} dx \right)^{\frac{1}{p}}$$

is topologically equivalent to the norm $\|u\|_{m,\,p}$ on $W_{0}^{m,\,p}(\Omega)$. If $p = 2$, then the inner product may be taken on $W_{0}^{m,\,2}(\Omega)$ as

$$(u, v) = \int_{\Omega} \sum_{|\alpha|=m} D^{\alpha}u(x)\, D^{\alpha}v(x)\, dx.$$

2.4. Denote the dual space of $W_{0}^{m,\,p}(\Omega)$ by $W^{-m,\,p'}(\Omega)$ where $\dfrac{1}{p} + \dfrac{1}{p'} = 1$.

Then

$$\mathscr{D}(\Omega) \subset W^{-m,\,p'}(\Omega) \subset \mathscr{D}'(\Omega).$$

THEOREM *If $p > 1$, then any functional $f \in W^{-m,\,p'}(\Omega)$ can be represented as*

$$f = \sum_{|\alpha|\leqslant m} D^{\alpha}f_{\alpha} \quad with \quad f_{\alpha} \in L^{p'}(\Omega).$$

This representation is not unique.

Proof: Since $W_{0}^{m,\,p}(\Omega) \subset \prod_{|\alpha|\leqslant m} L^{p}(\Omega)$, by the Hahn-Banach theorem we can extend any $f \in W^{-m,\,p'}$ to $\prod_{|\alpha|\leqslant m} L^{p}(\Omega)$. By virtue of the Riesz-Fréchet theorem, the dual space of $\prod_{|\alpha|\leqslant m} L^{p}(\Omega)$ is $\prod_{|\alpha|\leqslant m} L^{p'}(\Omega)$ and hence there exist $z_{\alpha} \in L^{p'}(\Omega)$ such that

$$(f, u) = \sum_{|\alpha|\leqslant m} \int_{\Omega} z_{\alpha}(x)\, u_{\alpha}(x)\, dx \quad (\forall)\, u \in \prod_{|\alpha|\leqslant m} L^{p}(\Omega).$$

For any $v \in W^{m,\,p}(\Omega)$,

$$(f, v) = \sum_{|\alpha|\leqslant m} \int_{\Omega} z_{\alpha}(x)\, D^{\alpha}v(x)\, dx.$$

In particular, if $v \in C_{0}^{\infty}(\Omega)$, then

$$(f, v) = \sum_{|\alpha|\leqslant m} (-1)^{\alpha}\, (D^{\alpha}z_{\alpha}, v).$$

The representation asserted by the theorem is valid for $f_{\alpha} = (-1)^{|\alpha|} z_{\alpha}$, which completes the proof.

COROLLARY *Let* $u \in W^{1,2}(\Omega)$ *and let* $\triangle = \sum\limits_{i=1}^{n} \dfrac{\partial^2}{\partial x_i^2}$ *be the Laplace operator.*
Then $\triangle u \in W^{-1,2}(\Omega)$ *and the Green formula*

$$(-\triangle u, v) = \sum_{i=1}^{n} \int_{\Omega} \frac{\partial u(x)}{\partial u_i} \frac{\partial v(x)}{\partial x_i} \, dx \quad (\forall) \, u, v \in W_0^{1,2}(\Omega)$$

is valid.

Proof: Since $\triangle u = \sum\limits_{i=1}^{n} \dfrac{\partial}{\partial x_i} \left(\dfrac{\partial u}{\partial x_i} \right)$ with $\dfrac{\partial u}{\partial x_i} \in L^2(\Omega)$, it follows that $\triangle u$ can be represented as in the theorem. Hence $\triangle u \in W^{-1,2}(\Omega)$ and

$$(\triangle u, \varphi) = -\left(\sum_{i=1}^{n} \frac{\partial u}{\partial x_i}, \frac{\partial \varphi}{\partial x_i} \right) = -\sum_{i=1}^{n} \int_{\Omega} \frac{\partial u}{\partial x_i} \frac{\partial \varphi}{\partial x_i} \, dx \quad (\forall) \, \varphi \in C_0^{\infty}(\Omega).$$

The corollary follows now by the density of $C_0^{\infty}(\Omega)$ in $W_0^{1,2}(\Omega)$.

2.5. Let us show that the elements of Sobolev spaces are obtained as limits of sequences of smooth function. Consider the vector space

$$C^{m,p}(\Omega) = \{ \varphi \in C^m(\Omega) \mid \|\varphi\|_{m,p} < \infty \}$$

and let $H^{m,p}(\Omega)$ be the completion of $C^{m,p}(\Omega)$ in the norm $\|\cdot\|_{m,p}$. As the generalized derivatives coincide with the classical ones when the latter exist and are continuous, it is clear that $C^{m,p}(\Omega)$ is contained in $W^{m,p}(\Omega)$. By the extension of this inclusion to the completion of $C^{m,p}$ in the norm $\|\cdot\|_{m,p}$ we obtain that $H^{m,p}(\Omega) \subseteq W^{m,p}(\Omega)$.

THEOREM *If* $1 < p < \infty$ *and* $m \in \mathbb{N}$, *then* $H^{m,p}(\Omega) = W^{m,p}(\Omega)$.

Proof: By the above inclusion it is sufficient to prove that $H^{m,p}(\Omega) \supseteq W^{m,p}(\Omega)$, i.e., $C^{m,p}(\Omega)$ is dense in $W^{m,p}(\Omega)$. Thus let $w \in W^{m,p}(\Omega)$ and $\varepsilon > 0$. We shall show that there exists a $u \in C^{\infty}(\Omega)$ such that $\|w - u\|_{m,p} < \varepsilon$.

For each natural number n set

$$U_n = \left\{ x \in \Omega \,\middle|\, |x| < n \text{ and } \operatorname{dist}(x, \partial\Omega) > \frac{1}{n} \right\}$$

and adopt the convention that $U_0 = U_{-1} = \emptyset$. Then the sets $\{\Omega_n\}$, where $\Omega_n = U_{n+1} - \overline{U}_{n-1}$, form an open covering of Ω. Let $\{\beta_n\}$ be a C^m partition of unity subordinated to this covering. As $\beta_n w \in W^{m,p}$ and supp $\beta_n w \subset \Omega_n$, we may choose a sequence of positive numbers $\{r_n\}$ such that

$$\operatorname{supp} \mathscr{M}_{r_n}(\beta_n w) \subset \Omega_{n-1} \cup \Omega_n \cup \Omega_{n+1}$$

and

$$\| \mathscr{M}_{r_n}(\beta_n w) - \beta_n w \|_{m,\,p} < \frac{\varepsilon}{3 \cdot 2^n}.$$

Here $\mathscr{M}_{r_n}$ is the molliffication defined in Section 2.1.

Let $u = \sum\limits_{n=1}^{\infty} \mathscr{M}_{r_n}(\beta_n w)$. For each $x \in \Omega_n$ we have $u(x) = \sum\limits_{j=-1}^{1} M_{r_{n+j}}(\beta_{n+j} w)$ and hence $u \in C^{\infty}(\Omega)$. Thus,

$$\| w - u \|_{m,\,p} \leqslant \sum_{n=1}^{\infty} \sum_{j=-1}^{1} \| \mathscr{M}_{r_{n+j}}(\beta_{n+j} w) - \beta_n w \|_{m,\,p} < \varepsilon,$$

as claimed.

COROLLARY *Let $\{\Omega_j'\}$ be a finite covering of a bounded domain Ω in $\mathbb{R}^N$ and set $\Omega_j = \Omega \cap \Omega_j'$. If a function u belongs to $W^{m,\,p}(\Omega_j)$ for every j, then $u \in W^{m,\,p}(\Omega)$.*

Proof: Let $\{\beta_j\}$ be the partition of unity subordinated to this covering. For a $u \in W^{m,\,p}(\Omega_j) \equiv H^{m,\,p}(\Omega_j)$ there exists a sequence $\{u_{n,\,j}\} \subset C^{m,\,p}(\Omega)_j$ such that $u_{n,\,j} \to u$ in $W^{m,\,p}(\Omega_j)$. Denote

$$u_n = \sum_j \beta_j u_{n,\,v}.$$

Then $\{u_n\} \subset C^{m,\,p}(\Omega)$ and $\| u_n - u \|_{m,\,p} \to 0$.

We remark that if for each $x \in \Omega$ there exists a neighbourhood in which u has α-th weak (strong) derivatives, then u has α-th (strong) derivatives in Ω.

Imbedding theorems

2.6. The study of Sobolev spaces is useful since the elements of these spaces possess special properties. These properties follow easily by the imbedding theorems.

We proceed by stating a negative property: *The functions of $W^{1,\,2}(\Omega)$ with $\Omega \subset \mathbb{R}^N$, $N \geqslant 2$, are generally non-continuous.* For instance, if $N = 2$ and Ω is the unit ball with 0 as its centre, then we can easily check that

$$u(x, y) = \left(\ln \frac{1}{r} \right)^k, \quad r = \sqrt{x^2 + y^2}, \quad k < \frac{1}{2},$$

belongs to $W^{1,\,2}(\Omega)$ and that it has a singularity at the origin.

5 — c. 1529

On the other hand, for $N = 1$ and $\Omega = (a, b)$, after an possibly a correction on a null-measure set, any function $u \in W^{1, 2}(\Omega)$ is absolutely continuous as we have stated at the end of Section 2.2.

We say that the Banach space X is *imbedded* into the Banach space Y provided that X is a subspace of Y and there exists a constant $K > 0$ such that

$$\|u\|_Y \leqslant K \|u\|_X \text{ for all } u \in X.$$

The space X is *compactly imbedded* in Y if the imbedding operator $I : X \to Y$ is compact.

Next let us discuss the representation (1) in Section 1.3. It remains also valid for the functions in $W^{1, p}(\Omega)$, considered as limits of sequences in $C^{1, p}(\Omega)$ (because of the continuity of the integral operators used in this representation).

LEMMA *Let Ω be a bounded convex domain in $\mathbb{R}^N$.*

(a) *If $p > N$, then $W^{1, p}(\Omega)$ is compactly imbedded in $C(\bar{\Omega})$ and*

$$\|u\| \leqslant K_1 \|u\|_{1, p} \text{ for all } u \in W^{1, p}(\Omega),$$

where $\| \cdot \|$ denotes the norm in $C(\bar{\Omega})$.

(b) *If $p \leqslant N$, then $W^{1, p}(\Omega)$ is compactly imbedded in $L^q(\Omega')$ and*

$$\|u\|_q \leqslant K_2 \|u\|_{1, p} \quad \text{for all} \quad u \in W^{1, p}(\Omega),$$

where

$$q < q^* = \frac{N'p}{N - p}, \quad N' > N - p$$

and Ω' is the N'-dimensional plane section of Ω.

Both constants K_1 and K_2 depend only on Ω.

Proof: (a) Set $\lambda = N - 1$ in Theorem 1.3. Then $N - 1 < \dfrac{N}{p'}$ implies that $p > N$ and thus $P_{N-1} : W^{1, p}(\Omega) \to C(\bar{\Omega})$ is compact. Moreover

$$\|u\| \leqslant K_1 \|u\|_{1, p},$$

where $K_1 = \max (1, \|P_{N-1}\|)$.

(b) Taking $\lambda = N - 1$ in Proposition 1.3, we obtain the assertion in a similar way. The lemma is proved.

Let us give a more accurate description of the above imbeddings. Strictly speaking, the elements of $W^{1, p}(\Omega)$ are equivalence classes of functions defined and equal up to the sets of null-measure. The imbedding $W^{1, p}(\Omega) \hookrightarrow C(\bar{\Omega})$ implies the existence of continuous functions in each equivalence class of

$W^{1,p}(\Omega)$. Thus each element of $W^{1,p}(\Omega)$ belongs to $C(\bar{\Omega})$, possibly after an alteration on a set of null-measure.

THEOREM *(Sobolev-Kondrashov) Let Ω be a bounded convex domain in $\mathbb{R}^N$.*

(a) *If $mp > N$, then $W^{m,p}(\Omega)$ is compactly imbedded in $C(\Omega)$ and*

$$\|u\| \leqslant C_1 \|u\|_{m,p} \text{ for all } u \in W^{m,p}(\Omega).$$

(b) *If $mp \leqslant N$, then $W^{m,p}(\Omega)$ is compactly imbedded in $L^q(\Omega')$ where*

$$q < q^* = \frac{N'p}{N - mp} \text{ and } N' \geqslant N - mp,$$

i.e., if $u \in W^{m,p}(\Omega)$, then $u \in L^q(\Omega')$ and

$$\|u\|_q \leqslant C_2 \|u\|_{m,p}.$$

The two positive constants C_1 and C_2 depend only on Ω.

Proof: We use an induction argument. The lemma proves the assertion for $m = 1$. Suppose that the conclusions remain valid for $m - 1$.

(a) If $mp > N$, then either $(m - 1)p > N$ or $(m - 1)p \leqslant N$. For $(m - 1)p > N$ it follows that $W^{m-1,p}(\Omega)$ is compactly imbedded in $C(\bar{\Omega})$. Then for $u \in W^{m,p}(\Omega)$ all $D_j u \in C(\bar{\Omega})$, hence $u \in C(\bar{\Omega})$ and

$$\|u\| \leqslant C_1 \|u\|_{m-1,p} \leqslant C_1 \|u\|_{m,p}.$$

If $(m - 1)p \leqslant N$, then $W^{m-1,p}(\Omega)$ is compactly imbedded in $L^{q_1}(\Omega)$ with $q_1 < \dfrac{Np}{N - (m - 1)p}$. Hence $D_j u \in L^{q_1}(\Omega)$ and

$$\|D_j u\|_{q_1} \leqslant C_2 \|D_j u\|_{m-1,p} \leqslant C_2 \|u\|_{m,p}.$$

Moreover $\|u\|_{q_1} \leqslant C_2 \|u\|_{m-1,p} \leqslant C_2 \|u\|_{m,p}$ which implies that $u \in W^{1,q_1}(\Omega)$ and

$$\|u\|_{1,q_1} \leqslant C_2 (1 + N) \|u\|_{m,p}.$$

We may apply the first part of the lemma because $q_1 > N$. Indeed, $\dfrac{Np}{N - (m - 1)p} > N$ is equivalent to $mp > N$. Therefore $u \in C(\bar{\Omega})$ and

$$\|u\| \leqslant K_1 \|u\|_{1,q_1} \leqslant C_1 \|u\|_{m,p},$$

where $C_1 = K_1 C_2 (1 + N)$. This proves (a).

(b) If $mp \leqslant N$, then $(m - 1)p < N$ and by induction $W^{m-1,p}(\Omega)$ is compactly imbedded in $L^{q_2}(\Omega')$, where

$$q_2 < \frac{N'_2 p}{N - (m - 1)p} \text{ and } N'_2 \geqslant N - (m - 1)p.$$

As above, $u \in W^{1, q_2}(\Omega')$ and $\|u\|_{1, q_2} \leqslant C_2 (1 + N) \|u\|_{m, p}$. Observe that

$$p \leqslant q_2 \leqslant \frac{Np}{N - (m - 1) p} .$$

Next, applying the lemma for $W^{1, q_2}(\Omega)$, which is possible because $q_2 \leqslant N$ is equivalent to $mp \geqslant N$, we obtain $u \in L^q(\Omega')$, and

$$\|u\|_q \leqslant K_2 \|u\|_{1, q_2} \leqslant K_2 C_2 (1 + N) \|u\|_{m, p}.$$

Here,

$$q < \frac{N' q_2}{N - q_2} < \frac{N' \dfrac{Np}{N - (m - 1) p}}{N - \dfrac{Np}{N - (m - 1) p}} = \frac{N'p}{N - mp}$$

and $N' > N'_2 - q_2 \geqslant N - (m - 1)p - p = N - mp$.

The compactness of the imbedding $W^{m, p}(\Omega)$ in $L^q(\Omega')$ follows from the compact imbedding of $W^{1, q_2}(\Omega)$ in $L^q(\Omega')$.

We note that the theorem remains true for $q = q^*$ but in this case, as it was mentioned at the end of Theorem 1.2, the imbedding is not compact.

Similarly, if we assume $u \in W^{m, p}(\Omega)$ and $|\alpha| = j$, then we can prove the

COROLLARY (a) *If* $mp > N$ *and* $0 \leqslant j \leqslant m - \dfrac{N}{p}$, *then*

$$W^{m, p}(\Omega) \hookrightarrow C^{m - \left[\frac{N}{p}\right] - 1}(\bar{\Omega}),$$

where $\left[\dfrac{N}{p}\right]$ *is the integral part of* $\dfrac{N}{p}$, *i.e., any derivative* $D^\alpha u$ *is equivalent to a continuous function and*

$$\|D^\alpha u\| \leqslant K \|u\|_{m, p}. \tag{1}$$

(b) *If* $j > 0$ *and* $j \geqslant m - \dfrac{N}{p}$, *then* $D^\alpha u$ *is equivalent to a function belonging to* $L^{q_j}(\Omega')$, *where*

$$q_j < \frac{N'p}{N - (m - j) p}, \quad N' > N - (m - j) p \tag{2}$$

and

$$\|D^\alpha u\|_{q_j} \leqslant K \|u\|_{m, p}.$$

The imbeddings are compact.

The condition $j \geqslant m - \dfrac{N}{p}$ is clearly redundant in (b). Solving the first inequality in (2) with respect to j we get

$$0 < j \leqslant m - \frac{N}{p} - \frac{N'}{q_j} = m - \frac{N - N'}{p} - \left(\frac{1}{p} - \frac{1}{q_j} \right) N'. \tag{3}$$

In this case $W^{m, p}(\Omega)$ is continuously imbedded in $W^{j, q_j}(\Omega')$ and

$$\|u\|_{j, q_j} \leqslant K \|u\|_{m, p}.$$

For $p = 2$, we denote the Hilbertean case of a Sobolev space $W^{m, 2}(\Omega)$ by $H^m(\Omega)$.

It is obvious by (2) that the imbedding $I: H^m(\Omega) \mapsto H^{m-1}(\Omega)$ is compact. As we can easily see in (3), if $N' = N - 1$, then $H^m(\Omega)$ is continuously imbedded in $H^j(\Omega')$ for all j with $0 \leqslant j \leqslant m - \dfrac{1}{2}$.

The Sobolev imbedding theorem remains valid for more general bounded domains, such as the so-called domains with the cone property. Given a point $x \in \mathbb{R}^N$, an open ball B_1 with center x, and an open ball B_2 not containing x, the set $C_x = \{x + t\,(y - x) \mid y \in B_2, t > 0\}$ is called a *finite cone* in $\mathbb{R}^N$ with the vertex at x. The domain Ω has the *cone property* if there exists a finite cone C such that each point $x \in \Omega$ is the vertex of a finite cone $C_x \subset \Omega$ obtained from C by a rigid motion (see Adams [1, p. 97]).

We shall use the following special case of the lemma:

PROPOSITION *Suppose that* $u \in W_0^{1, p}(\mathbb{R}^N)$, *where* $p > N$. *Then the inequality*

$$|u(x) - u(y)| \leqslant C\,|x - y|^{1 - \frac{N}{p}} \sum_{|\alpha| = 1} \|D^\alpha u\|_p$$

holds for any $x, y \in \mathbb{R}^N$.

Proof: There exists an $R > 0$ such that $K = \operatorname{supp} u \subseteq B(0, R)$. It is sufficient to prove the inequality for $u \in C_0^1(\mathbb{R}^N)$. Let x and y be two points in $\mathbb{R}^N$ and put $\rho = |x - y|$. Put $B_1 = B(x, \rho)$, $B_2 = B(y, \rho)$ and $A = B_1 \cap B_2$. Denote the Lebesgue measure of A by $\mu(A)$. Then

$$|u(x) - u(y)|\,\mu(A) = \int_A |u(x) - u(y)|\,dz \leqslant$$

$$\leqslant \int_A |u(x) - u(z)|\,dz + \int_A |u(z) - u(y)|\,dz. \tag{4}$$

Introduce the polar coordonates with pole x by writing $z = x + r\zeta$ where $r = |z - x|$ and $\zeta \in S = \partial B(x,1)$. The first integral on the right-hand of (4) is bounded by

$$\int_{B_1}\left(\int_0^r |u'(x + t\zeta)|\, dt\right) dz = \int_S \int_0^\rho r^{N-1}\int_0^r |u'(x + t\zeta)|\, dt\, dr\, d\zeta =$$

$$= \int_S \int_0^\rho |u'(x + t\zeta)|\int_t^\rho r^{N-1}\, dr\, dt\, d\zeta = \int_S d\zeta \int_0^\rho |u'(x + t\zeta)|\frac{\rho^N - t^N}{N}\, dt \leqslant$$

$$\leqslant \rho^N \int_{B_1} \frac{|u'(z)|}{r^{N-1}}\, dz.$$

Now, since $p' < \dfrac{N}{N-1}$, we have

$$\int_{B_1} \frac{dz}{r^{N-1}\,p'} \leqslant \int_S d\zeta \int_0^\rho r^{(N-1)(1-p')} dr = c\rho^{N-(N-1)p'}$$

and by Hölder's inequality,

$$\int_{B_2} |u(x) - u(z)|\, dz \leqslant c\rho^{1 + \frac{N}{p'}}\left[\int_{B_1} |u'(z)|^p\, dz\right]^{\frac{1}{p}}.$$

A similar estimate can be found for the second integral on the right side of (4). Since $\mu(A)$ is proportional to ρ^N,

$$|u(x) - u(y)| \leqslant c\rho^{\frac{N}{p'} - N + 1}\sum_{\alpha=1} \|D^\alpha u\|_p = c|x - y|^{1 - \frac{N}{p}}\sum_{|\alpha|=1} \|D^\alpha u\|_p.$$

In other words, if $p > N$, then $W_0^{1,p}(\Omega)$ is imbedded in $C^{1 - \frac{N}{p}}(\Omega)$, i.e. in the space of all Lipschitz continuous functions of exponent $1 - \dfrac{N}{p}$.

Generalities about the trace

2.7. If Ω is all of $\mathbb{R}^N$, then by using the Fourier transform, we can extend the definition of the Sobolev space to real values of m.

The Fourier transform of a function $u \in L^1(\mathbb{R}^N)$ is defined by

$$\hat{u}(\xi) = Fu(\xi) = \frac{1}{(2\pi)^{\frac{N}{2}}}\int_{\mathbb{R}^N} e^{-ix\xi}u(x)\, dx,$$

where $x\xi = x_1\xi_1 + \ldots + x_N\xi_N$. Consider the set

$$S = \{u \in C^\infty(\mathbb{R}^N) \mid \max_{x \in \mathbb{R}^N} |x^\alpha D^\beta u(x)| < \infty\},$$

for all multi-indices α and β. We can easily check that $F: S \mapsto S$ has the following properties

$$F(D^\alpha u) = (i\xi)^\alpha Fu,$$

$$D^\beta Fu = F((-ix)^\beta u)$$

$$u(x) = \overline{F}\hat{u}(\xi) = \frac{1}{(2\pi)^{\frac{N}{p}}} \int_{\mathbb{R}^N} e^{ix\xi}\hat{u}(\xi)\,d\xi$$

for $u \in S$, where α and β are arbitrary multi-indices and $\overline{F}$ denotes the inverse Fourier transform.

Moreover, we can prove that *if $u \in L^p(\mathbb{R}^N)$ for $1 \leqslant p \leqslant 2$, then*

$$\int_{|x|<r} e^{-i\xi x}u(x)\,dx$$

converges in $L^{p'}(\mathbb{R}^N)$ as $r \to \infty$. If Fu denotes the limit, then $\|Fu\|_{p'} \leqslant \|u\|_p$ *and* $(Fu, v) = (u, \overline{F}v)$ *for all $v \in S$. In particular,*

$$(Fu, Fv) = (u, v) \text{ and } \|Fu\|_2 = \|u\|_2 \text{ for all } u, v \in L^2(\mathbb{R}^N).$$

For any $u \in S$ and any two real numbers, s and $1 \leqslant p \leqslant \infty$, we define a norm in S as follows:

$$\|u\|_{W^{s,p}(\mathbb{R}^N)} = \|\overline{F}(1 + |\xi^2|)^{\frac{s}{2}} Fu\|_p.$$

Denote by $W^{s,p}(\mathbb{R}^N)$, or simply by $W^{s,p}$, the completion of S with respect to this norm. If s is a natural number then $W^{s,p}$ coincides with the Sobolev space defined in Section 2.3.

In order to introduce Sobolev spaces on the boundary of a bounded domain, we must view the boundary as a manifold.

A compact set $\overline{U}$ in $\mathbb{R}^N$ is said to be an *N-dimensional manifold of class C^m* with the boundary ∂U if there exists a finite family $\{U_j \mid 1 \leqslant j \leqslant n\}$ of open bounded sets with the following properties:

(i) $\overline{U} \subseteq \bigcup \{U_j \mid 1 \leqslant j \leqslant n\}$

(ii) Consider the cylinder

$$K_N = \{y \in \mathbb{R}^N \mid y = (y', y_N), \ y' \in \mathbb{R}^{N-1}, \ |y'| < 1, -1 < y_N < 1\}.$$

To each $j, 1 \leqslant j \leqslant n$, there corresponds a C^m-diffeomorphism $\varphi_j \colon \bar{U} \cap U_j \mapsto K_N$ such that

$$\varphi_j(U \cap U_j) = K_N^+ \equiv \{y \in K_N \mid y_N > 0\},$$

$$\varphi_j(\partial U \cap U_j) = K_N^0 \equiv \{y \in K_N \mid y_N = 0\},$$

$$\varphi_j(\complement \bar{U} \cap U_j) = K_N^- \equiv \{y \in K_N \mid y_N < 0\}.$$

(iii) If $U_i \cap U_j \neq \emptyset$ for some $1 \leqslant i, j \leqslant n$, then there exists a C^m-homeomorphism Φ_{ij} from $\varphi_i(U_i \cap U_j)$ onto $\varphi_j(U_i \cap U_j)$ with positive Jacobian such that

$$\varphi_j(x) = \Phi_{ij}(\varphi_i(x)) \text{ for all } x \in U_i \cap U_j \cap \bar{U}.$$

Any set $\{\varphi_j\}$ of such functions is called a *local chart* or *coordinate system* on $\bar{U}$.

The set ∂U of all points $x \in \bar{U}$ with the property that $\varphi_j(x) \in K_N^0$ can now be viewed as an $(N-1)$-dimensional manifold of class C^m, where the restriction of φ_j to the sets $\partial U_j = \partial U \cap U_j$ are taken as local charts.

Let Ω be a bounded domain in $\mathbb{R}^N$ whose boundary $\Gamma = \partial\Omega$ is an $(N-1)$-dimensional manifold of class C^m and let $\{\varphi_j\}$ be a family of local charts on Γ. Denote by $\{\beta_j\}$ a partition of unity on Γ such that $\beta_j \in C_0^m(\Gamma)$, $\operatorname{supp}\beta_j \subset \Omega_j \cap \Gamma$ for all $1 \leqslant j \leqslant n$ and $\sum\limits_{j=1}^{n} \beta_j = 1$ on Γ. Here $\bigcup \{\Omega_j \mid 1 \leqslant j \leqslant n\}$ is a finite covering of Γ and for any $u \in L^p(\Gamma)$ we can put $u = \sum\limits_{j=1}^{n} \beta_j u$. Denote $u_j = \varphi_j(\beta_j u)$, $1 \leqslant j \leqslant n$. For $s \geqslant 0$ and $1 < p < \infty$, we say that $u \in W^{s,p}(\Gamma)$ provided that $u_j \in W^{s,p}(\mathbb{R}^{N-1})$ for all j, $1 \leqslant j \leqslant n$.

It can be shown that $W^{s,p}(\Gamma)$ does not depend on the chosen family of local charts and the subordinated partition of unity.

The space $W^{s,p}(\Gamma)$ is a Banach space with respect to the norm

$$\|u\|_{s,p,\Gamma}^p = \sum_{j=1}^{n} \|u_j\|_{W^{s,p}(\mathbb{R}^{N-1})}^p.$$

It can also be checked that $C^\infty(\Gamma)$ is a dense subset of $W^{s,p}(\Gamma)$.

Now, for $u \in C_0^m(\mathbb{R}^N)$, we can define the *outward normal derivative of order* j *to* Γ

$$\gamma_j = \frac{\partial^j u}{\partial n_j}, \qquad 0 \leqslant j \leqslant m - 1.$$

The restrictions to Ω of functions of $C_0^m(\mathbb{R}^N)$ form a dense subset of $W^{m,p}(\Omega)$. They define the *trace* map

$$u \mapsto \gamma u = (\gamma_0 u, \ldots, \gamma_{m-1} u).$$

This mapping is linear and it can be extended by continuity to all u in $W^{m,p}(\Omega)$. More precisely, we have the following trace result extended (see Adams [1, p. 216]).

THEOREM *Let Ω be a bounded domain in $\mathbb{R}^N$ such that the boundary $\Gamma = \partial\Omega$ is an $(N-1)$-dimensional manifold of class C^m, and suppose the domain is located on one side of Γ. If $1 < p < \infty$, then the map $u \mapsto \gamma u$ from $C_0^\infty(\mathbb{R}^N)$ into $[C_0^\infty(\Gamma)]^m$ can be extended to a linear continuous operator γ from $W^{m,p}(\Omega)$ into*

$$\Pi \ \{W^{m-j-\frac{1}{2},p}(\Gamma) \mid 1 \leqslant j \leqslant m-1\}.$$

As a consequence of this theorem one can deduce that $W_0^{m,p}(\Omega)$ coincides with the kernel of γ, i.e.,

$$\ker \ \gamma = \left\{ u \in W^{m,p}(\Omega) \ \middle| \ \frac{\partial^j u}{\partial n_j} = 0, \ 1 \leqslant j \leqslant m-1 \right\}.$$

Further details on this subject can be found in Lions and Magenes [1, Chap. 1] or Hörmander [1, Chap. 1].

Imbeddings in the case of unbounded domains

2.8. Many mathematical models of hydrodynamics lead to boundary-value problems on unbounded domains. A functional approach to these problems requires some extensions of the Sobolev imbedding theorem to general domains which will be used in Section IV.3.2.

We denote

$$\nabla u = \left(\frac{\partial u}{\partial x_1}, \ldots, \frac{\partial u}{\partial x_N} \right) \quad \text{and} \quad |\nabla u| = \left(\sum_{j=1}^{N} \left| \frac{\partial u}{\partial x_j} \right|^2 \right)^{\frac{1}{2}}$$

for the sake of simplicity.

For an unbounded domain $\Omega \subseteq \mathbb{R}^N$ the imbedding of $W_0^{1,p}(\Omega)$ into $L^q(\Omega)$ is based on a particular estimate which is similar to the representation in 1.3.

LEMMA *Suppose that $u \in C_0^1(\Omega)$. Then*

$$|u(x)| \leqslant \frac{1}{|S|d} \int\limits_{\Omega \cap B(x,d)} [|u(y)| + d|\nabla u(y)|] |x - y|^{1-N} \, dy, \tag{1}$$

where $B(x,d)$ is the ball centred at $x \in \Omega$ with the radius d and $|S|$ is the measure of the surface of the unit sphere $S \subset \mathbb{R}^N$.

Proof: Let $h \in C^1([0,\infty))$ be such that $0 \leqslant h \leqslant 1$. For a $t \in \left(0, \dfrac{1}{3}\right)$, take

$$h(r) = \begin{cases} 1 & \text{if} \quad 0 \leqslant r \leqslant td, \\ 0 & \text{if} \quad (1-t)\,d \leqslant r \leqslant d, \end{cases}$$

with

$$|h'(r)| \leqslant \frac{1}{(1-3t)\,d}, \quad r \in [0, \infty),$$

Such a function can be obtained as a mollification of a polygonal function, which is possible for small d.

Using polar coordinates centred at x, we can write any point y in $B(x, d)$ as $y = x + r\zeta$, where $\zeta \in S$, i.e., $|\zeta| = 1$ and $0 \leqslant r \leqslant d$. Extend u to all of $\mathbb{R}^N$ with zero outside Ω. For $y \in B(x, d)$ we have $u(y) = u(x + r\zeta) = \varphi(r\zeta)$ and

$$\varphi(s\zeta) = - \int_d^s \frac{\partial}{\partial r} [\varphi(r\zeta)\, h(r)]\, dr,$$

where $0 \leqslant s \leqslant td$. Then

$$\left| \int_S \varphi(s, \zeta)\, d\zeta \right| \leqslant \int_S \int_s^d \left| \frac{\partial}{\partial r} [\varphi(r, \zeta)\, h(r)] \right| dr\, d\zeta \leqslant$$

$$\leqslant \int_S \int_0^d \left[\left| \frac{\partial \varphi(r, \zeta)}{\partial r} \right| + \frac{|\varphi(r, \zeta)|}{d(1 - 3t)} \right] dr\, d\zeta =\!=$$

$$= \int_{\Omega \cap B(x, d)} \left[|\zeta \nabla u(y)| + \frac{|u(y)|}{d(1 - 3t)} \right] |x - y|^{1-N}\, dy \leqslant$$

$$\leqslant \frac{1}{d} \int_{\Omega \cap B(x, d)} \left[d\,|\nabla u(y)| + \frac{|u(y)|}{1 - 3t} \right] |x - y|^{1-N}\, dy.$$

Multiplication of the above inequality by $|S|^{-1} s^{N-1}$ and integration with espect to s over the interval $[0, \rho]$, where $s \leqslant \rho \leqslant td$, yields

$$\frac{1}{|S|} \left| \int_0^\rho \int_S \varphi(s, \zeta)\, s^{N-1}\, ds\, d\zeta \right| \leqslant$$

$$\leqslant \frac{\rho^N}{Nd|S|} \int_{\Omega \cap B(x, d)} \left[d\,|\nabla u(y)| + \frac{|u(y)|}{1 - 3t} \right] |x - y|^{1-N}\, dy.$$

Since $\mu(B(x, \rho)) = \dfrac{\rho^N |S|}{N}$ and

$$\frac{1}{\mu(B(x, \rho))} \int_{B(x,\ \rho)} u(y)\, dy \to u(x) \quad \text{as} \quad \rho \to 0,$$

we have

$$|u(x)| \leqslant \frac{1}{d|S|} \int_{\Omega \cap B(x,\ d)} \left[d\,|\nabla u(y)| + \frac{|u(y)|}{1 - 3t} \right] |x - y|^{1-N}\, dy.$$

The lemma follows now since t is arbitrarily small.

Estimate (1) permits to establish the following

THEOREM *Let $p \leqslant N$ and $\alpha > 0$ be such that*

$$\frac{1}{p} - \frac{1}{N} < \frac{1}{q} \leqslant \frac{1}{p} \quad \text{and} \quad \frac{\alpha - N}{q} < 1 - \frac{N}{p}.$$

If $u \in W_0^{1,\ p}(\Omega)$ and $P \colon \Omega \to \mathbb{R}^N$ is a measurable function with the property that

$$M_\alpha(|P|^q) = \sup_{x \in \Omega} \int_\Omega |P(y)|^q \,|x - y|^{\alpha - N}\, dy < \infty,$$

then there exists a constant $K > 0$ such that

$$\|Pu\|_{0,\ q} \leqslant K[M_\alpha(|P|^q)]^{\frac{1}{q}} \|u\|_{1,\ p}.$$

Proof: Denote $v(y) = |u(y)| + d\,|\nabla u(y)|$. Then, by the lemma, for $\beta > 0$ we can write (1) in the form

$$|u(x)| \leqslant$$

$$\leqslant \frac{1}{|S|\,d} \int_{\Omega \cap B(x,\ d)} [|v(y)|^p\, |x - y|^{\beta r(1-N)}]^{\frac{1}{q}}\, |x - y|^{\frac{1}{p'}[\beta r + (1-\beta)\,p'](1-N)} |v(y)|^{\frac{p}{v}}\, dy,$$

where

$$\frac{1}{p} + \frac{1}{p'} = 1, \quad \frac{1}{v} = \frac{1}{p} - \frac{1}{q}, \quad \frac{1}{r} = 1 + \frac{1}{q} - \frac{1}{p} - \frac{1}{q} = \frac{1}{q} + \frac{1}{p'}.$$

By Hölder's generalized inequality applied for the exponents satisfying

$$\frac{1}{q} + \frac{1}{p'} + \frac{1}{v} = 1,$$

it follows that

$$|u(x)| \leqslant \frac{1}{|S| \, d} \left(\int\limits_{\Omega \cap B(x, \, d)} |v(y)|^p \, |x-y|^{\beta r(1-N)} \, dy \right)^{\frac{1}{q}} \times$$

$$\times \left(\int\limits_{\Omega \cap B(x, \, d)} |x-z|^{[\beta r + (1-\beta)\, p'](1-N)} \, dz \right)^{\frac{1}{p'}} \left(\int\limits_{\Omega \cap B(x, \, d)} |v(y)|^p \, dy \right)^{\frac{1}{v}}. \qquad (2)$$

Here the integral

$$I = \int\limits_{\Omega \cap B(x, \, d)} |x-z|^{[\beta r + (1-\beta)\, p'](1-N)} \, dz$$

is improper. A sufficient condition for the convergence of I is that

$$N + [\beta r + (1-\beta)\, p'] \, (1 - N) > 0.$$

As $p \leqslant N$, we may set $\beta r(1-N) = \alpha - N$ with $\alpha > 0$. Then this condition may by written as

$$\frac{N-\alpha}{q} + 1 - \frac{N}{p} > 0$$

since

$$0 < (1-N)\,[\beta r + (1-\beta)\, p'] + N = (\alpha - N) + (1-N)\, p' -$$

$$- \frac{\alpha - N}{r}\, p' + N = p'\left[(\alpha - N)\left(\frac{1}{p'} - \frac{1}{r} \right) + 1 - N + \frac{N}{p'} \right] =$$

$$= p'\left(\frac{N-\alpha}{q} + 1 - \frac{N}{p} \right).$$

Therefore I is convergent by hypothesis. Take the polar coordinates with pole at x and ζ on the unit sphere $S \subset \mathbb{R}^N$. Thus, $z = x + \rho\zeta$, $dz = \rho^{N-1} d\rho \, d\zeta$ and

$$I \leqslant \int_0^d \rho^{N-1+p'\left(\frac{N-\alpha}{q} + 1 - \frac{N}{p} \right) - N} \, d\rho \int_S d\zeta = \frac{1}{tp'} \, d^{tp'} |S|,$$

where $t = \dfrac{N-\alpha}{q} + 1 - \dfrac{N}{p} > 0$. Hence

$$|u(x)| \leqslant$$

$$\leqslant \frac{1}{|S| \, d} \left(\frac{1}{tp'} \, d^{tp'} |S| \right)^{\frac{1}{p'}} \left(\int\limits_{\Omega \cap B(x, \, d)} |v(y)|^p \, |x-y|^{\alpha - N} \, dy \right)^{\frac{1}{q}} \left(\int\limits_{\Omega \cap B(x, \, d)} |v(y)|^p \, dy \right)^{\frac{1}{v}}.$$

Multiply the above inequality by $|P(x)|$, raise to the power q and integrate it over Ω. One obtains

$$\int_\Omega |P(x)\,u(x)|^q\,dx \leqslant$$

$$\leqslant K^q\|v\|_v^{\frac{pq}{v}}\int_\Omega |P(x)|^q \int_{\Omega\cap B(x,d)} |v(y)|^p\,|x-y|^{a-N}\,dy\,dx, \tag{3}$$

where K is a positive constant which satisfies

$$K \geqslant \left(\frac{1}{tp'}\right)^{\frac{1}{p'}} d^{t-1}\,|S|^{-\frac{1}{p}} > 0.$$

The integral on the right-hand side can be estimated as follows:

$$\int_\Omega |P(x)|^q \int_{\Omega\cap B(x,d)} |v(y)|^p\,|x-y|^{a-N}\,dy\,dx =$$

$$= \int_{\Omega\cap B(x,d)} |v(y)|^p \int_\Omega |P(x)|^q\,|x-y|^{a-N}\,dx\,dy \leqslant$$

$$\leqslant \sup_{y\in\Omega}\int_\Omega |P(x)|^q\,|x-y|^{a-N}\,dx \int_{\Omega\cap B(x,d)} |v(y)|^p\,dy \leqslant M_\alpha(|P|^q)\,\|v\|_{0,p}^p.$$

Hence (3) becomes

$$\int_\Omega |P(x)\,u(x)|^q\,dx \leqslant K^q M_\alpha(|P|^q)\,\|v\|_{0,p}^{\frac{pq}{v}+p}.$$

Since $\dfrac{1}{v}+\dfrac{1}{q}=\dfrac{1}{p}$ and $\|v\|_{0,p}\leqslant\|u\|_{1,p}$, we obtain

$$\|Pu\|_{0,q} \leqslant K[M_\alpha(|P|)^q]^{\frac{1}{q}}\|u\|_{1,p}$$

as claimed.

To complete this result we shall prove the next

PROPOSITION *Suppose that $p > N$. Then*

$$W_0^{1,p}(\Omega) \subset C^{1-\frac{N}{P}}(\bar\Omega) \cap L^\infty(\Omega)$$

and

$$\|u\|_\infty = \operatorname*{ess\,sup}_{x \in \Omega} |u(x)| \leqslant K_0 \|u\|_{1,\,p}.$$

Proof: Take $\beta = 0$ in (2). Then

$$|u(x)| \leqslant$$

$$\leqslant \frac{1}{|S|\,d} \left(\int\limits_{\Omega \cap B(x,\,d)} |v(y)|^p \, dy \right)^{\frac{1}{q}} \left(\int\limits_{\Omega \cap B(x,\,d)} |x - z|^{p'(1-N)} \, dz \right)^{\frac{1}{p'}} \left(\int\limits_{\Omega \cap B(x,\,d)} |v(y)|^p \, dy \right)^{\frac{1}{v}}.$$

Since $p'(1 - N) + N = p'\left(1 - \dfrac{N}{p}\right) > 0$, the middle integral is convergent and the estimate

$$\int\limits_{\Omega \cap B(x,\,d)} |x - z|^{p'(1-N)} \, dz \leqslant \frac{1}{\left(1 - \dfrac{N}{p}\right) p'} \, |S| \, d^{\left(1 - \frac{N}{p}\right) p'}$$

holds. Therefore, if we denote by K_0 a positive constant such that

$$K_0 \geqslant \frac{1}{|S|\,d} \left[\frac{1}{\left(1 - \dfrac{N}{p}\right) p'} \right]^{\frac{1}{p'}} d^{\left(1 - \frac{N}{p}\right)} |S|^{\frac{1}{p'}},$$

then

$$|u(x)| \leqslant K_0 \|v\|_{0,\,p}^{\frac{p}{q} + \frac{p}{v}} \leqslant K_0 \|u\|_{1,\,p}.$$

The proposition follows now since every $u \in W_0^{1,\,p}(\Omega)$ can be identified with a function in the space $C^{1 - \frac{N}{p}}(\Omega)$ of all Lipschitz continuous functions with exponent $1 - \dfrac{N}{p}$, as we know by Proposition 2.6.

2.9. We shall give now a generalization of the compact imbedding of $W_0^{1,\,p}(\Omega)$ to unbounded domains.

THEOREM *Suppose that $p, q, \alpha, u(x)$ and $P(x)$ fulfil the conditions of theorem 2.8 and*

$$\int\limits_{\Omega \cap B(x,\,d)} |P(y)|^q \, dy \to 0 \quad \text{as } |x| \to \infty.$$

Then the multiplication by P is a compact operator from $W_0^{1,\,p}(\Omega)$ into $L^q(\Omega)$.

Proof: First, we remark that for $\varepsilon > 0$

$$\int_{\Omega \cap B(x, d)} |P(y)|^q \, |x - y|^{a-N+\varepsilon} \, dy \to 0 \quad \text{as} \quad |x| \to \infty. \tag{1}$$

If $\alpha + \varepsilon \geqslant N$, then (1) is obvious. If $\alpha + \varepsilon < N$, we denote $s = \dfrac{N-\alpha}{\varepsilon} > 1$,

and apply Hölder's inequality for the exponents satisfying $\dfrac{1}{s} + \dfrac{1}{s'} = 1$. Since

$(\alpha - N + \varepsilon) \, s' = \alpha - N,$

$$\int_{\Omega \cap B(x, d)} |P(y)|^q \, |x - y|^{a-N+\varepsilon} \, dy \leqslant$$

$$\leqslant \left(\int_{\Omega \cap B(x, d)} |P(y)|^q \, dy \right)^{\frac{1}{s}} \left(\int_{\Omega \cap B(x, d)} |P(y)|^q \, |x - y|^{a-N} \, dy \right)^{\frac{1}{s'}}.$$

The second term in the above inequality converges to zero because $M_\alpha(|P|^q) < \infty$. Hence assertion (1) is valid.

Let $\psi \in C_0^\infty(\Omega)$ be such that $0 \leqslant \psi(x) \leqslant 1$ for all $x \in \mathbb{R}^N$ and

$$\psi(x) = \begin{cases} 1 & \text{if } |x| \leqslant 1 \\ 0 & \text{if } |x| > 2. \end{cases}$$

Denote $\psi_R(x) = \psi\left(\dfrac{x}{R}\right)$ for $R > 0$ so that

$$Pu = P\psi_R u + P(1 - \psi_R)\, u.$$

As the support of $\psi_R u$ is bounded, by Theorem 2.6, $P\psi_R$ is a compact operator from $W_0^{1,\, p}(\Omega) \cap B(0, 2R))$ into $L^q(\Omega \cap B(0, 2R))$.

On the other hand,

$$\|Pu\|_{0,\, q} \leqslant \|P\psi_R u\|_{0,\, q} + \|P(1 - \psi_R)\, u\|_{0,\, q}.$$

Now, let $\beta > 0$ be such that $\alpha < \beta < q\left(1 - \dfrac{N}{p}\right) + N$. By Theorem 2.8 we have that

$$\|P(1 - \psi_R)\, u\|_{0,\, q} \leqslant K[M_\beta(|P(1 - \psi_R)|^q)]^{\frac{1}{q}} \|u\|_{1,\, p}$$

and by (1) we obtain that

$$\lim_{R \to \infty} M_\beta(|P(1 - \psi_R)|^q) = 0 \quad \text{for } \varepsilon = \beta - \alpha > 0.$$

Therefore, the operator of multiplication by P is the limit in the norm of $L^q(\Omega)$ of the compact operators $P\psi_R$. Thus, by Proposition 1.1, multiplication by P is compact from $W_0^{1,p}(\Omega)$ into $L^q(\Omega)$.

COROLLARY *Suppose that $\{u_n\} \subset W_0^{1,p}(\Omega)$ converges weakly to u in $W_0^{1,p}(\Omega)$. Then there exists a subsequence $\{u_j\} \subset \{u_n\}$ such that $u_j(x) \to u(x)$ for almost every $x \in \Omega$.*

Proof: Take $P(x) = e^{-|x|^2}$ for $x \in \Omega$ and denote

$$\Omega(R) = \{x \in \Omega \mid |x| > R\}.$$

Then

$$\int_{\Omega \cap B(x,\,d)} |P(y)|^q \, dy = \int_{\Omega(R) \cap B(x,\,d)} e^{-q|y|^2} \, dy \leqslant \mu(B(x,d)) e^{-qR^2} \to 0,$$

as $R \to \infty$. By the theorem, the mapping $u \mapsto Pu$ is compact from $W_0^{1,p}(\Omega)$ into $L^q(\Omega)$ and there exists a subsequence $\{u_j\} \subseteq \{u_n\}$ such that $Pu_j \to Pu$ in $L^q(\Omega)$. As P is a positive function, we deduce that $u_j(x) \to u(x)$ for almost every $x \in \Omega$.

3. Theory of topological degree

The use of compactness methods in solving nonlinear equations is fully exemplified by the fixed point theorems.

By a *fixed point* of a multivalued mapping T defined on a compact convex set K of a real separated topological vector space X, we mean an element x of K such that $x \in T(x)$. The main results in this field are the famous Brouwer and Borsuk theorems. The proofs of these theorems require the use of the topological degree on finite-dimensional spaces. This degree will be extended to general Banach spaces which allow to establish the Schauder fixed point theorem.

Degree of a mapping in N-dimensional spaces

3.1. Let Ω be a bounded domain in $\mathbb{R}^N$ and let $T: \Omega \mapsto \mathbb{R}^N$ be a continuous differentiable mapping. To discuss the solvability of functional equation

$$T(x) = f, \tag{1}$$

we shall define presently an integer deg (T, f, Ω) which depends continuously on T and f.

Denote by $J_T(x)$ the Jacobian matrix of $T = (T_1, \ldots, T_N)$,

$$J_T(x) = \left(\frac{\partial T_i}{\partial x_j} \right)_{N \times N},$$

and define the *critical set* as

$$\Gamma_T = \Gamma_T(\Omega) = \{T(x) \mid x \in \Omega, \, \det J_T(x) = 0\}.$$

The classical inverse function theorem asserts that *for each* $f_0 \notin \Gamma_T$, *the condition* $T(x_0) = f_0$ *implies the existence of two neighbourhoods* V *of* f_0 *and* U *of* x_0 *such that equation* (1) *has a unique solution* $x = T^{-1}(f)$ *in* U *for any* $f \in V$. *Moreover* T *is a one-to-one mapping of* U *onto* V (see Martin Jr. [1, p. 116]).

Let f be a point in $\mathbb{R}^N$, which belongs neither to $\Gamma_T(\Omega)$ nor to $T(\partial\Omega)$. Then the pre-image

$$T^{-1}(f) = \{x \in \Omega \mid T(x) = f\}$$

consists of isolated points in Ω. Since this set is bounded, it is finite and so we can write $T^{-1}(f) = \{x_1, \ldots, x_n\}$.

For an algebraic definition of topological degree denote the integers

$$p = \text{card } \{x \in T^{-1}(f) \cap \Omega \mid \det J_T(x) > 0\}$$

and

$$q = \text{card } \{x \in T^{-1}(f) \cap \Omega \mid \det J_T(x) < 0\}.$$

Then $n = p + q$ and the integer

$$\text{deg } (T, f, \Omega) = p - q$$

is called the *topological degree* of T on Ω with respect to the element f. This number can also be writen in the following form

$$\text{deg } (T, f, \Omega) = \sum_{k=1}^{n} \text{sign det } J_T(x_k) = \sum \{\text{sign det } J_T(x) \mid x \in T^{-1}(f)\}. \qquad (2)$$

Clearly, deg $(T, f, \Omega) = 0$ if $f \notin T(\bar\Omega)$ and deg $(I, f, \Omega) = 1$, where I is the identity map on $\mathbb{R}^N$.

This definition has the drawback that it applies only to $f \notin \Gamma^T \cup \cup T(\partial\Omega)$. To eliminate this restriction we shall give an integral evaluation of degree.

Denote by Λ_ε the set of all continuous functions $\Phi : \mathbb{R} \mapsto \mathbb{R}$ such that

$$\text{supp } \Phi \subset (0, \varepsilon) \quad \text{and} \quad \int_{\mathbb{R}^N} \Phi(|x|) \, dx = 1,$$

where $\varepsilon > 0$ is an arbitrary number. The mollifiers defined in Section 2.1 are examples of elements of this family.

It is obvious that the difference, say φ, of any two functions from Λ_ε satisfies the condition

$$\int_{\mathbb{R}^N} \varphi(|x|) \, dx = 0, \tag{3}$$

or, equivalently, in polar coordinates,

$$\int_0^\infty r^{N-1} \varphi(r) \, dr = 0.$$

We shall prove that a function $\varphi(|x|)$ with these properties may be put in divergence form. Indeed, define the map

$$\psi(r) = \begin{cases} r^{-N} \displaystyle\int_0^r \rho^{N-1} \varphi(\rho) \, d\rho & \text{for} \quad 0 < r < \infty, \\ 0 & \text{for} \quad r = 0. \end{cases}$$

This mapping belongs to $C^1((0, \varepsilon), \mathbb{R})$ and it satisfies the differential equation

$$r\psi'(r) + N\psi(r) = \varphi(r).$$

Consider the functions

$$P_j(x) = \psi(|x|) x_j, \quad 1 \leqslant j \leqslant N,$$

where $x \equiv (x_1, \ldots, x_N) \in \mathbb{R}^N$. Obviously $P_j \in C^1((0, \varepsilon), \mathbb{R})$ and

$$\sum_{j=1}^N D_j P_j = N\psi(|x|) + \psi'(|x|) |x| = \varphi(|x|), \quad D_j = \frac{\partial}{\partial x_j}.$$

Conversely, if the continuous function $\varphi(|x|)$ is the divergence of a function $P \in C^1(\mathbb{R}^N, \mathbb{R}^N)$ with compact support K, then a simple calculation leads to (3). Therefore $\varphi(|x|)$ *is the divergence of some function if and only if* $\varphi \in C((0, \varepsilon), \mathbb{R})$ *and it satisfies* (3). The connection between these functions and the topological degree is given by the following

PROPOSITION *Let* $\varphi(|x|)$ *be the divergence of a function* $P \in C^1(\mathbb{R}^N, \mathbb{R}^N)$ *with compact support* K. *If* $T \in C^1(\mathbb{R}^N, \mathbb{R}^N)$ *is such that*

$$T(\partial\Omega \cap)K = \emptyset,$$

then

$$\int_\Omega \varphi(|T(x)|) \det J_T(x) \, dx = 0 \tag{4}$$

Proof: First, we suppose that $T \in C^2(\mathbb{R}^N, \mathbb{R}^N)$, and define

$$Q_j(x) = \sum_{k=1}^{N} A_{jk}(x) \, P_k(T(x)), \quad x \in \Omega, \quad 1 \leqslant j \leqslant N,$$

where A_{jk} is the algebraic complement of the element $\dfrac{\partial T_k}{\partial x_j}$ in $J_T(x)$.

Obviously $Q_j(x)$, $1 \leqslant j \leqslant N$, are functions with compact supports contained in Ω. Moreover, denoting $Q = (Q_1, \dots, Q_N)$ we have

$$\operatorname{div} Q = \sum_{j=1}^{N} D_j Q_j = \sum_{k,j=1}^{N} [A_{jk}(x) \, D_j P_k(T(x)) + P_k(T(x)) \, D_j A_{jk}(x)] =$$

$$= \operatorname{div} P(T(x)) \det J_T(x) + \sum_{k=1}^{N} P_k(T(x)) \left[\sum_{k=1}^{N} D_j A_{jk}(x) \right].$$

Since for each k, $1 \leqslant k \leqslant N$, the sum

$$\sum_{j=1}^{N} D_j A_{jk}(x) = 0$$

for reasons of symmetry, we obtain

$$\operatorname{div} Q = \varphi(|T(x)|) \det J_T(x),$$

and by integrating this over Ω we get (4).

We extend now this conclusion to $T \in C^1(\mathbb{R}^N, \mathbb{R}^N)$. Since $T(\partial\Omega)$ and $K \equiv \operatorname{supp} P$ are disjoint compact sets, we may consider $d = \operatorname{dist}(T(\partial\Omega), K)$. Then, for any $x \in K$ and $\rho < d$, by the properties of the mollifiers, we have

$$\mathcal{M}_\rho(D_j \varphi(|T(x)|)) = D_j(\mathcal{M}_\rho \varphi(|T(x)|)), \quad 1 \leqslant j \leqslant N.$$

Hence for $\rho \to 0$ we get $\varphi(|\mathcal{M}_\rho T(x)|) \to \varphi(|T(x)|)$ and $J_{\mathcal{M}_\rho T}(x) \to J_T(x)$ uniformly for $x \in K$. Hence, by the above arguments,

$$\int_\Omega \varphi(|\mathcal{M}_\rho T(x)|) \det J_{\mathcal{M}_\rho T}(x) \, dx = 0.$$

Passing to limit as $\rho \to 0$ we easily get (4).

THEOREM *Let f be a regular value of $T \in C^1(\mathbb{R}^N, \mathbb{R}^N)$, i.e., $f \notin T(\partial\Omega) \cup \Gamma_T(\Omega)$. If $\Phi \in \Lambda_\varepsilon$, then for a small enough $\varepsilon > 0$*

$$\deg(T, f, \Omega) = \int_\Omega \Phi(|T(x) - f|) \det J_T(x) \, dx. \tag{5}$$

Proof: By the proposition, the integral in (5) does not depend on $\Phi \in \Lambda_\varepsilon$ provided that $\varepsilon < d$. Let us show the equivalence of both definitions.

Suppose that $\{x_1, \ldots, x_n\} \subset \Omega$ is the set of solutions of equation (1). Since $\det J_T(x_k) \neq 0$, $1 \leqslant k \leqslant n$, by the inverse function theorem to each x_k there corresponds a neighbourhood U_k such that equation (1) has a solution for every $f \in T(U_k)$. We can select for each k, $1 \leqslant k \leqslant n$, an open subset $\mathcal{O}_k \subset U_k$ such that $\det J_T(x)$ keeps a constant sign on $\mathcal{O}_k$ and $\mathcal{O}_k \cap \mathcal{O}_m = \emptyset$ for $k \neq m$, $1 \leqslant k$, $m \leqslant n$. Since,

$$\bigcap \{T(\mathcal{O}_k) \mid 1 \leqslant k \leqslant n\}$$

is a neighbourhood of f, we can choose $\varepsilon > 0$ such that

$$B(f, \varepsilon) \subset \bigcap \{T(\mathcal{O}_k) \mid 1 \leqslant k \leqslant n\}.$$

Denote

$$\mathscr{V}_k(\varepsilon) = \mathcal{O}_k \cap T^{-1}(B(f, \varepsilon)), \quad 1 \leqslant k \leqslant n.$$

Thus we obtain a system of open neighbourhoods $\mathscr{V}_k$ such that $\Phi(|T(x) - f|)$ vanishes on the complementary set of $\bigcap \{\mathscr{V}_k(\varepsilon) \mid 1 \leqslant k \leqslant n\}$. Now, we can write that

$$\int_\Omega \Phi(|T(x) - f|) \det J_T(x) \, dx = \int_{\{x \in \Omega \mid |T(x) - f| < \varepsilon\}} \Phi(|T(x) - f|) \det J_T(x) \, dx =$$

$$= \sum_{k=1}^n \int_{\mathscr{V}_k(\varepsilon)} \Phi(|T(x) - f|) \det J_T(x) \, dx.$$

Observe that

$$\int_{\mathscr{V}_k(\varepsilon)} \Phi(|T(x) - f|) \det J_T(x) \, dx = \operatorname{sign} \det J_T(x_k) \int_{B(f, \varepsilon)} \Phi(|z|) \, dz =$$

$$= \operatorname{sign} \det J_T(x_k).$$

Hence,

$$\int_\Omega \Phi(|T(x) - f|) \det J_T(x) \, dx = \sum \{\operatorname{sign} \det J_T(x) \mid x \in T^{-1}(f)\},$$

which proves the theorem.

3.2. To extend the definition of degree to the elements $f \in \Gamma_T(\Omega)$, we have to prove first the following special case of the Stoilow-Sard lemma:

LEMMA *If* $T \in C^1(\Omega, \mathbb{R}^N)$, *then the critical set* $\Gamma_T(\Omega)$ *is of measure null.*

Proof: The set $A = \{x \in \Omega | \det J_T(x) = 0\}$ is bounded and closed. We shall show that $\Gamma_T(\Omega) = T(A)$ is contained in a set of arbitrarily small measure. Consider an N-dimensional closed cube $C \subset \Omega$ with d as side and let $\varepsilon > 0$ be small enough. Since the map $x \mapsto J_T(x)$ is uniformly continuous on C, there exists a $\delta > 0$ such that $|x - z| < \delta$ implies that $|J_T(x) - J_T(z)| < \varepsilon$ for all $x, z \in C$. Divide C into m^N subcubes $C_m \subset C$ of side-length $\dfrac{d}{m}$,

each with the diameter $\sqrt{N} \dfrac{d}{m} < \delta$. Then on each of these subcubes we have

$$|T(z) - T(x) - J_T(x)(z - x)| = \left| \int_0^1 [J_T(x + t(z - x)) - J_T(x)](z - x)\, dt \right| \leqslant$$

$$\leqslant \varepsilon |z - x| \leqslant \varepsilon \sqrt{N} \dfrac{d}{m} \quad (\forall)\ x, y \in C_m.$$

Consider now only those subcubes C_m for which $C_m \cap A \neq \varnothing$. For any fixed $x \in C_m \cap A$ it follows that the set

$$\{J_T(x)(z - x) \mid z \in C_m\}$$

belongs to an $(N-1)$-dimensional subspace V of $\mathbb{R}^N$. Thus, the set $\{T(z) \mid z \in C_m\}$ is in an $\varepsilon \sqrt{N} \dfrac{d}{m}$-neighbourhood of the subspace $V + T(x)$. Now, we have

$$T_i(z) - T_i(x) = \sum_{j=1}^N T_i(z_1, \ldots, z_j, x_{j+1}, \ldots, x_N) - T_i(z_1, \ldots, z_{j-1}, x_j, \ldots, x_N) =$$

$$= \sum_{j=1}^N D_j T_i(y)(z_j - x_j), \quad 1 \leqslant i \leqslant N,$$

where $y = x + \theta(z - x)$, $0 < \theta < 1$. Hence, if we put

$$M = \sup \{|J_T(x)| \mid x \in C\},$$

where $|J_T(x)|$ denotes the norm of $J(T)$, the above equality yields

$$|T(z) - T(x)| \leqslant M|z - x| \leqslant MN^{\frac{3}{2}} \dfrac{d}{m} \quad (\forall)\ z, x \in C_m.$$

Therefore the set $\{T(z) | z \in C_m\}$ is contained in a cylinder of height smaller than $2\varepsilon\sqrt{N} \dfrac{d}{m}$ and base lying in an $(N-1)$-dimensional sphere of radius $MN^{\frac{3}{2}} \dfrac{d}{m}$. The volume of this cylinder is $C \left(\dfrac{d}{m} \right)^N \varepsilon$, where C is a positive constant which does not depend on ε. Since the number of subcubes which intersect A is less than m^N, the image $T(C \cap A)$ is contained in a set whose measure is

$$C \left(\frac{d}{m} \right)^N \varepsilon m^N = C d^N \varepsilon.$$

Thus $T(C \cap A)$ is of null measure and so is the measure of $T(A)$ because the cube C is arbitrarily chosen. The proof is complete.

In other words, the above lemma asserts that $T(\Gamma_T(\Omega))$ has empty interior. This lemma allows us to view the elements $f \in \Gamma_T$ as limits of sequences of elements $\{f_n\}$, with $f_n \notin \Gamma_T \cap T(\partial \Omega)$ and we define the degree as

$$\deg (T, f, \Omega) = \lim_{n \to \infty} \deg (T, f_n, \Omega).$$

To prove that this definition is independent of the particular sequences $\{f_n\}$ we need the following

PROPOSITION *Consider* $\Phi \in C^1(\mathbb{R}, \mathbb{R})$ *with compact support* $K = \operatorname{supp} \Phi(|x|)$ *contained in* Ω. *If* y *and* z *are two points in* $\mathbb{R}^N$ *such that the convex hull of the set* $(K - y) \cup (K - z)$ *is contained in* Ω, *then*

$$\varphi(x) = \Phi(|x + y|) - \Phi(|x + z|)$$

is the divergence of a function $P \in C^1(\mathbb{R}^N, \mathbb{R}^N)$ *with compact support contained in* Ω.

Proof: As the support of φ is contained in $(K - y) \cup (K - z)$, for any fixed y, we may consider the function

$$\psi(x) = \int_{-\infty}^{0} \varphi(x + ty) \, dt$$

and define

$$P_j(x) = \psi(x) \, y_j, \quad 1 \leqslant j \leqslant N.$$

These functions are with compact supports contained in Ω and they satisfy

$$\operatorname{div} P(x) = \sum_{j=1}^{N} D_j P_j = \sum_{j=1}^{N} y_j D_j \psi = \left[\frac{d}{dt} \psi(x + ty) \right]_{t=0}.$$

We may write

$$\left[\frac{d}{dt} \psi(x + ty) \right]_{t=0} = \left[\int_{-\infty}^{0} \frac{d}{dt} \varphi(x + ty + sy) \, ds \right]_{t=0} =$$

$$= \int_{-\infty}^{0} \frac{d}{ds} \varphi(x + sy) \, dy = \varphi(x) = \Phi(|x + y|) - \Phi(|x + z|),$$

which proves the assertion.

COROLLARY *If $T \in C^1(\Omega, \mathbb{R}^N)$ and the points $f, g \notin \Gamma_T(\Omega) \cup T(\partial\Omega)$ belong to the same connected component of $\mathbb{R}^N - T(\partial\Omega)$, then*

$$\deg(T, f, \Omega) = \deg(T, g, \Omega).$$

Proof: Choose $\varepsilon > 0$ small enough and such that the spheres $B_\varepsilon + y$ and $B_\varepsilon + z$ do not intersect $T(\partial\Omega)$. By the definition of degree, $\deg(T, f, \Omega) -$

$$- \deg(T, g, \Omega) = \int_\Omega [\Phi(|Tx - f|) - \Phi(|Tx - g|)] \det J_T(x) \, dx = \int_\Omega \operatorname{div} P \, dx = 0,$$

in virtue of the proposition. This proves the assertion.

In other words, this corollary asserts that *for any $T \in C^1(\Omega, \mathbb{R}^N)$ and $f \notin \Gamma_T \cup T(\partial\Omega)$, the degree $\deg(T, f, \Omega)$ is constant on each connected component of $\mathbb{R}^N - T(\partial\Omega)$.* This shows the independence of $\deg(T, f, \Omega)$, for $f \in \Gamma_T$, from the particular sequence approximating f.

3.3. The definition of degree for the functions $T \in C(\bar\Omega, \mathbb{R}^N)$ utilizes the uniform topology norm

$$\|T\| = \sup\{|T(x)| \mid x \in \bar\Omega\}$$

on $C(\bar\Omega, \mathbb{R}^N)$ and the Čebyšev norm

$$\|T\|_{C^1} = \sup_{x \in \bar\Omega}\left\{|T(x)| + \sum_{i=1}^{N}\left|\frac{\partial T}{\partial x_i}\right|\right\}$$

on $C^1(\bar\Omega, \mathbb{R}^N)$.

We shall use the

PROPOSITION *Consider the family $\{T_t\} \subset C^1(\Omega \times [0, 1], \mathbb{R}^N)$. If for any $t \in [0, 1]$, $f \notin T_t(\partial\Omega)$, then*

$$\deg(T_{t_1}, f, \Omega) = \deg(T_{t_2}, f, \Omega), \quad (\forall)\, t_1, t_2 \in [0, 1]$$

Proof: Since the map $t \mapsto \deg(T_t, f, \Omega)$ from $[0, 1]$ into $\mathbb{Z}$ is continuous, the image of $[0, 1]$ is a connected and discrete set. Hence it reduces to only one point.

Now, for any $f \notin T(\partial\Omega)$ we denote $d = \operatorname{dist}(f, T(\partial\Omega))$ and consider the neighbourhood of T in Čebyšev norm

$$\mathscr{V} = \left\{P \in C^1(\Omega, \mathbb{R}^N) \mid \|P - T\|_{C^1} \leqslant \frac{d}{2}\right\}.$$

If the maps $P_0, P_1 \in \mathscr{V}$, then $P_t = (1 - t)P_0 + tP_1$ belongs to $\mathscr{V}$ for all $t \in [0, 1]$ and

$$|P_t(x) - f| \geqslant \frac{d}{2} \qquad (\forall)\, x \in \partial\Omega.$$

Hence, by the proposition,

$$\deg (P_0, f, \Omega) = \deg (P_1, f, \Omega) = \deg (T, f, \Omega).$$

Since $C^1(\bar{\Omega}, \mathbb{R}^N)$ is a dense subset of $C(\bar{\Omega}, \mathbb{R}^N)$, we can define the degree of a map $T \in C(\bar{\Omega}, \mathbb{R}^N)$ as follows: *If* $\{T_n\} \subset C^1(\bar{\Omega}, \mathbb{R}^N)$ *converges to* $T \in C(\bar{\Omega}, \mathbb{R}^N)$ *uniformly on* Ω, *then for any* $f \notin T(\partial \Omega)$ *we put*

$$\deg (T, f, \Omega) = \lim_{n \to \infty} \deg (T_n, f, \Omega).$$

The independence of degree on the sequence $\{T_n\}$ chosen is shown as above by taking the neighbourhood of T,

$$\mathcal{U} = \left\{ P \in C^1(\bar{\Omega}, \mathbb{R}^N) \mid \|P - T\|_{C^1} \leqslant \frac{d}{2} \right\}.$$

Thus the topological degree is well defined in the case of continuous mappings.

Properties of the degree

3.4. Let Ω be a bounded domain in $\mathbb{R}^N$ and let $C(\bar{\Omega}, \mathbb{R}^N)$ be the space of all continuous functions endowed with the uniform norm topology. The topological degree deg (T, f, Ω) of $T \in C(\bar{\Omega}, \mathbb{R}^N)$ has the following properties:

(d1) Continuity with respect to the function: *If* $T \in C(\bar{\Omega}, \mathbb{R}^N)$ *and* $f \notin T(\partial \Omega)$, *then there exists a neighbourhood* $\mathscr{V}$ *of* T *in* $C(\bar{\Omega}, \mathbb{R}^N)$ *such that* $f \notin P(\partial \Omega)$ *and* deg (P, f, Ω) = deg (T, f, Ω) *for all* $P \in \mathscr{V}$.

(d2) Homotopy invariance: *If the family* $\{T_t\} \subset C(\bar{\Omega}, \mathbb{R}^N)$ *depends continuously on* $t \in [0, 1]$ *and* $f \notin T_t(\partial \Omega)$ *for all* $t \in [0, 1]$, *then* deg (T_t, f, Ω) *is constant for t in* $[0, 1]$.

Indeed, this property follows easily by the definition and by Proposition 3.3.

(d3) Dependence on boundary-values; *If*

$$T|_{\partial \Omega} = P|_{\partial \Omega} \quad and \quad f \notin T(\partial \Omega) \equiv P(\partial \Omega),$$

then

$$\deg (T, f, \Omega) = \deg (P, f, \Omega).$$

In fact, the map

$$T_t = tP + (1 - t) T \qquad (\forall) \ t \in [0, 1]$$

satisfies all the conditions of (d2), and deg (T_t, f, Ω) does not depend on $t \in [0, 1]$, i.e.,

$$\deg (T_0, f, \Omega) = \deg (T, f, \Omega) = \deg (T_1, f, \Omega) = \deg (P, f, \Omega)$$

(d4) Additivity with respect to the domain: *If $\Omega = \Omega_1 \bigcup \Omega_2$, where Ω_1, Ω_2 are both bounded domains in $\mathbb{R}^N$ and $f \notin T(\Omega')$, $\Omega' = (\Omega_1 \bigcap \Omega_2) \bigcup \bigcup \partial\Omega_1 \bigcup \partial\Omega_2$, then*

$$\deg(T, f, \Omega) = \deg(T, f, \Omega_1) + \deg(T, f, \Omega_2),$$

and the assertion remains true if Ω is a finite union of disjoint open subsets.

In fact, by definition, if $T \in C^1(\Omega, \mathbb{R}^N)$ and $f \notin \Gamma_T \bigcup T(\Omega')$, then we have

$$\deg(T, f, \Omega) = \sum \{\text{sign det } J_T(x) | x \in T^{-1}(f)\} =$$

$$= \sum_{i=1}^{2} \sum \{\text{sign det } J_T(x) | x \in T^{-1}(f) \bigcap \Omega_i\} = \sum_{i=1}^{2} \deg(T, f, \Omega_i)$$

and we may extend this, as previously, to the case $T \in C(\Omega, \mathbb{R}^N)$ and $f \notin T(\Omega').$,

(d5) *If $T \in C(\bar{\Omega}, \mathbb{R}^N)$, $f \notin T(\partial\Omega)$ and $\deg(T, f, \Omega) \neq 0$, then there exists at least one point $x \in \Omega$ such that $T(x) = f$.*

Indeed, if such a point does not exist, then $f \notin T(\bar{\Omega})$ and we can find a $d > 0$ such that

$$|T(x) - f| \geqslant d \quad (\forall) x \in \bar{\Omega}.$$

Since, by definition, $\deg(T, f, \Omega) = \lim \deg(T_n, f, \Omega)$, where $\{T_n\} \subset C^1(\Omega, \mathbb{R}^N)$ and $T_n \to T$ uniformly on Ω, there exists an $n_1 \geqslant 1$ such that

$$|T_n(x) - f| \geqslant \frac{d}{2} \quad (\forall) x \in \bar{\Omega}$$

for all $n \geqslant n_1$. Hence,

$$\deg(T_n, f, \Omega) = 0 \quad (\forall) n \geqslant n_1$$

and thus $\deg(T, f, \Omega) = 0$, which contradicts the assumption. Therefore, there exists an $x \in \Omega$ such that $T(x) = f$.

We give now some immediate applications of these properties:

Denote by $\bar{B}_1$ the closed unit ball in $\mathbb{R}^N$ centred at the origin and let $T: \bar{B}_1 \mapsto \mathbb{R}^N$ be a continuous function.

PROPOSITION *Suppose that $T(x)$ never points opposite to x for $x \in \partial B_1$, i.e,*

$$T(x) + \lambda x \neq 0 \quad (\forall) x \in \partial B_1, \quad \lambda \geqslant 0.$$

Then the equation $T(x) = 0$ has a solution inside B_1.

Proof: By hypothesis, $0 \notin T(\partial B_1)$ so that $\deg(T, 0, B_1)$ is well-defined. Define the homotopy

$$T_t = tT + (1 - t)I, \quad t \in [0, 1],$$

where I is the identity map. Then $0 \notin T_t(\partial B_1)$ and by (d2)

$$\deg(T, 0, B_1) = \deg(T_t, 0, B_1) = \deg(I, 0, B_1) = 1.$$

By (d5), there exists at least one point x_0 inside B_1 such that $T(x_0) = 0$.

A direct consequence of this proposition, which will be used in the next chapter, is the following

THEOREM *If* $T \in C(\mathbb{R}^N, \mathbb{R}^N)$ *is such that*

$$(T(x), x) \geq 0 \quad (\forall) \ x \in \partial B(0, R),$$

then the equation $T(x) = 0$ *has a solution* $x \in \bar{B}(0, R)$.

Proof: Indeed, suppose that $T(x) \neq 0$ for $|x| = R$. Then, we have

$$(T(x), x) > 0 \qquad \text{for } |x| = R,$$

which implies that $T(x)$ never points oposite to x for all $|x| = R$, i.e.,

$$T(x) + \lambda x \neq 0 \text{ for } \lambda \geq 0 \text{ and } |x| = R.$$

The theorem follows now by the previous proposition.

3.5. By property (d5), it is natural to find the conditions which make the degree odd. In this respect we shall need some results concerning the extensions of continuous odd functions.

LEMMA *Let* $\bar{\Omega}$ *be a compact set in* $\mathbb{R}^N$, *let* $T : \bar{\Omega} \to \mathbb{R}^{N+1}$ *be a continuous function with* $0 \notin T(\bar{\Omega})$ *and let* $K \subset \mathbb{R}^N$ *be any compact set containing* $\bar{\Omega}$. *Then there exists a continuous extension* $\tilde{T}$ *of* T *such that* $0 \notin \tilde{T}(K)$.

Proof: Identify $\mathbb{R}^N$ with the hyperplane $\{x \in \mathbb{R}^{N+1} \mid x_{N+1} = 0\}$ and set $a = \text{dist}(0, T(\bar{\Omega})) > 0$. Let P be a continuous extension of T (a mollifier) such that

$$m = \sup_{x \in \bar{\Omega}} |T(x) - P(x)| < \frac{a}{4}.$$

We shall choose successively the functions P satisfying certain conditions.

First we determine $P : \mathbb{R}^N \to \mathbb{R}^{N+1} - \{0\}$. Since $P(\mathbb{R}^N)$ has null measure in $\mathbb{R}^{N+1}$, the origin is not an interior point of $P(K)$. Hence there exists a $z \in \mathbb{R}^{N+1}$ arbitrarily close to 0 and not in the range of P. We choose $|z| < \frac{a}{2} - m$.

Taking $P_1(x) = P(x) - z$, we have $0 \notin P_1(K)$ and $\sup_{x \in \bar{\Omega}} |T(x) - P_1(x)| < \frac{a}{4}$.

Consider now the continuous function

$$P_2(x) = \frac{P_1(x)}{\eta(|P_1(x)|)}, \qquad \text{where} \qquad \eta(t) = \begin{cases} \dfrac{2t}{a} & \text{if } 0 \leqslant t < \dfrac{a}{2} \\[2mm] 1 & \text{if } t \geqslant \dfrac{a}{2} \end{cases}.$$

We easily verify that $\inf\limits_{x \in K} |P_2(x)| \geqslant \dfrac{a}{2}$, $P_2 = P_1$ on $\bar{\Omega}$ and

$$|P_2(x) - T(x)| < \frac{a}{4} \quad \text{on } \bar{\Omega}.$$

Denote by P_3 the continuous extensions to K of the restriction $P_2|_{\partial\Omega} - T$ obtained by means of the Tietze theorem (see I.3.1.). Note that $\sup\limits_{x \in K} |P_3(x)| \leqslant \dfrac{a}{4}$ and define $\tilde{T} = P_2 - P_3$. We can see that

$$\tilde{T}(x) = P_2(x) - (P_2(x) - T(x)) = T(x) \quad \text{for all } x \in \bar{\Omega}$$

and

$$|\tilde{T}(x)| \geqslant |P_2(x)| - |P_3(x)| \geqslant \frac{a}{4} \quad \text{for all } x \in K.$$

The proof is complete.

PROPOSITION *Let Ω be a bounded open set in $\mathbb{R}^N$ and symmetric about the origin and with $0 \notin \Omega$. Let $T : \partial\Omega \to \mathbb{R}^{N+1}$ be a continuous odd function such that $0 \notin T(\partial\Omega)$. Then T can be extended to all of Ω as a continuous odd map into $\mathbb{R}^{N+1} - \{0\}$.*

Proof: We use an induction argument. When $N = 1$, there exist two positive numbers $\varepsilon < \delta$ such that $\bar{\Omega} \subseteq [-\delta, -\varepsilon] \cup [\varepsilon, \delta]$. By the lemma, we can extend T to $\tilde{T}$ as a continuous map of $[\varepsilon, \delta]$ into $\mathbb{R}^2 - \{0\}$. Define $\tilde{T}(x) = -\tilde{T}(-x)$ for $x \in [-\delta, -\varepsilon]$; we deduce the proposition for $N = 1$.

Let Ω be a set in $\mathbb{R}^N$. Identify $\mathbb{R}^{N-1}$ with the hyperplane $\{x \in \mathbb{R}^N | x_N = 0\}$ and denote $\Omega_0 = \Omega \cap \mathbb{R}^{N-1}$. By induction, T can be extended to a continuous function of $\partial\Omega \cap \Omega_0$ into $\mathbb{R}^N - \{0\}$. Divide Ω into two disjoint parts

$$\Omega_\pm = \Omega \cap \{x \in \mathbb{R}^N | x_N \lessgtr 0\}$$

and, by the lemma, extend T to $\tilde{T}$ on $\bar{\Omega}_+$. Define $\tilde{T}$ on $\bar{\Omega}_-$ by $\tilde{T}(x) = -\tilde{T}(-x)$; we obtain the desired continuous odd extension of T.

COROLLARY *Let Ω be a bounded domain in $\mathbb{R}^N$ with $0 \notin \Omega$ which is symmetric about the origin. If $T : \partial\Omega \to \mathbb{R}^N$ is a continuous odd map such that $0 \notin T(\partial\Omega)$, then $\deg(T, 0, \Omega)$ is an even integer.*

Proof: By the proposition, there exists a continuous extension $T_1 : \bar{\Omega}_0 \to \mathbb{R}^N - \{0\}$ of $T|_{\partial\Omega_0}$. Thus the map

$$T_2 = \begin{cases} T_1 & \text{on } \bar{\Omega}_0, \\ T & \text{on } \partial\Omega_+ \bigcup \partial\Omega_- \end{cases}$$

is continuous odd and nonvanishing on $\Omega' = \bar{\Omega}_0 \bigcup \partial\Omega_+ \bigcup \partial\Omega_-$. We apply again the proposition to T_2 and obtain a new map $\tilde{T} : \Omega \bigcup \Omega' \to \mathbb{R}^N - \{0\}$ with the same properties. By (d4), we have

$$\deg(T, 0, \Omega) = \deg(T, 0, \Omega_+) + \deg(T, 0, \Omega_-).$$

Since T is odd and the domains Ω_+ and Ω_- are symmetric, it follows that $\deg(T, 0, \Omega_+) = \deg(T, 0, \Omega_-)$. As $\tilde{T}|_{\partial\Omega} = T|_{\partial\Omega}$, by (d3), we conclude that deg $(T, 0, \Omega) = \deg(T, 0, \Omega)$ is an even integer.

Based on these facts we can prove now one of the main results in this area:

THEOREM *(Borsuk) Let Ω be a bounded domain in $\mathbb{R}^N$ with $0 \in \Omega$, which is symmetric about the origin. If $T: \bar{\Omega} \to \mathbb{R}^N$ is a continuous map, odd on $\partial\Omega$, and $0 \notin T(\partial\Omega)$, then $\deg(T, 0, \Omega)$ is an odd integer (in particular different from zero).*

Proof: Let $r > 0$ be small enough such that $\bar{B}(0, r) \bigcap \partial\Omega = \emptyset$. According to the proposition there exists a continuous odd extension P of $T|_{\partial\Omega}$ such that P is the identity map on $\bar{B}(0, r)$. As $T|_{\partial\Omega} = P|_{\partial\Omega}$, by (d3), we have

$$\deg(P, 0, \Omega) = \deg(T, 0, \Omega) = \deg(T, 0, \Omega - B(0, r)) + \deg(T, 0, B(0, r)).$$

By the corollary, deg $(T, 0, \Omega - B(0, r))$ is even, while

$$\deg(T, 0, B,(0, r)) = \deg(I, O, B(0, r)) = 1.$$

Thus deg $(T, 0, \Omega)$ is an odd integer.

We remark that $\deg(T, 0, \Omega)$ is defined independently of the extension of T inside Ω.

Fixed point theorems

3.6. We shall give now some results concerning the existence of solutions of nonlinear equations of the form

$$T(x) = x \tag{1}$$

via the theory of finite-dimensional degree. These results are known as fixed point theorems and the best known is the following

THEOREM *(Brouwer) Let K be a compact convex set in $\mathbb{R}^N$ and let $T \in C(K, K)$ such that $T(K) \subset K$. Then T has at least one fixed point. The assertion remains also valid when K is homeomorphic to a convex compact set in $\mathbb{R}^N$.*

Proof: Let $K = \bar{B}(0, r)$. If there exists an $x \in \partial K$ such that $x = T(x)$, then x is a fixed point. Otherwise, we have that $x - T(x) \neq 0$ on ∂K and we define the homotopy

$$H(t, x) = x - t\, T(x).$$

Then $0 \neq H(t, x)$ for all $[t, x] \in [0, 1] \times \partial K$ because

$$|x - tT(x)| \geqslant |x| - t|T(x)| \geqslant (1 - t)r > 0$$

and by (d2) we conclude that

$$\deg\,(I - T, 0, K) = \deg\,(H(1, .), 0, K) = \deg\,(H(0, .), 0, K) = \deg\,(I, 0, K) = 1.$$

Hence, by (d5), there exists an $x \in K$ such that $x - T(x) = 0$.

When K is an arbitrary convex compact set in $\mathbb{R}^N$ we choose a ball $B(0, r)$ which contains K and a continuous extension $\tilde{T}$ of T to $\bar{B}(0, r)$ such that

$$\tilde{T}(\bar{B}(0, r)) \subseteq \mathrm{conv}(T(K)) \subseteq K \subseteq \bar{B}(0, r).$$

The Tietze extension theorem (see I.3.1) assures that this can be done. Thus, by the first part of the proof, there exists an $x \in \bar{B}(0, r)$ which is a fixed point for $\tilde{T}$. Since $\tilde{T}(x) \in K$ we deduce that $x \in K$ and $x = T(x)$.

Finally, let K^* be a convex compact set in $\mathbb{R}^N$ and let H be a homeomorphic map of K^* on an arbitrary set $K \subset \mathbb{R}^N$. Then, by the above proof, $H \circ T \circ H^{-1}$ has at least one fixed point $x \in K^*$ and thus $H^{-1}(x) \in K$ is a fixed point for T. The proof is now complete.

COROLLARY *Any continuous mapping from an N-dimensional simplex into itself has a fixed point.*

Proof: The corollary follows by the theorem because an N-dimensional simplex is homeomorphic to a closed ball in $\mathbb{R}^N$.

An extension of the Brouwer theorem to the case of multivalued mappings is given by the following

PROPOSITION *Let K be a non-empty, compact convex set of a separated topological vector space X, and let $T: K \mapsto 2^K$ be a mapping such that for each $x \in K$ the image $T(x)$ is a non-empty convex subset of K. If the set*

$$T^{-1}(f) = \{x \in K | f \in T(x)\} \qquad (\forall)\ f \in K$$

is open in K, then T has a fixed point, i.e., there exists an $x_0 \in K$ with the property that $x_0 \in T(x_0)$.

Proof: For each $f \in K$, the set $T^{-1}(f)$ is open in K and each point of K lies in at least one of these open subsets. Since K is a compact set, there exists a finite family $\{f_0, f_1, \ldots, f_n\} \subset K$ such that

$$K = \bigcup \{T^{-1}(f_j) | 0 \leqslant j \leqslant n\}.$$

Let $\{\beta_0, \beta_1, \cdots, \beta_n\}$ be a partition of unity subordinated to this covering. Consider the function

$$P(x) = \sum_{j=0}^{n} \beta_j(x) f_j. \tag{1}$$

For any $x \in K$, $P(x)$ is contained in the closed convex hull of $\{f_0, f_1, \cdots, f_n\}$. Thus P maps the n-dimensional simplex σ-conv$(f_0, f_1, \cdots, f_n)$ into itself. Further, since $\beta_j(x) \neq 0$ for some $j, 0 \leqslant j \leqslant n$, we have that $x \in T^{-1}(f_j)$, i.e. $f_j \in T(x)$. As $P(x)$ is a convex combination of elements of $T(x)$, we have $P(x) \in T(x)$ for all $x \in K$.

Since the topology induced by X on each finite-dimensional subspace $X_n \subset X$ coincides with the Euclidian topology on $\mathbb{R}^n$, the function $P : \sigma \mapsto \sigma$ is continuous. By the corollary, it has a fixed point in σ, i.e., $x_0 = P(x_0) \in T(x_0)$.

Topological degree in Banach spaces

3.7. Let T be a continuous mapping from a Banach space X into itself. The Brouwer theorem in infinite-dimensional spaces is not generally true. This fact can be readily seen from the following

COUNTEREXAMPLE Let $X = l_2$ be the Hilbert space of all square summable real sequences $x = (x_1, x_2, \ldots, x_n, \ldots)$ with the norm

$$\|x\|^2 = \sum_{i=1}^{\infty} x_i^2.$$

Let $B(0, 1)$ be the closed unit ball in X and let $T : B(0, 1) \mapsto B(0, 1)$ be defined by

$$T(x) = (\sqrt{1 - \|x\|^2}, x_1, x_2, \ldots).$$

The map $T(x)$ is continuous and we can write

$$T(x) = (\sqrt{1 - \|x\|^2}, 0, 0, \ldots) + (0, x_1, x_2, \ldots).$$

T maps $\bar{B}(0, 1)$ into itself because $\|x\| \leqslant 1$ implies that

$$\|T(x)\|^2 = 1 - \|x\|^2 + \sum_{i=1}^{\infty} x_i^2 = 1.$$

This map has no fixed points. In fact, if $x_0 = (x_1^0, x_2^0, \ldots)$ is a fixed point of T, i.e., $T(x_0) = x_0$, then necessarily $\|x_0\| = 1$ and so $T(x_0) = (0, x_1^0, x_2^0, \ldots)$. This implies that $x_1^0 = 0, x_2^0 = x_1^0, x_3^0 = x_2^0$, etc., that is, $x^0 = (0, 0, \ldots)$ which contradicts $\|x_0\| = 1$.

In infinite-dimensional spaces more than mere continuity of T is required because a closed bounded set need not be compact. Therefore in this case we shall have to use straight compactness.

By means of the properties of compact mappings we can define the degree $\deg(T, f, \Omega)$ of a continuous map T of a Banach space X into itself provided that T may be written as

$$T = I + P,$$

where I is the identity map and P is a compact mapping. Here Ω is a bounded domain of X and $f \notin T(\partial\Omega)$.

First, remark that *if $\bar{\Omega}$ is a closed, bounded subset in the Banach space X, then $T(\bar{\Omega}) = (I + P)(\Omega)$ is closed in X.* Indeed, let $\{x_n\} \subset \bar{\Omega}$ be such that $T(x_n) - x_n + P(x_n) \to y$ in X. Since $\{x_n\}$ is bounded and P is compact, there exists a subsequence $\{x_j\} \subset \{x_n\}$ such that $P(x_j) \to z$ in X. Therefore, $x_j \to y - z = x$ and by the continuity of T, $x + P(x) = y$, i.e., $y \in T(\bar{\Omega})$.

Thus we may conclude that $T(\partial\Omega)$ is closed in X and hence $f \notin T(\partial\Omega)$ implies that

$$d = \mathrm{dist}\,(T(\partial\Omega), f) > 0.$$

Now let P_ε be an ε-approximation of P, as in Theorem 1.1, such that $\varepsilon < \dfrac{d}{2}$ and let X_n be the subspace of X spanned by the elements of the range of P_ε and which contains f. Choose a basis in X_n and identify X_n with $\mathbb{R}^n$. We have that $T_\varepsilon(x) \equiv (I + P_\varepsilon)(x) \neq f$ on $\partial(\Omega \cap X_n)$ because one has

$$\|T_\varepsilon(x) - f\| = \|T(x) - P(x) + P_\varepsilon(x) - f\| \geq$$

$$\geq \|T(x) - f\| - \|P(x) - P_\varepsilon(x)\| \geq d - \frac{d}{2} > 0$$

for all $x \in \partial(\Omega \cap X_n)$. If we consider the restriction

$$T_\varepsilon|_{\bar{\Omega} \cap X_n} : \bar{\Omega} \cap X_n \mapsto X_n.$$

then $\deg(T_\varepsilon, f, \Omega \cap X_n)$ can be defined as in Section 3.3.

We introduce the topological degree in arbitrary Banach spaces by means of the topological degree corresponding to the appropriate finite-dimensional Banach spaces. So, by definition,

$$\deg(T, f, \Omega) = \deg(T_\varepsilon, f, \Omega \cap X_n). \tag{1}$$

To prove that this degree is well defined we need the following

LEMMA *Let $P : \bar{\Omega} \mapsto \mathbb{R}^N$ be a continuous map, where Ω is a bounded domain in $\mathbb{R}^{N+M}(M > 0)$. Then $\deg(I + P, f, \Omega) = \deg(I + P, f, \Omega) \cap \mathbb{R}^N)$ for every $f \in \mathbb{R}^N$ with $f \notin (I + P)(\partial\Omega)$.*

Proof: It is sufficient to prove the result for $P \in C^1(\Omega, \mathbb{R}^N)$. Denote by J_{N+M} the Jacobian matrix in $\mathbb{R}^{N+M}$ of the map $I + P$, where $P = (P_1, \ldots, P_N, 0, \ldots, 0)$ is continuously differentiable from Ω into $\mathbb{R}^{N+M}$. By the algebraic definition of the degree,

$$\deg (I + P, f, \Omega) = \sum \{\text{sign det } J_{N+M}(x) \mid x \in (I + P)^{-1}(f)\}$$

$$= \sum_{x \in (J+P)^{-1}(f)} \text{sign det} \begin{bmatrix} J_N(x) & 0 \\ 0 & I_M \end{bmatrix} = \deg (I + P, f, \Omega \cap \mathbb{R}^N).$$

where I_M is the identity matrix in $\mathbb{R}^M$.

PROPOSITION $\deg (T_\varepsilon, f, \Omega \cap X_n)$ *is independent of the ε-approximation* T_ε, *chosen in* (1).

Proof: First, we shall prove that $\deg(T_\varepsilon, f, \Omega \cap X_n)$ is independent of the basis chosen in X_n. In fact, if T_ε is C^1, the sign of the Jacobian determinant at one point is invariant under a change of basis in X_n. Thus, the assertion follows by the algebraic definition of the degree. For generally continuous T_ε, we take the limit as in Section 3.3

By the lemma, $\deg(T_\varepsilon, f, \Omega X_n \cap)$ *remains unchanged if n increases, i.e.,*

$$\deg (T_\varepsilon, f, \Omega \cap X_n) = \deg(T, f, \Omega \cap X_{n+m}), \quad m > 0.$$

Indeed, let $T_n = I + P_n$ be an η-approximation of T and let X_m be the subspace of X generated by the elements of the range of P_n. Denote by F a finite-dimensional subspace of X containing X_n and X_m. Then, by the lemma, we have

$$\deg (T_\varepsilon, f, \Omega \cap X_n) = \deg (T_\varepsilon, f, \Omega \cap F)$$

and

$$\deg (T_\varepsilon, f, \Omega \cap X_m) = \deg (T_\eta, f, \Omega \cap F).$$

Consider the homotopy $H_t = t T_\varepsilon + (1 - t) T_\eta$; then, by the homotopy invariance, we get that

$$\deg T_\varepsilon, f, \Omega \cap F) = \deg (T_\eta, f, \Omega \cap F),$$

which proves the proposition.

We note that if $f \notin T(\bar{\Omega},)$ then, $\deg (T, f, \Omega) = 0$.

3.8. All properties of finite-dimensional degree and their consequences can be extended to topological degree in Banach spaces by applying these results to the ε-approximations of the maps. We mention here the following instance of such extension.

Let Ω be an open bounded set in the Banach space X and let $T : \bar{\Omega} \to X$ be such that $T = I + P$, with P, compact.

PROPOSITION *(Leray–Schauder principle) Suppose that for each* $t \in [0, 1]$, *the equation*

$$(I + tP)x = f \tag{1}$$

has no solution on $\partial\Omega$. *Then there exists an* $x_0 \in \Omega$ *such that* $Tx_0 = (I + P)x_0 = f$.

Proof: Let $\{P_n\}$ be a sequence of mappings with finite-dimensional ranges contained in X_n which approximate the compact map $P.$,i.e.,

$$\|Px - P_n x\| \leqslant \frac{1}{n} \qquad (\forall)\, x \in \bar{\Omega},$$

and let $\gamma_n = \deg(I + P_n, f, \Omega \cap X_n)$ be their associated finite-dimensional degrees. By the homotopy invariance of the finite-dimensional degree, it follows that

$$\gamma_n(t) = \deg(I + tP_n, f, \Omega \cap X_n)$$

is constant for all t in $[0, 1]$. Since $\gamma_n(0) = \deg(I, f, \Omega \cap X_n) = 1$ we obtain $\gamma_n(1) = 1$ for all $n \in \mathbb{N}$. Thus, by (d5), to each n there corresponds a point $x_n \in \Omega \cap X_n$ which satisfies the equation

$$x_n + P_n x_n = f.$$

As P is compact, the sequence of images $\{Px_n\}$ contains a Cauchy subsequence in X, say $\{Px_m\}$, and

$$\|x_m + Px_m - f\| \leqslant \|Px_m - P_m x_m\| + \|x_m + P_m x_m - f\| \leqslant \frac{1}{m}.$$

Hence, the sequence $\{x_m\}$ with $x_m = f - P_m x_m$ converges in X to a limit x_0 and $(I + P)x_0 = f$. The proof is complete.

We note that *the assumption of this proposition holds whenever the solutions* x_t *of the equation* (1) *fulfil the a priori estimate*

$$\|x_t\| \leqslant M \quad \text{with an } M > 0.$$

Indeed, if this inequality is true, then there exists a constant $N > 0$ such that $\|Px_t\| \leqslant N$. Consider the open set

$$\Omega = \{x \in X \mid \|x\| < M + N + \|f\|\}.$$

If the element $x_0 \in \partial\Omega$ is a solution of equation (1), then we have

$$M + N + \|f\| = \|x_0\| \leqslant N + \|f\|$$

which is impossible.

In the same way we can prove an extension of Brouwer's theorem.

THEOREM *(Schauder)* *Suppose that Ω is a bounded closed and convex set of a Banach space X. If P is a compact map and $P(\Omega) \subseteq \Omega$, then the equation $Px = x$ has at least one solution in Ω.*

Proof: Let P_ε be the ε-approximation of P with range contained in the n-dimensional subspace X_n. Then P_ε maps the compact set $\Omega \cap X_n$ into itself. Thus, by Brouwer's theorem for each $\varepsilon > 0$, there exists an $x_\varepsilon \in \Omega \cap X_n$ such that

$$P_\varepsilon x_\varepsilon = x_\varepsilon, \quad x_\varepsilon \in \Omega.$$

Now, by the compactness of P_ε, the family $\{P_\varepsilon x_\varepsilon\}$ contains a Cauchy sequence, say $\{P_n x_n\}$. Hence,

$$x_n = P_n x_n \to x_0 \quad \text{as} \quad n \to \infty.$$

On the other hand, $\|x_n - Px_n\| = \|P_n x_n - Px_n\| \to 0$ and therefore $P_n x_n \to Px_0$. Thus $Px_0 = x_0$ as claimed.

4. Related topics and exercises

4.1. Let $\theta : \mathbb{R} \mapsto \mathbb{R}_+$ be the Heaviside function

$$\theta(x) = \begin{cases} 1 & \text{if } x > 0 \\ 0 & \text{otherwise.} \end{cases}$$

Prove that the generalized derivative $\theta'(x) = \delta(x)$, where $\delta(x)$ is the Dirac function

$$\delta(x) = \begin{cases} \infty & \text{if } x = 0 \\ 0 & \text{otherwise.} \end{cases} \quad \text{and} \quad \int_{-\infty}^{+\infty} \delta(x)\, dx = 1.$$

SOLUTION By the definition of the weak derivative we have

$$\int_{-\infty}^{+\infty} \theta'(x)\, \varphi(x)\, dx = -\int_{-\infty}^{+\infty} \theta(x)\, \varphi'(x)\, dx = -\int_0^{+\infty} \varphi'(x)\, dx = \varphi(0) = \int_{-\infty}^{+\infty} \delta(x)\, \varphi(x)\, dx$$

for all $\varphi \in C_0^\infty$, as required.

4.2. Evaluate $\varDelta \dfrac{1}{r}$ in $\mathbb{R}^3$, where $r = |x|$ and $\varDelta = \sum_{i=1}^{3} \dfrac{\partial^2}{\partial x_i^2}$ is the Laplace operator.

SOLUTION We have

$$\int_{\mathbb{R}^3} \varDelta \frac{1}{r}\, \varphi(x)\, dx = \int_{\mathbb{R}^3} \frac{\varDelta \varphi}{r}\, dx = \lim_{\varepsilon \to 0} \int_{r \geqslant \varepsilon} \frac{\varDelta \varphi}{r}\, dx.$$

Suppose that supp $\varphi \subset B(0, a)$. By Green's formula we obtain

$$\int_{r \geqslant \varepsilon} \frac{\varDelta \varphi}{r}\, dx = \int_{r \geqslant \varepsilon} \varphi \varDelta \frac{1}{r}\, dx - \int_{r = \varepsilon} \frac{1}{r} \frac{\partial \varphi}{\partial r}\, d\sigma + \int_{r = \varepsilon} \varphi \frac{\partial}{\partial r} \frac{1}{r}\, d\sigma,$$

where the last two integrals are evaluated over $\partial B(0, \varepsilon)$. Since for $r \geqslant \varepsilon$ the function $\dfrac{1}{r}$ is harmonic, i.e. $\Delta \dfrac{1}{r} = 0$, it follows that the first integral in the second term vanishes. We evaluate the remaining two integrals. First, we note that

$$\left| \int_{r=\varepsilon} \frac{1}{r} \frac{\partial \varphi}{\partial r} \, \mathrm{d}\sigma \right| = \frac{1}{\varepsilon} \left| \int_{r=\varepsilon} \frac{\partial \varphi}{\partial r} \, \mathrm{d}\sigma \right| \leqslant 4\pi\varepsilon \max \left| \frac{\partial \varphi}{\partial r} \right|,$$

where the maximum of $\dfrac{\partial \varphi}{\partial r}$ is taken over $\partial B(0, \varepsilon)$. For the second integral we have

$$\int_{r=\varepsilon} \varphi \frac{\partial}{\partial r} \frac{1}{r} \, \mathrm{d}\sigma = -\int_{r=\varepsilon} \varphi \frac{1}{r^2} \, \mathrm{d}\sigma = -\frac{1}{\varepsilon^2} \int_{r=\varepsilon} \varphi \, \mathrm{d}\sigma.$$

Hence $\displaystyle\lim_{\varepsilon\to 0} \int_{r=\varepsilon} \varphi \frac{\partial}{\partial r} \frac{1}{r} \, \mathrm{d}\sigma = -4\pi\varphi(0)$ and therefore

$$\int_{\mathbb{R}^3} \Delta\frac{1}{r} \varphi \, \mathrm{d}x = \lim_{\varepsilon\to 0} \int_{r=\varepsilon} \frac{\Delta\varphi}{r} \, \mathrm{d}x = -4\pi\varphi(0) = -4\pi \int_{\mathbb{R}^3} \delta(x) \varphi(x) \, \mathrm{d}x,$$

that is, $\Delta \dfrac{1}{r} = -4\pi\delta(x)$.

Similarly, we prove that in the general case $\mathbb{R}^N$, $N > 3$

$$\Delta \frac{1}{r^{N-2}} = -(N-2)\,\omega_n\,\delta(x),$$

where ω_n is the surface area of $B(0, 1)$.

4.3. Prove that if $\partial\Omega$ is of class C^1, then Ω satisfies the cone property (Friedman [1, p. 18]).

4.4. Let $a : H^1(\Omega) \times H^1(\Omega) \mapsto \mathbb{R}$ be defined by

$$a(u, v) = \int_\Omega a_0(x)\, uv \, \mathrm{d}x + \sum_{i,j=1}^N \int_\Omega a_{ij}(x) \frac{\partial u}{\partial x_i} \frac{\partial v}{\partial x_j} \, \mathrm{d}x.$$

where $a_0, a_{ij} = a_{ji}$, $1 \leqslant i, j \leqslant N$, are bounded continuous functions on Ω which satisfy

$$a_0(x) \geqslant c_0 > 0 \quad \text{and} \quad \sum_{i,j=1}^N a_{ij}(x)\, \xi_i\xi_j \geqslant \alpha |\xi|^2$$

for all $x \in \Omega$ and all $\xi \in \mathbb{R}^N$. Prove that $a(u, v)$ defines a new inner product on $H^1(\Omega)$ and the corresponding norm $|u| = [a(u, u)]^{\frac{1}{2}}$ is equivalent to the norm $\|u\|_{1,2}$.

HINT One verifies easily that

$$a(u, v) = a(v, u) \quad \text{and} \quad a(u, u) \geqslant \min (c_0, \alpha)\, \|u\|_{1,2}^2 .$$

These relations and the boundedness of a_0 and a_{ij} on $\bar\Omega$ prove the assertion.

4.5. Let X_1, X_2, X_3 be three Banach spaces with the norms $\|\cdot\|_i$, $1 \leqslant i \leqslant 3$. Assume that $X_1 \subset X_2 \subset X_3$, the imbedding $I_{12} : X_1 \mapsto X_2$ is compact and the imbedding $I_{23} : X_2 \mapsto X$

is continuous. Prove that for any $\varepsilon \geqslant 0$ there exists a positive constant $c(\varepsilon)$ such that the inequality $\|u\|_2 \leqslant \varepsilon\|u\|_1 + c(\varepsilon)\|u\|_3$ holds for all $\mu \in X_1$. Apply the result to the Sobolev spaces $H^m(\Omega)$, $H^{m-1}(\Omega)$ and $L^2(\Omega)$

SOLUTION Assume that, on the contrary, there exists an $\varepsilon > 0$ such that to each $n \in \mathbb{N}$ there corresponds some u_n with the property $\|u_n\|_2 \geqslant \varepsilon\|u_n\|_1 + n\|u_n\|_3$. Take $v_n = \dfrac{u_n}{\|u_n\|_1}$ whence $\|v_n\|_2 > \varepsilon + n\|v_n\|_3$. As I_{12} is continuous, $\|v_n\|_2 \leqslant M$ for some $M > 0$, and consequently $\|v_n\|_3 \to 0$. The compactness of I_{12} implies that $\{v_n\}$ contains a subsequence $\{v_j\}$ converging to v in X_2. Then $v_j \to v$ in X_3 since I_{23} is continuous, and hence $v = 0$. This contradicts the fact that $\|v_j\|_2 > \varepsilon$.

Now, $H^m(\Omega) \subset H^{m-1}(\Omega) \subset L^2(\Omega)$ with compact imbeddings provided that Ω satisfies the conditions of Theorem 2.5. Therefore, as above, for $\varepsilon > 0$ there exists a positive constant $c(\varepsilon)$ such that the *Ehrling inequality*

$$\|u\|_{m-1,2} \leqslant \varepsilon\|u\|_{m,2} + c(\varepsilon)\|u\|_2 \quad (\forall)\, u \in H^m(\Omega),$$

holds (see Fichera [1]).

4.6. Let $U = \left\{ u \in L^p(0, t_0; X_1) \mid u' = \dfrac{du}{dt} \in L^r(0, t_0; X_3) \right\}$ be the Banach space with respect to the norm $\|u\|_U = \|u\|_{L^p(0, t_0; X_1)} + \|u'\|_{L^r(0, t_0; X_1)}$, where X_1, X_2, X_3 are three Banach spaces as in Problem 4.5 and $1 < p, r < \infty$. Prove that the injection of U in $L^p(0, t_0; X_2)$ is compact.

HINT Clearly, $U \subset L^p(0, t_0; X_2)$. Let $\{u_n\} \subset U$ with $\|u_n\|_U \leqslant M$ be such that $u_n \to u$ in U. To prove that $u_n \to u$ in $L^p(0, t_0; X_2)$ apply first the conclusion of Problem 4.5 for $(u_m - u_n) \in L^p(0, t_0; X_2)$ and then use the Ascoli-Arzelà theorem to show the compactness of $\{u_n\}$ in $L^p(0, t_0; X_2)$. Observe that the equicontinuity of $\{u_n\}$ follows by the inequality

$$\|u_n(t) - u_n(s)\|_{L^p(0, t_0; X_3)} = \left\| \int_s^t \frac{du_n}{d\tau}\, d\tau \right\|_{L^p(0, t_0; X_3)}$$

$$\leqslant |t - s|^{\frac{1}{p'}} \left(\int_s^t \left\| \frac{du_n}{d\tau} \right\|_{L^p(0, t_0; X_3)}^p d\tau \right)^{\frac{1}{p}} = M|t - s|, \ (\forall)\, s,\, t \in [0, t_0]$$

(see Aubin [1]).

4.7. Let Ω be a bounded domain in $\mathbb{R}^N$ and let $T \in C(\bar{\Omega})$ and $f \notin T(\partial\Omega)$. Then $\deg(T, f, \Omega) \neq 0$ implies by virtue of (d5) that the equation $T(x) = f$ has at least one solution. Moreover $|\deg(T, f, \Omega)|$ is a lower bound for the number of solutions of the above equation in Ω. Clearly this equation could have other solutions outside Ω. The case $\deg(T, f, \Omega) = 0$ is extremely unfavourable because we can say nothing of the solutions in Ω of the equation $T(x) = f$. In this case, it is better to consider the card $T^{-1}(f)$ instead of $\deg(T, f, \Omega)$ as we can easily see from the following example:

Let Ω be the interval $(-1,1)$ and let $T(x) = x^2$. Then $J_T(x) = 2x$, $\Gamma_T = \{0\}$ and $T(\partial\Omega) = \{1\}$. Hence, for $f \notin \Gamma_T \cup T(\partial\Omega)$, we have $\deg(T, f, \Omega) = 1 - 1 = 0$ if $f > 0$ and $\deg(T, f, \Omega) = 0$ if $f < 0$. On the other hand,

$$\text{card } T^{-1}(f) = \begin{cases} 2 & \text{if } f > 0, \\ 0 & \text{if } f < 0. \end{cases}$$

4.8. Let $T \in C(\bar{\Omega}, \mathbb{R}^N)$ and assume deg $(T, f, \Omega) \neq 0$. Prove that $T(\Omega)$ is a neighbourhood of in $\mathbb{R}^N$.

HINT By (d5), there exists an $x \in \Omega$ such that $T(x) = f$. Hence, $C_f \subset T(\Omega)$, where $C_f \subset \mathbb{R}^N - T\partial\Omega)$ is the connected component containing f. Since C_f is an open set and T is a continuous map, it follows that $T(\Omega)$ is a neighbourhood of f in $\mathbb{R}^N$.

4.9. Suppose that Ω is a bounded domain in $\mathbb{R}^N$, which contains the origin and is symmetric about it. If $T \in C(\bar{\Omega}, \mathbb{R}^N)$ is odd on $\partial\Omega$ and $T(\partial\Omega)$ lies in a proper subspace of $\mathbb{R}^N$, i.e., $T(\partial\Omega) \subset \mathbb{R}^k$, $k < N$, then there exists an $x_0 \in \partial\Omega$ such that $T(x_0) = 0$ (Nirenberg [1, p. 25]).

4.10. Let Ω be as in 4.9. Let $T \in C(\partial\Omega, \mathbb{R}^N)$ be such that $T(\partial\Omega)$ lies in a proper subspace of $\mathbb{R}^N$. Then there exists a point $x_0 \in \partial\Omega$ such that $T(x_0) = T(-x_0)$ (Borsuk-Ulam theorem).

HINT Apply 4.12 to the mapping $P(x) = T(x) - T(-x)$.

4.11. Let Ω be a bounded domain in $\mathbb{R}^N$, containing the origin and symmetric about it. Suppose that $\partial\Omega$ is covered by N closed sets $A_1, A_2, \dots, A_N$. Prove that one of these sets contains a pair of antipodal points x_0 and $-x_0$.

SOLUTION If this is not the case, then $\cap \{A_i | 1 \leqslant i \leqslant N\}$ is empty. For $x \in \partial\Omega$, put $d_i(x) = \text{dist } (x, A_i)$, $1 \leqslant i \leqslant N$. Then $d(x) = \sum_{i=1}^{N} d_i(x) > 0$. Define the map $T(x) = \left\{ \dfrac{d_1(x)}{d(x)}, \dots, \dfrac{d_{N-1}(x)}{d(x)} \right\}$ which maps $\partial\Omega$ into $\mathbb{R}^{N-1}$. By 4.10, there is a point x_0 such that $T(x_0) = T(-x_0)$. If $x_0 \in A_i$ for an $i < N$, then $d_i(x_0) = 0$ and since $T(x_0) = T(-x_0)$, it follows that $d_i(-x_0) = 0$, i.e., $-x_0 \in A_i$. If $x_0 \notin A_i$ with $i < N$, then $d_i(x_0) > 0$ and hence $d_i(-x_0) > 0$. Thus x_0 and $-x_0$ belong both to A_N.

4.12. Let Ω be as in 4.11. If $T \in C(\bar{\Omega}, \mathbb{R}^N)$ is such that $0 \notin T(\partial\Omega)$ and

$$\frac{T(x)}{|T(x)|} \neq \frac{T(-x)}{|T(-x)|} \qquad (\forall) \; x \in \partial\Omega,$$

prove that deg $(T, 0, \Omega)$ is odd.

SOLUTION Define the homotopy $H(x, t) = T(x) - tT(-x)$, $t \in [0, 1]$. Then $H \in C(\Omega \times [0, 1], \mathbb{R}^N)$ and $0 \notin H(\partial\Omega \times [0, 1])$. By (d2), deg $(H(., t), 0, \Omega)$ is constant for all $t \in [0, 1]$. If $t = 1$, then $H(x, 1) = T(x) - T(-x)$ is an odd map. By Theorem 2.5, deg $(H(., 1), 0, \Omega)$ is odd and so is deg $(H(., 0), 0, \Omega) \equiv$ deg $(T, 0, \Omega)$.

4.13. (*The Perron and Frobenius theorem*) Let $A = (a_{ij})_{N \times N}$ be a matrix with $a_{ij} \geqslant 0$ for $1 \leqslant i, j \leqslant N$. Then there exist a $\lambda > 0$ and an $x \neq 0$ with coordinates $x_i \geqslant 0$ such that $Ax = \lambda x$.

Proof: The set $K = \left\{ x \in \mathbb{R}^N | x_i \geqslant 0, \sum_{i=1}^{N} x_i = 1 \right\}$ is closed and convex. If $Ax = 0$ at least for an $x \in K$ then $Ax = \lambda x$ with $\lambda = 0$. Suppose that $Ax \neq 0$ for all $x \in K$. Then there exists a constant $a > 0$ such that $\sum_{i=1}^{N} (Ax)_i \geqslant a$ for all $x \in K$. The function

$$T(x) = \frac{Ax}{\sum_{i=1}^{N} (Ax)_i}$$

is continuous on K, $(Tx)_i \geqslant 0$ and $\sum_{i=1}^{N} (Tx)_i = 1$, i.e., $T(K) \subseteq K$. By the Brouwer theorem, there exists a fixed point for T, that is, $Ax = \lambda x$ with $\lambda = \sum_{i=1}^{N} (Ax)_i > 0$.

Bibliographical comments

The characterization of compact mappings given in section 1.1 follows Nirenberg [1]. The linear integral operators are presented as in Kantorovich — Akilov [1]. For integral operators of potential type we have used Smirnov's book [1].

The initial Russian approach to Sobolev spaces proceeds from linear compact integral operators and we prefer it because it seems to be simpler and more coherent. For the theory of Sobolev spaces we use the books of Agmon [1], Friedmann [1], Nečas [1] and Sobolev [1]—[2]. The Poincaré inequality in Section 2.3 follows Morrey Jr. [1] and the approximant description of Sobolev spaces (Theorem 2.5) is due to Meyer — Serrin [1]. The proof by induction of the compactness of imbedding in Section 2.6 is due to Kantorovich and the Lipschitz continuous character of the functions in $W_0^{1,p}(\Omega)$ is pointed out by Shinbrot [1]. We have considered that it is necessary to give (in Section 2.7) some outline on the construction of $W^{m,p}(\Omega)$ with fractional values of m and of the trace. For proofs of these and other results on Sobolev spaces we refer to the excellent book of Adams [1] and to Aizicovici [1]. Studies of imbeddings in the case of unbounded domains were initiated by Berger — Schecter [1] and they have been carried on mainly by the English group of D. E. Edmunds. The matter contents of Sections 2.8 and 2.9 give an easier variant of the results presented by Edmunds — Evans [1] and Edmunds — Webb [1]. For the proof of imbedding in this case we use Schecter's method [2]. A simple proof of Corollary 2.9 can be found in Simader's. paper [2].

The theory of topological degree in Section 3 proceeds from the analytical definition of Brouwer degree due to Heinz [1]. The new idea used here is the introduction of mollifiers which allows to define the degree for continuous differentiable functions. The integral evaluation of degree was given by Schwartz [1]. The proof of Sard's lemma in Section 3.2 follows Spivak's book [1]. For the properties of degree we have used the books of Berger — Berger [1], and Deimling [1]. Borsuk's theorem is given in Rabinovitz [1] and Fučik-Nečas-Souček-Souček [1]. Brouwer's theorem and its extension to multivalued mappings are presented as in Schwartz [1] a   Browder [7]. The extension of Brouwer degree to infinite dimensions is due to Leray and Schauder [1], Schauder [1], and is presented following Nirenberg [1] and Amann [7]. An interesting specification of topological degree in Banach spaces for nonlinear differential equations is the coincidence degree, studied by Gaines—Mawhin [1].

Among other Romanian contributions to the fixed point theory we mention the books of Istrățescu [1] and Rus [1].

Chapter III

Nonlinear mappings
of monotone type

This chapter deals with the structure of the most important classes of nonlinear mappings, which have been intensively studied of late. The origins of these types of operators are rooted in the calculus of variations and it is for the study of nonlinear functional equations that these operators provide a framework broader than the compact or weakly continuous operators.

Our present aim is to give a unitary approach to the theory of these mappings and to point out those results required in applications.

Considering the highly technical character of the first section we suggest the reader might first read the section on monotone mappings where many intuitive and well-known examples are given.

1. Pseudo-monotone mappings

The monotone mappings are interesting in applications especially when they are also maximal. The axiomatization of certain topological properties of monotone operators has allowed the introduction of wider classes of nonlinear mappings for which the basic results of monotone operators remain valid.

For two locally convex linear spaces, dual to each other, these mappings were introduced by Brézis in [1] by means of nets. For a Banach space X, a single-valued map $T : X \mapsto X^*$ is said to be *pseudo-monotone* if the following conditions are satisfied:

(i) For any bounded net $\{x_\alpha\} \subset X$ such that $x_\alpha \rightharpoonup x$ in X and

$$\limsup(T x_\alpha, x_\alpha - x) \leqslant 0,$$

the relation

$$(Tx, x - y) \leqslant \liminf (Tx_\alpha, x_\alpha - y)$$

holds for all $y \in X$.

(ii) For any fixed $y \in X$, the function $\varphi_y(x) = (Tx, x - y)$ is bounded from below on X, uniformly on each bounded subset of X.

In the case of a reflexive Banach space, we have Proposition I.1.4 (due to Browder [10]), which characterizes the weak closure of bounded sets in X. We consider ordinary sequences instead of nets and extend pseudo-mono-tonicity to multivalued mappings.

A general survey of pseudo-monotone mappings permits a more direct approach to the discussion of certain properties of the sum of monotone maps.

Mappings of variational calculus type

1.1. We proceed by giving a justification of the notion of pseudo-mono-tonicity.

Let X be a reflexive Banach space and let X^* be its dual space. A map $T: X \mapsto X^*$ is *monotone* provided that

$$(Tx - Ty, x - y) \geqslant 0 \quad (\forall) \, x, y \in D(T).$$

The operator T is said to be *hemicontinuous* at $x \in D(T)$ if ($D(T)$ is convex and) for any $y \in D(T)$ the map $t \mapsto T[(1 - t)x + ty]$ is continuous from $[0, 1]$ into the weak topology of X^*. Obviously, the hemicontinuity of T in $x \in D(T)$ is equivalent to the continuity of the map $t \mapsto T(x + ty)$ from the neighbourhood $[0, t_0)$ into the weak topology of X^*.

The notions of monotone operator and convex function are closely connected due to the following

PROPOSITION *Let* $\varphi: X \mapsto \overline{\mathbb{R}}$ *be a Gâteaux differentiable function. Then* φ *is convex if and only if its gradient* $\varphi': X \mapsto X^*$ *is a hemicontinuous monotone operator.*

Proof: For any $x, y \in X$, the real-valued function

$$\Phi(t) = \varphi[(1 - t)x + ty]$$

is convex and differentiable for $t \in [0, 1]$ and

$$\Phi'(t) = (\varphi'[(1 - t)x + ty], \, y - x).$$

Since $\Phi'(t)$ is a continuous increasing function (see I.2.7), we have that $\Phi'(0) \leqslant \Phi'(1)$ and thus

$$(\varphi'(x) - \varphi'(y), \, x - y) \geqslant 0 \quad (\forall) \, x, y \in X.$$

Conversely, the continuity of the increasing function Φ' implies easily the convexity of Φ and φ.

We give now an example of a pseudo-monotone mapping. An operator $P: X \mapsto X^*$ is said to be *semi-monotone* or *of the variational calculus type* provided that

$$Px = A\,(x, x),$$

where the map $A: X \times X \mapsto X^*$ has the following properties :

(j) For fixed $x \in X$, the mapping $A(x, .): X \mapsto X^*$ is hemicontinuous and

$$(A(x,x) - A\,(x,y),\ x - y) \geqslant 0 \qquad (\forall)\,y \in X;$$

(jj) For fixed $y \in X$, $A\,(.,y): X \mapsto X^*$ is a bounded hemicontinuous operator;

(jjj) If $x_n \to x$ in X and $(A(x_n, x_n) - A(x_n, x),\ x_n - x) \to 0$, then $A(x_n, y) \to A(x, y)$ in X^*, for all $y \in X$;

(jv) If $x_n \to x$ in X and $A\,(x_n, y) \to f$ in X^*, then $(A\,(x_n, y),\ x_n) \to (f, x)$.

THEOREM *Any semi-monotone operator is pseudo-monotone.*

Proof: Let $\{x_n\}$ such that $x_n \to x$ in X and

$$\lim \sup\,(Px_n, x_n - x) \leqslant 0.$$

First, we show that there exists a subsequence $\{x_k\} \subseteq \{x_n\}$ such that

$$r_k = (A\,(x_k, x_k) - A\,(x_k, x),\ x_k - x) \to 0 \text{ as } k \to \infty. \tag{1}$$

In fact, by (jj) the sequence $\{A(x_n, x)\}$ is bounded and we may choose a sequence $\{x_k\}$ such that $A(x_k, x) \to f$ in X^*. Then, by (jv), we have that $(A(x_k, x), x_k) \to (f, x)$ and thus $(A(x_k, x), x_k - x) \to 0$. Consequently $r_k \to 0$.

Now (jjj) implies that

$$A(x_k, y) \to A(x, y) \text{ in } X^* \quad (\forall)\,y \in X, \tag{2}$$

and (jv) guarantees

$$(A(x_k, y),\ x_k - x) \to 0 \qquad (\forall)\,y \in X. \tag{3}$$

By (j),

$$(Px_k, x_k - x) \geqslant (A\,(x_k, x),\ x_k - x) \to 0,$$

and this together with (1) yields

$$(Px_k, x_k - x) \to 0 \text{ as } k \to \infty. \tag{4}$$

Next we consider the monotonicity relation

$$(Px_k - A\,(x_k, z),\ x_k - z) \geqslant 0$$

with $z = (1 - t) x + ty$ and $t \in [0,1]$, and deduce that

$$t(Px_k, x - y) \geqslant - (Px_k, x_k - x) + (A(x_k, z), x_k - x) + t(A(x_k, z), x - y).$$

By using (2)–(4), we have

$$t \lim \inf (Px_k, x - y) \geqslant t \lim \inf (A(x_k, z), x - y) = t(A(x, z), x - y),$$

whence by division by $t \neq 0$,

$$\lim \inf (Px_k, x - y) \geqslant (A (x, z), x - y).$$

Since $(Px_k, x_k - y) = (Px_k, x_k - x) + (Px_k, x - y)$, by (jj) and (4), we conclude that

$$\lim \inf (Px_k, x_k - y) \geqslant (Px, x - y), \qquad (\forall) \, y \in X,$$

as $t \to 0$. This contradicts (i) in the definition of pseudo-monotonicity.

Since $(Px, x - y) \geqslant (A(x, y), x - y)$, the condition (ii) holds by the boundedness of $A(x, .)$ in (jj).

The class of pseudo-monotone mappings

1.2. Let C be a closed convex set in a reflexive Banach space X and let $T : C \mapsto 2^{X^*}$ be a multivalued mapping (generally nonlinear). The mapping T is said to be *pseudo-monotone* provided that

(P_1) For each $x \in C$, the image Tx is a non-empty closed and convex subset of X^*;

(P_2) If $\{x_n\}$ is a sequence in C converging weakly to $x \in C$ and if $f_n \in Tx_n$ is such that

$$\lim \sup (f_n, x_n - x) \leqslant 0,$$

then to each element $y \in C$ there corresponds an $f(y) \in Tx$ with the property that

$$(f(y), x - y) \leqslant \lim \inf (f_n, x_n - x);$$

(P_3) For each finite-dimensional subspace F of X, the map T is continuous from $C \cap F$ to X^* in the weak topology.

In particular, an operator $T : C \mapsto X^*$ is pseudo-monotone if $D(T) = C$, its restrictions to finite-dimensional subspaces are demicontinuous, and for every sequence $\{x_n\} \subset C, x_n \to x$ in X, the inequality $\lim \sup (Tx_n, x_n - x) \leqslant 0$ implies that

$$(Tx, x - y) \leqslant \lim \inf (Tx_n, x_n - y) \qquad (\forall) \, y \in C.$$

One can easily check that a *completely continuous operator* $T : C \mapsto X^*$ is *pseudo-monotone. In the case of a finite-dimensional space X, any continuous operator is also pseudo-monotone.*

Generally, we are concerned with the mappings that are defined on the space X, i.e., satisfying $D(T) = X$. In this case, *for any $x \in X$ the image Tx is a bounded subset of X*. Indeed, for any sequence $\{[x_n, f_n]\} \subset G(T)$ such that $x_n \to x$ in X and $\limsup(f_n, x_n - x) \leqslant 0$, by virtue of (P_2) we have

$$\liminf(f_n, x_n - y) \geqslant (f(y), x - y) \qquad (\forall) \, y \in X.$$

Since $(f_n, x - y) = (f_n, x_n - y) - (f_n, x_n - x)$, it follows that

$$\liminf(f_n, x - y) \geqslant (f(y), x - y).$$

Set $u = x - y$. Then $\liminf(f_n, u) \geqslant (f(x - u), u)$ and replacing u by $(-u)$ we obtain $\limsup(f_n, u) \leqslant (f(x + u), u)$. Hence $|(f_n, u)|$ is uniformly bounded for each $u \in X$ and, by the uniform boundedness principle I.1.1, the sequence $\{f_n\}$ is bounded in X.

PROPOSITION *If T_1 and T_2 are two pseudo-monotone mappings from X into 2^{X^*}, then their sum $T_1 + T_2$ is also pseudo-monotone.*

Proof: Clearly for $x \in X$ the subset $(T_1 + T_2)\,x = T_1 x + T_2 x$ is non-empty, convex and closed in X, and the sum $T_1 + T_2$ is continuous from any finite-dimensional subspace F of X to X^* in the weak topology.

Consider the sequence $\{x_n\} \subset X$, $x_n \to x$ and $h_n \in (T_1 + T_2)\,x_n$ such that $\limsup(h_n, x_n - x) \leqslant 0$. Since $h_n = f_n + g_n$, where $f_n \in T_1 x_n$ and $g_n \in T_2 x_n$, we have $\limsup[(f_n, x_n - x) + (g_n, x_n - x)] \leqslant 0$.

We shall prove that this inequality implies that

$$\limsup(f_n, x_n - x) \leqslant 0 \quad \text{and} \quad \limsup(g_n, x_n - x) \leqslant 0. \tag{1}$$

Indeed, if this is so then for reasons of symmetry we have $\limsup(g_n, x_n - x) = d > 0$ or, for a subsequence, $\lim(g_n, x_n - x) = d$. Then $\lim(f_n, x_n - x) = -d < 0$. As T_1 is pseudo-monotone, to each $y \in X$ there corresponds an element $f(y) \in T_1 x$ such that

$$\liminf(f_n, x_n - y) \geqslant (f(y), x - y).$$

In particular we may take $y = x$. We then obtain $\liminf(f_n, x_n - x) \geqslant 0$ which contradicts our assumption.

According to (1), the pseudo-monotonicity of T_2 implies that for any $y \in X$, there exist $f(y) \in T_1 x$ and $g(y) \in T_2 x$ such that

$$\liminf(f_n, x_n - y) \geqslant (f(y), x - y) \quad \text{and} \quad \liminf(g_n, x_n - y) \geqslant (g(y), x - y).$$

Adding these inequalities we obtain $\liminf(h_n, x_n - y) \geqslant (h(y), x - y)$, where $f(y) + g(y) = h(y) \in (T_1 + T_2)x$. This shows that $T_1 + T_2$ is pseudo-monotone.

Mappings with generalized pseudo-monotone property

1.3. A multivalued mapping T from a reflexive Banach space X into X^* is said to have the *generalized pseudo-monotone property* if for any sequence $\{[x_n, f_n]\} \subset G(T)$ with $x_n \rightharpoonup x$ in X and $f_n \rightharpoonup f$ in X^* such that $\limsup (f_n, x_n - x) \leq 0$, we have $f \in Tx$ and $(f_n, x_n) \to (f, x)$.

It follows from this definition that T has the generalized pseudo-monotone property if $T^{-1}: X^* \mapsto 2^X$ has the same property.

The relation between pseudo-monotonicity and generalized pseudo-monotonicity is given by the following

PROPOSITION *Any pseudo-monotone mapping has the generalized pseudo-monotone property. Moreover, a bounded mapping with generalized pseudo-monotone property which satisfies condition* (P_1) *is pseudo-monotone.*

Proof: Let $\{[x_n, f_n]\} \subset G(T)$ be a sequence such that $[x_n, f_n] \rightharpoonup [x, f]$ in $X \times X^*$, while $\limsup (f_n, x_n - x) \leq 0$. If T is pseudo-monotone, then to any $y \in X$ there corresponds an element $f(y) \in Tx$ such that $(f(y), x-y) \leq \liminf (f_n, x_n - y)$. Since $\limsup (f_n, x_n) \leq (f, x)$, we have

$$(f_n, x_n - y) \leq \limsup (f_n, x_n) - (f, y) \leq (f, x) - (f, y),$$

i.e.,

$$(f(y), x - y) \leq (f, x - y) \qquad (\forall)\ y \in X. \tag{1}$$

We show that $f \in Tx$. By (P_1), the image Tx is a closed convex subset of X^*. If $f \notin Tx$, there exists an element $u \in X$ such that

$$(f, u) < \inf \{(g, u)|\ g \in Tx\}.$$

Choosing in (1) $y = x - u$, we get a contradiction. Hence, $f \in Tx$. Since $\liminf (f_n, x_n - x) \geq (f(x), x_n - x) = 0$ and

$$(f, x) \leq \liminf (f_n, x_n) \leq \limsup (f_n, x_n) \leq (f, x),$$

we conclude that $[x_n, f_n] \to [x, f]$ in $X \times X^*$.

For the converse implication, we show that conditions (P_2) and (P_3) hold. Consider a sequence $\{x_n\} \subset X$ such that $x_n \rightharpoonup x$ in X and $\limsup(f_n, x_n - x) \leq 0$ for $f_n \in Tx_n$. The sequence $\{f_n\}$ is bounded by hypothesis. If T is not pseudo-monotone, then there exists an element $y \in X$ with the property that

$$\liminf (f_n, x_n - y) < \inf \{(g, x - y)|\ g \in Tx\},$$

or, for a subsequence, also denoted by $\{f_n\}$, one has

$$\lim (f_n, x_n - y) < \inf \{(g, x - y)\ |\ g \in Tx\}.$$

By the reflexivity of X^* we obtain that $f_n \rightharpoonup f$ in X^*. As T has the generalized pseudo-monotone property, we have $f \in Tx$ and $(f_n, x_n) \to (f, x)$. Therefore we get the following contradiction

$$\lim \ (f_n, x_n - y) = (f, x - y) < \inf \ \{(g, x - y) \mid g \in Tx\}.$$

Hence condition (P_2) is fulfilled.

To verify condition (P_3) let $\{x_n\} \subset X$ be a sequence such that $x_n \to x$ and let $f_n \in Tx_n$ for each $n \in \mathbb{N}$. Consider an open weak neighbourhood V of Tx in X^*. We prove that $f_n \in V$ for n sufficiently large. If this is not the case, there exists a subsequence, denoted again by $\{f_n\}$, such that $f_n \notin X^* - V$ for every $n \in \mathbb{N}$. Since $\{f_n\}$ is bounded, there exists a subsequence weakly convergent to an element f of the weakly closed set $X^* - V$. But $(f_n, x_n - x) \to 0$ and as condition (P_2) holds, to each $y \in X$ there corresponds an $f(y) \in Tx$ such that

$$(f, x - y) = \lim \ (f_n, x_n - y) \geqslant (f(y), x - y).$$

Since Tx is a closed convex set and $f \notin Tx$, we can find a $y_0 \in X$ such that

$$(f, x - y_0) < \inf \ \{(g, x - y_0) \mid g \in Tx\},$$

which contradicts the previous inequality. Therefore the continuity of T from X in the strong topology to X^* in the weak topology is established.

In our further considerations we shall use the following

LEMMA *Let T be a mapping with generalized pseudo-monotone property and let $\{[x_n, f_n]\} \subset G(T)$ be such that $[x_n, f_n] \to [x, f]$ in $X \times X^*$. If*

$$\limsup_{m,\, n \to \infty} \ (f_m - f_n, x_m - x_n) \leqslant 0,$$

then $[x, f] \in G(T)$ and $(f_n, x_n) \to (f, x)$.

Proof: It is sufficient to prove that $\limsup \ (f_n, x_n) \leqslant (f, x)$. Passing if necessary to a subsequence, we can assume that $(f_n, x_n) \to a$. We shall show that $a \leqslant (f, x)$. By the hypothesis, to any $\varepsilon > 0$ there corresponds an $N(\varepsilon)$ such that $m, n \geqslant N(\varepsilon)$ implies $(f_m - f_n, x_m - x_n) \leqslant \varepsilon$, i.e.

$$(f_m, x_n) + (f_n, x_n) \leqslant (f_m, x_n) + (f_n, x_m) + \varepsilon.$$

If we fix $m \geqslant N$ and pas to the limit as $n \to \infty$ we obtain $(f_m, x_n) + a \leqslant (f_m, x) + (f, x_m) + \varepsilon$. Now, for $m \to \infty$ one gets $2a \leqslant 2(f, x) + \varepsilon$. As this inequality holds for any $\varepsilon > 0$, it follows that $a \leqslant (f, x)$.

Let us examine another question: When does the sum of two maps T_1 and T_2 have the generalized pseudo-monotone property? Here some restrictions must be imposed on the summands; the most natural ones can be derived from a detailed study of Brézis' condition (ii) above. If the mapping $T : X \mapsto X^*$ satisfies condition (ii), then there exists a continuous non-decreasing function $\rho : \mathbb{R}_+ \mapsto \mathbb{R}_+$ such that

$$(Tx, x - y) \geqslant - \rho(\|y\|), \ \text{for all} \ y \in X, \ \text{i.e.,} \ (Tx, y) \leqslant (Tx, x) + \rho(\|y\|).$$

Hence,

$$\|Tx\| = \sup \ \{(Tx, y) \mid \|y\| \leqslant 1\} \leqslant (Tx, x) + \rho(1).$$

A mapping $T : X \mapsto 2^{X^*}$ is *quasi-bounded* if to each $M > 0$ there corresponds a constant C such that whenever $[x, f] \in G(T)$, $(f, x) \leqslant M \|x\|$ and $\|x\| \leqslant M$, it follows that $\|f\| \leqslant C$. The quasi-boundedness is a generalization of boundedness which is meaningful in the setting of Lemma 3.6 as well as of Section VI.3.2.

THEOREM *Let X be a reflexive Banach space and let T_1 and T_2 be two multi-valued mappings from X into X^*, both with the generalized pseudo-monotone property. Suppose that T_1 is quasi-bounded and that there exists a continuous function ρ from $\mathbb{R}_+$ into $\mathbb{R}_+$ such that*

$$(g, x) \geqslant - \rho (\|x\|) \|x\| \quad (\forall) \, [x, g] \in G \, (T_2).$$

Then $(T_1 + T_2)$ has the generalized pseudo-monotone property.

Proof: Consider $\{[x_n, h_n]\} \subset G(T_1 + T_2)$ such that $[x_n, h_n] \rightharpoonup [x, h]$ in $X \times X^*$ and $\lim \sup (h_n, x_n) \leqslant (h, x)$. We have $h_n = f_n + g_n$ with $f_n \in T_1 x_n$ and $g_n \in T_2 x_n$ for any $n \in \mathbb{N}$. As the sequences $\{x_n\}$ and $\{h_n\}$ are both bounded, there exists an $M > 0$ such that $\|x_n\| + \|h_n\| \leqslant M$ for all $n \in \mathbb{N}$. The assumption on T_2 yield

$$(g_n, x_n) \geqslant - \rho (\|x_n\|) \|x_n\| \geqslant - M_1 \|x_n\| \quad (\forall) \, n \in \mathbb{N},$$

where M_1 depends on M and ρ. Then,

$$(f_n, x_n) = (h_n, x_n) - (g_n, x_n) \leqslant (M + M_1) \|x_n\| \quad (\forall) \, n \in \mathbb{N},$$

and by the quasi-boundedness of T_1, there exists an M_2, which depends on M and M_1 such that $\|f_n\| \leqslant M_2$ for all $n \in \mathbb{N}$. Finally,

$$\|g_n\| \leqslant \|h_n\| + \|f_n\| \leqslant M + M_2 = M_3.$$

Now, as $\{[x_n, h_n]\}$ is bounded in $X \times X^*$, we may assume that $(h_n, x_n) \to a$. It remains to show that $a = (h, x)$ and $[x, h] \in G(T_1 + T_2)$. By choosing again subsequences, we may assume that $f_n \rightharpoonup f$ and $g_n \rightharpoonup g$ in X^* and that the limits are such that $h = f + g$ and $(f_n, x_n) \to a_1$, $(g_n, x_n) \to a_2$, where $a = a_1 + a_2$. Since

$$\lim \sup (h_n, x_n - x) = \lim \sup (f_n, x_n - x) + (g_n, x_n - x) \leqslant 0,$$

$(f_n, x_n - x) \to a_1 - (f, x)$ and $(g_n, x_n - x) \to a_2 - (g, x)$, one gets $a_1 + a_2 - (f + g, x) = a - (h, x) \leqslant 0$. Let us prove that

$$a_1 - (f, x) \leqslant 0 \quad \text{and} \quad a_2 - (g, x) \leqslant 0.$$

Indeed, if this is not the case, we have $a_1 - (f, x) = p > 0$ and from $a_1 - (f, x) + a_2 - (g, x) \leqslant 0$ one derives that $a_2 - (g, x) \leqslant - p < 0$. Since T_2 has the generalized pseudo-monotone property, we deduce that $(g_n, x_n) \to (g, x)$. Hence, $a_2 = (g, x)$ and $0 = a_2 - (g, x) \leqslant - p < 0$. This is a contradiction. Therefore $a_1 \leqslant (f, x)$, and similarly $a_2 \leqslant (g, x)$. This implies that

$$\lim (f_n, x_n - x) = a_1 - (f, x) \leqslant 0 \quad \text{and} \quad \lim (g_n, x_n - x) = a_2 - (g, x) \leqslant 0.$$

As both T_1 and T_2 have the generalized pseudo-monotone property, $[x, f] \in G(T_1)$, $[x, g] \in G(T_2)$, $(f_n, x_n) \to (f, x)$ and $(g_n, x_n) \to (g, x)$. Hence, $[x, h] = [x, f + g]$ lies in $G(T_1 + T_2)$ and

$$(h_n, x_n) = (f_n, x_n) + (g_n, x_n) \to (f + g, x) = (h, x),$$

i.e., the sum $T_1 + T_2$ has the generalized pseudo-monotone property.

2. Monotone mappings

The extension of the monotonicity definition to operators from a Banach space into its dual has been the starting point for the development of nonliniar functional analysis some fifteen years ago. The monotone maps constitute the most manageable class, because of the very simple structure of the monotonicity condition. The monotone mappings appear in a rather wide variety of contexts, since they can be found in many functional equations. Many of them appear also in the calculus of variations as subdifferentials of convex functions and their investigation will be useful as a proper means in applications.

Definitions and examples

2.1. Let X be a real Banach space, let X^* be its dual space and let $T : X \to 2^{X^*}$ be a mapping (possibly nonlinear). The mapping T is said to be *monotone*, if for any $x, y \in D(T)$, the inequality

$$(f - g, x - y) \geqslant 0 \tag{1}$$

holds for all $f \in Tx$ and $g \in Ty$. The mapping T is *strictly monotone* if equality in (1) implies that $x = y$. In particular, $T : X \mapsto X^*$ is a monotone operator if

$$(Tx - Ty, x - y) \geqslant 0 \qquad (\forall)\ x, y \in D(T).$$

In the case of a linear operator $L : X \to X^*$, the monotonicity of L is equivalent to its non-negativity

$$(Lx, x) \geqslant 0,\ (\forall)\ x \in D(L).$$

The set of all monotone mappings is closed under addition.

Example 1. Let $\varphi : \mathbb{R} \mapsto \bar{\mathbb{R}}$ be a monotone increasing function. Then the multivalued map $T : \mathbb{R} \mapsto 2^{\mathbb{R}}$, given by

$$Tx = [\varphi(x - 0),\ \varphi(x + 0)],\ x \in D(\varphi),$$

is monotone.

Example 2. Let $\varphi : X \mapsto \overline{\mathbb{R}}$ be a subdifferentiable proper function. The map $\partial\varphi : X \mapsto 2^{X^*}$ is monotone since for any $f \in \partial\varphi(x)$, $g \in \partial\varphi(y)$, we have

$$\varphi(y) - \varphi(x) \geqslant (f, \ y - x), \qquad \varphi(x) - \varphi(y) \geqslant (g, \ x - y),$$

and addition of these inequalities yields

$$(f - g, \ x - y) \geqslant 0 \qquad (\forall) \ x, y \in D\,(\partial\varphi).$$

Example 3. Let Ω be a bounded domain in $\mathbb{R}^N$. The *pseudo-laplacian* map $T : W_0^{1,p}(\Omega) \mapsto W^{-1,p'}(\Omega)$, $(p \geqslant 2)$ defined by

$$(Tu, v) = \int_\Omega \sum_{i=1}^N |D_i u|^{p-2} \, D_i u \, D_i v \ dx,$$

is strictly monotone. Applying Theorem II.2.3, we consider in $W_0^{1,p}(\Omega)$ the equivalent norm

$$\|u\|_{1,p} = \left(\sum_{i=1}^N \|D_i u\|_p^p \right)^{1/p}.$$

For all $u, v \in W_0^{1,p}(\Omega)$ we have

$$(Tu, u) = \sum_{i=1}^N \|D_i u\|_p^p = \|u\|_{1,p}^p,$$

and

$$(Tu, v) = \sum_{i=1}^N (|D_i u|^{p-2} \, D_i u, \, D_i v) \leqslant \sum_{i=1}^N \|D_i u\|_p^{p-1} \|D_i v\|_p \leqslant$$

$$\leqslant \sum_{i=1}^N \|D_i u\|_p^p \Big)^{\frac{1}{p'}} \left(\sum_{i=1}^N \|D_i v\|^p \right)^{\frac{1}{p}} \leqslant \|u\|_{1,p}^{p-1} \|v\|_{i,p}.$$

Then

$$(Tu - Tv, \ u - v) \geqslant (\|u\|_{1,p}^{p-1} - \|v\|_{1,p}^{p-1}) \, (\|u\|_{1,p} - \|v\|_{1,p}) \geqslant 0.$$

Example 4. Let H be a Hilbert space, whose dual is identified with H. If $A : H \mapsto H$ is a contraction i.e.,

$$\|Ax - Ay\| \leqslant \|x - y\| \qquad (\forall) \ x, y \in D\,(A),$$

then the map $T = I - A$, where I is the identity operator, is monotone.

Monotonicity in Hilbert spaces can be characterized by the following

PROPOSITION *Let H be a Hilbert space. Then a mapping $T : H \mapsto 2^H$ is monotone if and only if*

$$\|x - y + t\,(f - g)\| \geqslant \|x - y\|$$

for all $x, \quad y \in H, f \in Tx, g \in Ty$ and $t \geqslant 0$.

Proof: Suppose T is monotone. Then the equality

$$\|(x - y) + t(f - g)\|^2 = \|x - y\|^2 + 2t(f - g, x - y) + t^2\|f - g\|^2$$

yields

$$\|(x - y) + t(f - g)\| \geqslant \|x - y\|.$$

Conversely, the last inequality implies that

$$2(f - g, x - y) + t\|f - g\| \geqslant 0,$$

and letting $t \to 0$, we obtain the monotonicity of T.

Local boundedness and continuity

2.2. A basic property of monotone mappings is local boundedness. A mapping $T: X \to 2^{X^*}$ is *locally bounded* at $x \in X$ if there exists a neighbourhood U of x such that the set

$$T(U) = \{Ty \mid y \in U \cap D(T)\}$$

is bounded in X^*.

A simple proof of the local boundedness of monotone mappings may be obtained by means of the following

LEMMA *Let $\{x_n\} \subset X$ and $\{f_n\} \subset X^*$ be two sequences such that $x_n \to 0$ in X and $\|f_n\| \to \infty$. Then, for every $\rho > 0$, there exist an element $z \in \bar{B}(0, \rho)$ and subsequences $\{f_j\}$ and $\{x_j\}$ of $\{f_n\}$ and $\{x_n\}$ respectively, such that*

$$\lim (f_j, x_j - z) = -\infty.$$

Proof: Suppose that the conclusion of the lemma is not true. Then to each $z \in \bar{B}(0, \rho)$ there corresponds a constant c_z such that $(f_n, x_n - z) \geqslant c_z$ for all $n \in \mathbb{N}$. For any $k \in \mathbb{N}$ the set

$$E_k = \{u \in \bar{B}(0, \rho) \mid (f_n, x_n - u) \geqslant -k, \ n \in \mathbb{N}\}$$

is closed and $\bar{B}(0, \rho) = \bigcup_{k \in \mathbb{N}} E_k$. According to the Baire category theorem (see Yosida [1, p. 11]), there exist $r > 0$, $y \in \bar{B}(0, \rho)$ and $k_0 \in \mathbb{N}$ such that $\bar{B}(y, r) \subseteq E_{k_0}$. We denote $c_{-y} = c$ and applying addition we obtain

$$(f_n, 2x_n + y - u) \geqslant c - k_0 \quad (\forall) \ n \in \mathbb{N}, \ u \in \bar{B}(y, r).$$

Now, we choose n_0 so that $\|x_n\| \leqslant \dfrac{r}{4}$ for all $n \geqslant n_0$. For $n \geqslant n_0$ and

$\|v\| \leqslant \dfrac{r}{2}$, we have $u = 2x_n + y - v \in \bar{B}(y, r)$ and $(f_n, v) \geqslant c - k_0$. Hence the sequence $\{(f_n, v)\}$ is bounded from below on $\bar{B}\left(0, \dfrac{r}{2}\right)$. Replace v by $-v$ and the boundedness of $\{\|f_n\|\}$ follows. This fact contradicts one of our hypotheses.

THEOREM *A monotone mapping $T: X \mapsto 2^{X^*}$ is locally bounded at the interior points of $D(T)$.*

Proof: Assume the existence of a point $x_0 \in \operatorname{Int} D(T)$ in which T is not locally bounded. Since the monotonicity is invariant under translations, we may suppose that $x_0 = 0$. Then there exists a sequence $\{x_n\} \subset D(T)$ with $x_n \to 0$ in X and a sequence $\{f_n\}, f_n \in Tx_n$ such that $\|f_n\| \to \infty$. By the lemma, passing if necessary to subsequences, we conclude that there exists a $z \in \bar{B}(0, \rho) \subset D(T)$ such that $\lim (f_n, x_n - z) = -\infty$. For any $g \in Tz$ the monotonicity of T implies that

$$(f_n, x_n - z) \geqslant (g, x_n - z) \qquad (\forall) \; n \in \mathbb{N}.$$

We have arrived to a contradiction since the second term in the above inequality is bounded from below. The proof is now complete.

By virtue of this theorem, the image Tx is a bounded subset of X^* for any $x \in \operatorname{Int} D(T)$. As we shall see in Section 6.9, it is possible that T is not locally bounded at any point of the boundary of $D(T)$.

Recall that for any two sequences $\{x_n\} \subset X$, $\{f_n\} \subset X^*$ such that $x_n \to x$ in X and $f_n \to f$ in X^*, we have $(f_n, x_n) \to (f, x)$ because of the boundedness of weakly convergent sequences and the Cauchy-Bunjakowski inequality.

In general, any demicontinuous operator is hemicontinuous. To prove the converse assertion we have to restrict our discussion to monotone operators defined on a reflexive Banach space.

PROPOSITION *Any monotone hemicontinuous operator $T: X \mapsto X^*$ is demicontinuous on $\operatorname{Int} D(T)$.*

Proof: Suppose that T is hemicontinuous on $\operatorname{Int} D(T)$. Let $x \in \operatorname{Int} D(T)$, and $\{x_n\} \subset \operatorname{Int} D(T)$ be such that $x_n \to x$ in X. By the theorem, the sequence $\{Tx_n\}$ is bounded (for n large enough), and by the reflexivity of X, we can assume that $Tx_n \to f$ in X^*. Passing to the limit in the monotonicity relation $(Tx_n - Ty, x_n - y) \geqslant 0$, we obtain

$$(f - Ty, x - y) \geqslant 0 \qquad (\forall) \; y \in D(T). \tag{1}$$

Since $\operatorname{Int} D(T)$ is open, to any $u \in X$ there corresponds a $t_u > 0$ such that $y_t = x + tu \in \operatorname{Int} D(T)$ for all t with $0 < t \leqslant t_u$. Take in (1) the element y_t instead of y; then we have $(f - Ty_t, x - y_t) \geqslant 0$. Set $t \to 0$, and we obtain by the hemicontinuity of T that $(f - Tx, u) \geqslant 0$ for all $u \in X$; i.e. $Tx = f$. Therefore $Tx_n \to Tx$, that is, T is demicontinuous.

COROLLARY *If* $T: X \mapsto X^*$ *is a monotone hemicontinuous operator with* $D(T) = X$, *then* T *is continuous on finite-dimensional subspaces of* X.

The proof is obvious because in this case the weak convergence is equivalent to the strong one.

Any monotone linear operator $L: X \mapsto X^*$ with $D(L) = X$ is continuous. Indeed, its hemicontinuity implies its continuity, in virtue of Proposition I.1.7.

Maximal monotone mappings

2.3. In the investigation by means of the monotonicity methods of the simplest functional equations one makes use of some properties of the corresponding mappings such as the demiclosedness of their graphs and the convexity of the closure of their effective domains.

A set M of $X \times X^*$ is said to be *demiclosed* if $x_n \to x$ in $X, f_n \rightharpoonup f$ in X^* or $x_n \rightharpoonup x$ in $X, f_n \to f$ in X^* and $[x_n, f_n] \in M$ imply that $[x, f] \in M$. These properties are seen to be as consequences of the maximality conditions. It is therefore important to determine whether various given monotone operators are maximal and to be able to generate maximal monotone operators satisfying given conditions.

A set $M \subseteq X \times X^*$ is *monotone* provided that

$$(f - g, x - y) \geqslant 0$$

for any pair $[x, f], [y, g] \in M$. A monotone set M is *maximal* if it is not a proper subset of a monotone set in $X \times X^*$.

The mapping $T: X \mapsto 2^{X^*}$ is said to be *maximal monotone* if its graph $G(T)$ is a maximal monotone set of $X \times X^*$. If follows from this definition that T is maximal monotone if and only if the inequality

$$(f - g, x - y) \geqslant 0 \qquad (\forall) \ [y, g] \in G(T)$$

implies $x \in D(T)$ and $f \in Tx$.

The element $[x, f] \in X \times X^*$ lies in $G(T)$ if and only if $[f, x] \in X \times X^*$ lies in $G(T^{-1})$. Since the monotonicity is invariant under transposition of the domain and the range of a map, T is (maximal) monotone if and only if T^{-1} has this property.

If T *is a maximal monotone mapping then, for any* $x \in D(T)$, *the image* Tx *is a closed convex subset of* X^*. Indeed, if $f_1, f_2 \in Tx$, then for any $t \in [0,1]$ and any $[y, g] \in G(T)$ we have the inequality

$$(f - g, x - y) = (1 - t)(f_1 - g, x - y) + t(f_2 - g, x - y) \geqslant 0$$

for all $f = (1 - t)f_1 + tf_2$. If follows by the maximality of $G(T)$ that $f \in Tx$. Now, if the sequence $\{f_n\} \subset Tx$ converges weakly to f in X^*, then $(f_n - g, x - y) \geqslant 0$, and passing to limit we obtain $(f - g, x - y) \geqslant 0$ and $f \in Tx$.

Similarly, we may conclude that $G(T)$ is demiclosed in $X \times X^*$.

THEOREM *Let $T: X \mapsto 2^{X^*}$ be a monotone mapping such that for each $x \in X$ the image Tx is a non-empty closed convex subset of X^*. If T is continuous from the line segments in X to X^* in the weak topology, then T is maximal monotone.*

Proof: We prove that for any $[x, f] \in X \times X^*$, the inequality

$$(f - g, x - y) \geqslant 0 \qquad (\forall) \, [y, g] \in G(T) \tag{1}$$

implies that $[x, f] \in G(T)$. Suppose that, on the contrary, $f \notin Tx$. Since Tx is convex and closed in X^*, we conclude from the second separation theorem that there exists a $z \in X$ such that

$$(f, z) > \sup \, \{(g, z) \mid g \in Tx\}.$$

Let us take in (1) elements of the form $z_t = x + tz$ with $t > 0$ and $g_t \in Tz_t$, and hence $(g_t - f, z) \geqslant 0$. Passing to limit as $t \to 0$, we have $(g - f, z) \geqslant 0$ by the continuity of T, and this contradicts the choice of the element z. Hence $f \in Tx$.

COROLLARY *Any monotone hemicontinuous operator $T: X \mapsto X^*$ with $D(T) = X$ is maximal monotone.*

We note that *if $T: X \mapsto 2^{X^*}$ is a monotone mapping with $R(T) = X^*$ and T^{-1} is hemicontinuous, then T is a maximal monotone mapping.*

2.4. As was mentioned at the beginning of this chapter, we may study more conveniently the maximal monotone mappings by means of pseudo-monotone ones.

THEOREM *Any maximal monotone mapping $T: X \mapsto 2^{X^*}$ with $D(T) = X$ is a pseudo-monotone one.*

Proof: We prove that T verifies the conditions of pseudo-monotonicity.

(P_1) For each $x \in D(T) = X$, the image Tx is a nonempty closed convex set of X^* (by Theorem 2.2, it is also bounded).

(P_2) Let $\{x_n\} \subset X$ be a sequence which converges weakly to x in X and let $f_n \in Tx_n$ be such that $\lim \sup \, (f_n, x_n - x) \leqslant 0$. For an $f \in Tx$, the monotonicity of T yields $(f, x_n - x) \leqslant (f_n, x_n - x)$ and as $n \to \infty$ we obtain $\lim (f_n, x_n - x) = 0$. Taking $[z, g] \in G(T)$, we can write

$$(f_n, x_n - z) = (f_n, x_n - x) + (f_n, x - z),$$

which implies that

$$\lim \inf \, (f_n, x_n - z) = \lim \inf \, (f_n, x - z).$$

Since $(g, x_n - z) \leqslant (f_n, x_n - z)$, we get

$$(g, x - z) \leqslant \lim \inf \, (f_n, x - z). \tag{2}$$

For $y \in X$ and $t > 0$ we let $z_t = (1 - t) x + ty$ and $g_t \in Tz_t$. Take $[z_t, g_t]$ instead of $[z, g]$ in (2). We obtain

$$(g_t, x - y) \leqslant \lim \inf \, (f_n, x - y).$$

Now, since T is locally bounded at x, we may assume that there exists a sequence $\{t_j\}$, $t_j \to 0$, such that $z_{t_j} \to x$ in X and $g_{t_j} \to f(y)$ in X^*. As T is maximal monotone, we have $f(y) \in Tx$ and

$$(f(y), x - y) \leqslant \lim \inf (f_n, x - y) \leqslant \lim \inf (f_n, x_n - y).$$

(P_3) If this property does not hold, then for a given weak neighbourhood V of Tx in X^*, there exist a sequence $\{x_n\} \subset X$, $x_n \to x$ and $f_n \in Tx_n$ such that $f_n \notin V$ for each $n \in \mathbb{N}$. By the local boundedness of T in x, the sequence $\{f_n\}$ is bounded; hence, we may assume that $f_n \to f$ in X^*. As T is maximal monotone we get $f \in Tx$ but, on the other hand, by considering the weak closure of $X^* - V$ we see that $f \notin V$, which is a contradiction.

REMARK The continuity of maximal monotone mappings from the strong topology of X into the weak topology of X^* follows by the above proof of (P_3).

For further applications we have to discuss the restriction of the preceding results to the case of single-valued mappings defined on a closed convex subset C of a reflexive Banach space X.

PROPOSITION *Any hemicontinuous monotone operator $T: C \mapsto X^*$ is pseudo-monotone.*

Proof: Let $\{x_n\} \subset C$ be a sequence such that $x_n \to x$ in X and

$$\lim \sup (Tx_n, x_n - x) \leqslant 0.$$

By the monotonicity of T, we have $(Tx, x_n - x) \leqslant (Tx_n, x_n - x)$ and thus

$$\lim (Tx_n, x_n - x) = 0.$$

Let $y \in C$ be arbitrary and set $z_t = (1 - t) x + ty$, $t \in (0, 1)$. The monotonicity relation

$$(Tx_n - Tz_t, x_n - z_t) \geqslant 0$$

implies that

$$- (Tx_n, x_n - x) + (Tz_t, x_n - x + t(x - y)) \leqslant t (Tx_n, x - y).$$

Letting $n \to \infty$ and dividing by t we obtain

$$(Tz_t, x - y) \leqslant \lim (Tx_n, x - y).$$

This yields

$$(Tx, x - y) \leqslant \lim (Tx_n, x - y)$$

by the hemicontinuity of T.

Since $\lim \inf (Tx_n, x - y) = \lim \inf (Tx_n, x - x_n) + \lim \inf (Tx_n, x_n - y) = \lim \inf (Tx_n, x_n - y)$, the above inequality may be put in the form

$$(Tx, x - y) \leqslant \lim \inf (Tx_n, x_n - y) \quad (\forall) \ y \in C.$$

This inequality together with the hemicontinuity of T suffice to establish the assertion.

2.5. Given a finite number of mappings $T_j: X \mapsto 2^{X^*}$, $1 \leqslant j \leqslant n$ we define their sum $T: X \to 2^{X^*}$ as follows

$$Tx = \begin{cases} \sum_{j=1}^{n} f_j, & f_j \in T_j(x) \text{ for } x \in \bigcap_{j=1}^{n} D(T_j), \\ \varnothing & \text{otherwise.} \end{cases}$$

Obviously, the sum of a finite number of monotone mappings is also monotone. The maximality of the sum of maximal monotone maps requires some additional assumptions since the graph of $T_1 + T_2$ is empty when $D(T_1) \cap D(T_2) = \varnothing$.

We first give some auxiliary results:

THEOREM *Let $T_1: X \mapsto 2^{X^*}$ be a monotone mapping and let $T_2: X \mapsto X^*$ be a monotone operator with $D(T_2) = X$. If the sum $T = T_1 + T_2$ is a maximal monotone mapping, then T_1 is also maximal monotone.*

Proof: Let $[x_0, f_0] \in X \times X^*$ be such that

$$(f_1 - f_0, x - x_0) \geqslant 0 \qquad (\forall) [x, f_1] \in G(T_1).$$

We denote $g_0 = T_2 x_0$ and $f_2 = T_2 x$. Since f_1 is an arbitrary element in $T_1 x$, the elements of Tx have the form $f = f_1 + f_2$ and therefore

$$(f - g_0 - f_0, x - x_0) = (f_1 - f_0, x - x_0) + (f_2 - g_0, x - x_0) \geqslant 0.$$

By the maximality of T, $x_0 \in D(T)$ and $f_0 + g_0 \in Tx_0$. Hence,

$$f_0 \in Tx_0 - T_2 x_0 = T_1 x_0,$$

which proves the maximality of T_1.

PROPOSITION *Let $T: C \mapsto X^*$ be a hemicontinuous monotone operator and let $P: C \mapsto X^*$ be a pseudo-monotone mapping. Then the sum $S = T + P$ is pseudo-monotone.*

Proof: Let $\{x_n\} \subset C$ be such that $x_n \to x$ and

$$\lim \inf (Sx_n, x_n - x) \leqslant 0.$$

By the monotonicity of T we have $(Px_n, x_n - x) \leqslant (Sx_n, x_n - x) - (Tx, x_n - x)$ and thus

$$\lim \sup (Px_n, x_n - x) \leqslant \lim \sup (Sx_n, x_n - x) \leqslant 0. \tag{1}$$

Since P is pseudo-monotone, the inequality

$$(Px, x - y) \leqslant \lim \inf (Px_n, x - y) \qquad (\forall) \ y \in C \tag{2}$$

holds. By (1) we deduce that $\limsup (Tx_n, x_n - x) \leqslant 0$ and by Proposition 2.4,

$$(Tx, x - y) \leqslant \liminf (Tx_n, x_n - y) \qquad (\forall) \; y \in C. \tag{3}$$

The pseudo-monotonicity of S follows now easily by adding (2) and (3).

Normalized duality map

2.6. The discussion of the maximality of monotone mappings and their surjectivity properties requires the introduction of duality maps.

Let X be a normed space and X^* its dual space. The map $J: X \mapsto 2^{X^*}$ given by

$$Jx = \{ f \in X \mid (f, x) = \|x\|^2 = \|f\|^2 \}$$

is called the *normalized duality map* of X.

According to the Hahn-Banach theorem, Jx is a non-empty set of X^* for every $x \in X$. To each x there corresponds an $f \in X^*$ satisfying $\|f\| = \|x\|$ and such that $(f, x) = \|x\|^2$. Hence $D(J) = X$. It follows also from the definition that J *is odd* $(J(-x) = -J(x))$, *positive homogeneous* $(J(\lambda x) = \lambda J(x)$ for all $\lambda > 0)$ and *bounded*.

The duality map can be equivalently defined as the subdifferential of the convex function $\varphi(x) = \dfrac{1}{2} \|x\|^2$. Indeed,

$$\partial\varphi(x) = \left\{ f \in X^* \left| \; \frac{1}{2}\|z\|^2 - \frac{1}{2}\|x\|^2 \geqslant (f, z - x) \qquad (\forall) \; z \in X \right. \right\}.$$

If $f \in Jx$, then

$$(f, z - x) \leqslant \|f\| \, \|z\| - \|x\|^2 \leqslant \frac{1}{2}(\|x\|^2 + \|z\|^2) - \|x\|^2 =$$

$$= \frac{1}{2}\|z\|^2 - \frac{1}{2}\|x\|^2,$$

i.e., $f \in \partial\varphi(x)$.

Conversely, if $f \in \partial\varphi(x)$ and $z + ty$ with $t > 0$, we have

$$2t(f, y) \leqslant t^2 \|y\|^2 + 2t \, \|x\| \, \|y\|,$$

and $(f, y) \leqslant \|x\| \, \|y\|$ as $t \to 0$. Therefore, $\|f\| \leqslant \|x\|$ and $(f, x) \leqslant \|x\|^2$. Taking now $z = (1 + t) x$ with $t < 0$ we obtain similarly $\|x\|^2 \leqslant (f, x)$. Therefore $f \in J(x)$.

In virtue of the first property of the subdifferential, it follows by this definition that for any $x \in X$, *the image Jx is a closed convex subset of X^**.

When the dual space X^* is strictly convex, J is a single-valued map. In what follows, the reflexive Banach spaces will be considered equivalently renormed as strictly convex spaces and we may assume that J is an operator.

THEOREM *If X is a reflexive Banach space, the operator $J : X \mapsto X^*$ is strictly monotone and demicontinuous.*

Proof: For all $x, y \in X$, $x \neq y$ we have

$$(Jx - Jy, x - y) = (Jx, x) - (Jx, y) - (Jy, x) + (Jy, y) \geqslant$$

$$\geqslant \|x\|^2 - 2\|x\|\,\|y\| + \|y\|^2 = (\|x\| - \|y\|)^2 \geqslant 0.$$

Moreover, if

$$0 = (Jx - Jy, x - y) = (\|x\| - \|y\|)^2 + [\|x\|\,\|y\| - (Jx, y)] +$$

$$+ [\|x\|\,\|y\| - (Jy, x)], \tag{1}$$

and X^* is strictly convex, then Jx and Jy assume their minimum in the points $\dfrac{x}{\|x\|}$ and $\dfrac{y}{\|y\|}$ respectively (see Proposition I.1.5). Therefore

$$(Jx, y) < \|x\|\,\|y\| \quad \text{and} \quad (Jy, x) < \|x\|\,\|y\|$$

and this contradicts (1). Hence, J is strictly monotone on X.

Now let $\{x_n\} \subset X$ be such that $x_n \to x$. Since $\{Jx_n\}$ is a bounded sequence, we can assume, passing, if needed, to a subsequence, that $Jx_n \to f$ in X^*. Then $\|x_n\|^2 = (Jx_n, x_n) \to (f, x) = \|x\|^2$ and

$$\|f\| \leqslant \lim \inf \|Jx_n\| = \lim \inf \|x_n\| = \|x\|.$$

Hence $\|f\| = \|x\|$ and $f = Jx$, which proves the demicontinuity of J.

COROLLARY *If X is a (locally uniformly convex) reflexive Banach space, then $J : X \mapsto X^*$ is continuous. If further X^* is a uniformly convex space, then T is uniformly continuous on bounded sets of X.*

Proof: Since J is single-valued and demicontinuous, $x_n \to x$ in X implies that $Jx_n \to Jx$ in X^*. But $\|Jx_n\| = \|x_n\| \to \|x\| = \|Jx\|$ and the locally uniform convexity of X implies that $Jx_n \to Jx$ in X^*.

Suppose that to the contrary of what has been asserted for some uniformly convex space X^*, the operator J is not uniformly continuous on a bounded set in X. After a suitable homotety has been performed, which is possible because of the positive homogeneity of J, we can assume that the intersection of this set and the unit sphere is nonempty. Hence for any x, y in this intersection, i.e. satisfying $\|x\| = \|y\| = 1$, with the property that $\|x - y\| \to 0$ and $\|Jx - Jy\| \geqslant \varepsilon$ for some $\varepsilon > 0$, we have

$$(Jx + Jy, x) = (Jx, x) + (Jy, y) + (Jy, x - y) \geqslant$$

$$\geqslant \|x\|^2 + \|y\|^2 - \|x - y\| = 2 - \|x - y\|$$

or

$$\left\| \frac{Jx + Jy}{2} \right\| \geqslant 1 - \frac{1}{2} \|x - y\|.$$

This contradicts the uniform convexity of X^*. Therefore, J is a uniformly continuous operator on the bounded sets of X.

We use the duality map to characterize the *projection* P_K on a closed convex set K of a reflexive Banach space X:
For any $x \in X, u = P_K x$ *if and only if*

$$(J(x - u), u - v) \geqslant 0 \qquad (\forall) \, v \in K. \tag{2}$$

By the definition of the projection $P_K : X \mapsto K$,

$$\|x - u\| \leqslant \|x - (1 - t) \, u - tv\| \qquad (\forall) \, v \in K, \; t \in (0,1)$$

and by the definition of the duality map,

$$(J(x - u - t(v - u)), t(v - u)) \leqslant \frac{1}{2} \|x - u\|^2 - \frac{1}{2} \|x - (1 - t) \, u - tv\|^2 \leqslant 0$$

Dividing by $t > 0$ and letting $t \to 0$; we deduce (2).
Conversely, if we assume that (2) holds, then

$$\frac{1}{2} \|x - v\|^2 - \frac{1}{2} \|x - u\|^2 \geqslant (J(x - u), u - v) \geqslant 0 \qquad (\forall) \, v \in K.$$

Hence $\|x - u\| \leqslant \|x - v\|$ and $P_K x = u$. Therefore we have

$$(J(x - P_K x), P_K x - v) \geqslant 0 \qquad (\forall) \, v \in K$$

for each $x \in X$.

In particular, *if* $X = H$ *is a Hilbert space, then* $J = I$ *is the identity map and* $P_K: H \mapsto K$ *is a monotone operator.* Indeed, for $x, y \in K$ we have

$$(x - P_K x, P_K x - v) \geqslant 0 \text{ and } (y - P_K y, P_K y - w) \geqslant 0 \qquad (\forall)\, v, w \in K.$$

Set $v = P_K y$ and $w = P_K x$. We obtain by adding the above inequalities that

$$(P_K x - P_K y, x - y) \geqslant \| P_K x - P_K y \|^2.$$

Linear maximal monotone mappings

2.7. Let X be a reflexive Banach space and X^* its dual space. A mapping $L : X \mapsto 2^{X^*}$ is said to be *linear* if its graph $G(L)$ is a linear subspace of $X \times X^*$, that is, if $f \in Lx$ and $g \in Ly$ imply that $\lambda f + \mu g \in L(\lambda x + \mu y)$ for all real numbers λ and μ. As we have mentioned above the monotonicity of L is equivalent to its non-negativity, i.e.,

$$(f, x) \geqslant 0 \qquad (\forall)\, f \in Lx.$$

As in the case of single-valued maps, we define the *kernel* of L by

$$N(L) = \ker L = \{ x \in D(L) \mid 0 \in Lx \} = L^{-1}(0),$$

and denote by $\bar{L}$ the map which corresponds to the closure of $G(L)$. Then L is *closed* if $L = \bar{L}$.

The mapping $L^*: X \mapsto 2^{X^*}$ is said to be the *adjoint* of L provided that $g \in L^* y$ implies

$$(g, x) = (f, y) \qquad (\forall)\, [x, f] \in G(L).$$

It is clear that L^* is a linear closed map because $G(L^*) = V(G(L)^\perp)$, where $V([u, h]) = [-h, u]$. Indeed, $[y, g] \in G(L^*)$ is equivalent to $[y, g] \perp [-f, x]$ for all $[x, f] \in G(L)$.

Moreover, $L^{**} = \bar{L}$ and $(L^*)^{-1} = (L^{-1})^*$.

By the definition of the adjoint map it follows that

$$L^*(0) = D(L)^\perp. \tag{1}$$

Unlike the case of single-valued maps, L^{-1}, $\bar{L}$ and L^* exist without other restriction on L. We remark that L is single-valued if and only if $L(0) = \{0\}$, and L^* is single-valued if an only if $\overline{D(L)} = X$. If $\overline{D(L^*)} = X$, then L is closable, i.e., $\bar{L}$ is single-valued.

By (1) for L^{-1} we have $N(L^*) = (L^*)^{-1}(0) = D(L^{-1})^\perp = (R(L))^\perp$ and thus

$$N(L^*)^\perp = \overline{R(L)}. \tag{2}$$

PROPOSITION *If L is a linear monotone mapping, then*

$$N(L) \subseteq R(L)^\perp = N(L^*) \quad and \quad L(0) \subseteq D(L)^\perp = L^*(0).$$

Proof: For $x \in N(L)$, by the monotonicity of L we have

$$(tg, ty - x) \geqslant 0 \qquad (\forall) [y, g] \in G(L), \qquad t \in \mathbb{R}.$$

Thus $(g, ty - x) \geqslant 0$ if $t > 0$ and $(g, ty - x) \leqslant 0$ if $t < 0$. As $t \to 0$, we get $(g, x) = 0$ for all $g \in R(L)$, i.e., $N(L) \subseteq R(L)^\perp$. Replacing L by L^{-1} we get

$$N(L^{-1}) = L(0) \subseteq R(L^{-1})^\perp = D(L)^\perp = L^*(0),$$

as claimed.

In particular, any linear monotone map L is single-valued provided that $\overline{D(L)} = X$.

The maps L^{-1} and $\bar{L}$ are monotone whenever L is monotone, but generally L^* is not necessarily monotone. The monotonicity of L^* may be used to give a very simple maximality criterion.

THEOREM *A linear monotone mapping $L: X \mapsto 2^{X^*}$ is maximal if and only if L is closed and L^* is monotone.*

Proof: It is obvious that a linear maximal monotone map is closed. We show that in this case L^* *is monotone.* For $[x, f] \in G(L)$ and $[y, g] \in G(L^*)$, by the definition of the adjoint map, we have

$$(f+g, x-y) = (f, x) - (f, y) + (g, x) - (g, y) = (f, x) - (g, y) \geqslant -(g, y).$$

If $(g, y) < 0$, then by the maximality of L we obtain $[y, -g] \in G(L)$. Taking here $x = y$ and $y = -g$ we arrive to a contradiction. Hence $(g, y) \geqslant 0$, i.e., L^* is monotone.

Conversely, suppose that L is closed and L^* is monotone. Then for any $[z, h] \in X \times X^*$ we can show that

$$(f - h, x - z) \geqslant 0 \qquad (\forall) [x, f] \in G(L)$$

implies $[z, h] \in G(L)$. To do that, let us consider the continuous convex function

$$\varphi(x, f) = \frac{1}{2} \|f - h\|^2 + \frac{1}{2} \|x - z\|^2 + (f - h, x - z), \quad [x, f] \in G(L).$$

The subspace $G(L)$ with the norm induced by $X \times X^*$, i.e.,

$$|[x, f]| = \|x\| + \|f\|, \qquad [x, f] \in G(L)$$

is a (strictly convex) reflexive Banach space. Since $\varphi(x, f) \to \infty$ as $\|[x, f]\| \to \infty$, Corollary I.2.4 implies that the function φ assumes its minimum value at a point $[z_0, h_0] \in G(L)$. In this case φ is Gâteaux differentiable and we can write $\varphi'(z_0, h_0) = 0$ in the form

$$(g, J^{-1}(h_0 - h)) + (J(z - z_0), y) + (h_0 - h, y) + (g, z_0 - z) = 0$$
$$\forall \, [y, g] \in G(L),$$

or equivalently

$$h - h_0 + J(z - z_0) \in L^*[z_0 - z + J^{-1}(h_0 - h)].$$

By the monotonicity of L^*, we deduce that

$$(h - h_0 + J(z - z_0), z_0 - z + J^{-1}(h_0 - h)) \geq 0.$$

Thus

$$\|h - h_0\|^2 + \|z - z_0\|^2 + (h - h_0, z - z_0) \leq \|h - h_0\| \, \|z - z_0\|,$$

and this implies that $z = z_0$ and $h = h_0 \in Lz$, that is, the maximality of $G(L)$ is shown. The proof is complete.

It is obvious that L^* is also a maximal monotone map because of its closedness.

If L is a linear maximal monotone mapping then by the theorem, L^* is monotone and in virtue of the proposition we have

$$N(L) = N(L^*) \text{ and } L(0) = L^*(0).$$

By relations (1) and (2) we can deduce further that

$$D(L) = \overline{D(L^*)} \text{ and } \overline{R(L)} = \overline{R(L^*)}.$$

It is well known that there exist linear operators which are not densely defined and are maximal among all linear single-valued monotone mappings. These operators have no single-valued maximal monotone extensions. However, each such operator has a unique multivalued maximal monotone extension given by $Tu = Lu + D(L)^\perp$.

Surjectivity of coercive mappings

2.8. Let X be a Banach space. A mapping $T : X \mapsto 2^{X^*}$ is called *surjective* if for each element $f \in X^*$ there exists an element $x \in D(T)$ such that $f \in Tx$, i.e., $R(T) = X^*$. The mapping T is said to be *coercive* if there is a function $c : \mathbb{R}_+ \mapsto \mathbb{R}$ with $c(r) \to \infty$ as $r \to \infty$ and such that

$$(f, x) \geq c(\|x\|) \, \|x\| \qquad (\forall) \, [x, f] \in G(T).$$

A mapping T is *coercive with respect to the element* $h \in X^*$ if there exists a number $r > 0$ such that $x \in D(T)$ and $\|x\| \geqslant r$ imply that

$$(f - h, x) > 0 \qquad (\forall) f \in Tx.$$

The coercivity condition may be also expressed as

$$\lim_{\|x\| \to \infty} \frac{(f, x)}{\|x\|} = \infty, \text{ for each selection } f \in Tx.$$

Indeed, if one considers the function

$$c(r) = \inf \left\{ \frac{|(f, x)|}{\|x\|} \quad \|x\| = r, \ f \in Tx \right\}$$

and adopts the convention $\inf \emptyset = \infty$, we have $\lim_{r \to \infty} c(r) = \infty$.

A coercive mapping is coercive with respect to *each element* $h \in X^*$. Indeed, for any h, the properties of $c(r)$ imply the existence of a positive number r such that for all $\|x\| \geqslant r$ the inequalities

$$(f - h, x) = (f, x) - (h, x) \geqslant (c(\|x\|) - \|h\|) \|x\| > 0 \qquad (\forall) f \in Tx$$

hold. Moreover, the coercivity condition is satisfied whenever the mapping is defined on a bounded domain.

The following proposition will emphasize better than Theorem II.3.4 the role of coercivity on finite-dimensional spaces.

PROPOSITION *Let X be a finite-dimensional Banach space. If $T : X \to 2^{X^*}$ is a continuous mapping which is coercive with respect to the element $h \in X^*$, then there exists at least one element $x \in X$, such that $h \in Tx$.*

Proof: Without loss of generality we may consider single-valued maps. For any $t \in [0, 1]$ define the homotopy

$$T_t x = t(Tx - h) + (1 - t)x.$$

As T is coercive with respect to h, there exists an $r > 0$ so that

$$(Tx - h, x) > 0 \qquad (\forall) \|x\| \geqslant r,$$

and therefore

$$(T_t x, x) = t(Tx - h, x) + (1 - t)(x, x) > 0,$$

i.e., $0 \notin T_t(S(0, r))$ for all $t \in [0, 1]$. By property (d2) of the topological degree, $\deg(T_t, 0, B(0, r))$ is constant for all $t \in [0, 1]$ and since it is equal to one for $t = 0$, there exists at least one element $x \in B(0, r)$ such that $0 \in T_1 x$, i.e., $h \in Tx$. The proof is complete.

The above proposition implies that *in any finite-dimensional Banach space X, a continuous coercive mapping $T : X \to 2^{X^*}$ is surjective.*

In order to extend such results to general Banach spaces we need to note some more facts. First, let us observe that *for any fixed element y in a Banach space X, the function*

$$\varphi_y(x) = (Px, y - y) : K \mapsto \overline{\mathbb{R}}$$

is continuous provided that $P : X \mapsto X^*$ *is a continuous operator and K a bounded set of X.* Indeed, let $x_1 \in K$ and $\varepsilon > 0$ be given. We may choose a neighbourhood V of x_1 in X such that

$$|(Px - Px_1, y - x)| < \frac{\varepsilon}{2} \quad \text{for all} \quad x \in K \cap V.$$

Moreover, there exists a neighbourhood U of x_1 such that

$$|(Px_1, x - x_1)| < \frac{\varepsilon}{2} \quad \text{for any } x \in U.$$

Then for any $x \in K \cap V \cap U$, we have

$$|\varphi_y(x) - \varphi_y(x_1)| \leqslant |(Px - Px_1, y - x)| + |(Px_1, x_1 - x)| < \varepsilon,$$

i.e., φ_y is continuous.

LEMMA *(Debrunner-Flor) Let K be a compact convex set of a Banach space X and let G be a monotone set of the product* $K \times X^*$. *If* $P: K \mapsto X^*$ *is a continuous operator and* $h \in X^*$, *then there exists an element* $u \in K$ *with the property that*

$$(f + Pu - h, x - u) \geqslant 0 \quad (\forall) [x, f] \in G. \tag{1}$$

Proof: If one considers the operator $Tx = Px - h$ instead of P we can assume that $h = 0$. If (1) does not hold, then to $[x, f]$ there corresponds a $u \in K$ such that $(f + Pu, x - u) < 0$. For any such pair consider the set

$$N(x, f) = \{y \in K \mid (f + Py, x - y) < 0\}.$$

The function $(f + Py, x - y) : K \mapsto \overline{\mathbb{R}}$ is continuous in y as a sum of continuous functions: $(f + Py, x - y) = (f, x - y) + (Py, x - y)$. Hence the family $\{N(x, f) \mid [x, f] \in G\}$ forms an open covering of the compact set K, and thus there exists a finite subfamily $\{[x_j, f_j] \in G \mid 0 \leqslant j \leqslant n\}$ with the property

$$K = \bigcup_{j=0}^{n} N(x_j, f_j).$$

Let $\{\beta_0, \ldots, \beta_n\}$, where $\beta_j : N(x_j, f_j) \mapsto [0, 1]$ are continuous, be a partition of unity subordinated to this finite covering. Define the maps

$$p(y) = \sum_{j=0}^{n} \beta_j(y) x_j \quad \text{and} \quad q(y) = \sum_{j=0}^{n} \beta_j(y) f_j.$$

As $x_j \in K$, $0 \leqslant j \leqslant n$, we have that $p : K \mapsto K$ due to the convexity of K. Consider the n-dimensional simplex $\sigma = \mathrm{conv}\,(x_0, \ldots, x_n)$. Since $p : \sigma \mapsto \sigma$ is continuous, we conclude from Brouwer's theorem that p has a fixed point $y_0 \in \sigma \subseteq K$, that is, $p(y_0) = y_0$.

On the other hand, let us consider

$$k(y) = (q(y) + Py, p(y) - y) = \sum_{i,j=0}^{n} \beta_i(y)\beta_j(y)(f_i + Py, x_j - y) =$$

$$= k_1(y) + k_2(y),$$

where

$$k_1(y) = \sum_{j=0}^{n} \beta_i^2(y)\,(f_j + Py, x_j - y)$$

and

$$k_2(y) = \sum_{0 \leqslant i < j \leqslant n} \beta_i(y)\beta_j(y)\,[(f_i + Py, x_j - y) + (f_j + Py, x_i - y)].$$

For $y \in K$ there exists at least one m such that $\beta_m(y) \neq 0$. Then $y \in N(x_m, f_m)$ and $(f_m + Py, x_m - y) < 0$. This implies that $k_1(y) < 0$. For the second term, $k_2(y)$, one has

$$(f_i + Py, x_j - y) + (f_j + Py, x_i - y) = (f_i + Py, x_i - y) +$$

$$+ (f_j + Py, x_j - y) + (f_i - f_j, x_j - x_i) < 0.$$

Since $\beta_i(y)\,\beta_j(y) \neq 0$ implies $y \in N(x_i, f_i) \cap N(x_j, f_j)$, i.e.,

$$(f_i + Py, x_i - y) < 0 \text{ and } (f_j + Py, x_j - y) < 0,$$

the monotonicity of G yields

$$(f_i - f_j, x_j - x_i) = -(f_i - f_j, x_i - x_j) \leqslant 0.$$

Hence $k(y) < 0$ for any $y \in K$, which contradicts

$$k(y_0) = (q(y_0) + Py_0, p(y_0) - y_0) = 0.$$

Thus *Debrunner-Flor's inequality* (1) holds.

Next, for a map $T : X \mapsto 2^{X^*}$ and a subset K in X we consider the restriction

$$G(T|_K) = \{[x, f] \in G(T) \mid x \in K \cap D(T)\}.$$

We shall apply the lemma to closed balls with centers at the origin in a finite-dimensional space.

THEOREM *Let F be a finite-dimensional Banach space, let $T : F \mapsto 2^{F^*}$ be a monotone mapping with $0 \in D(T)$ and let $P : F \mapsto F^*$ be a continuous operator* which is coercive with respect to the element $h \in X^*$. Then there exists an element $u \in \bar{B}(0, R)$ with the property

$$(f + Pu - h, x - u) \geqslant 0 \quad (\forall) \, [x, f] \in G(T).$$

Proof: As above, we assume $h = 0$. Since the monotonicity is invariant under translations, one can also assume without loss of generality that $[0, 0] \in G(T)$. Let K_r be the compact ball $K_r = \{x \in F \,|\, \|x\| \leqslant r, r > 0\}$. By the lemma, there exists an element $u_r \in K_r$ such that

$$(f + Pu_r, x - u_r) \geqslant 0 \quad (\forall) \, [x, f] \in G(T|_{K_r}),$$

hence

$$(Pu_r, u_r) \leqslant (f, x) + (Pu_r, x) - (f, u_r), \quad [x, f] \in G(T|_{K_r}).$$

In particular, taking the pair $[0, 0] \in G(T|_{K_r})$ one obtains $(Pu_r, u_r) \leqslant 0$. This implies that $\|u_r\| \leqslant R$ for any $r > 0$, in virtue of the coercivity of P. Consider now the sets

$$S_r = \{u \in F \,|\, (f + Pu, x - u) \geqslant 0, \, [x, f] \in G(T|_{K_r})\}.$$

$S_r \cap K_r$ are non-empty compact sets for any $r > R$. These sets are decreasing with increasing r and

$$\bigcap_{r \geqslant R} (S_r \cap K_r) \neq \varnothing.$$

Let u be an element of this intersection. Since each $y \in D(T)$ belongs to at least one ball K_r, it follows that

$$(f + Pu, x - u) \geqslant 0 \quad (\forall) \, [x, f] \in G(T).$$

2.9. We shall extend Debrunner-Flor's inequality to infinite-dimensional spaces.

PROPOSITION *Let C be a closed convex set in a reflexive Banach space X. If $T : X \mapsto 2^{X^*}$ is a monotone mapping with $0 \in D(T) \cap C$ and $P : C \mapsto X^*$ is a pseudo-monotone bounded operator which is coercive with respect to $h \in X^*$, then there exists at least one element $u \in C$ such that*

$$(f + Pu - h, x - u) \geqslant 0 \quad (\forall) \, [x, f] \in G(T|_C).$$

Proof: Denote $S = T + P$, with $D(S) = D(T) \cap D(P) \subseteq C$ and $Sx = \{f + Px \,|\, f \in Tx\}$. As above, we assume for simplicity that $h = 0$ and write T instead of $T|_C$.

Let Λ be the partially ordered set of all finite-dimensional subspaces F of X with $\|x\|_F = \|x\|_X = \|x\|$ for all $x \in F$. To each $F \in \Lambda$ we associate the injection (inclusion) $j_F : F \mapsto X$, its dual map (the surjection) $j_F^* : X^* \mapsto F^*$ and the mappings

$$T_F = j_F^* T : F \mapsto F^* \quad \text{and} \quad P_F = j_F^* P : F \cap C \mapsto F^*.$$

For any y in F we have

$$(P_F x, y) = (j_F^* P x, y) = (Px, j_F y) = (Px, y),$$

and similarly,

$$(f_F, y) = (f, y) \quad (\forall)\, x \in D(T) \cap F,\ f \in T_F x.$$

Thus T_F is monotone and P_F is pseudo-monotone which implies that P_F is continuous. Moreover, due to the coercivity of P, there exists an $r > 0$ such that

$$(P_F x, y) = (Px, x) \geqslant 0 \quad (\forall)\, x \in B(0, r) \cap C \cap F,$$

and this is the condition of uniform coercivity of P_F. According to Theorem 2.8, there exists an element $u_F \in B(0, r)$ such that

$$(f_F + P_F u_F, x - u_F) \geqslant 0 \quad (\forall)\, [x, f_F] \in G(T_F),$$

and recalling the definitions of T_F and P_F, we get

$$(f + P u_F, x - u_F) \geqslant 0 \quad (\forall)\, [x, f] \in G(T),\ x \in F, \tag{1}$$

where $f = j_F f_F$. Since P is a bounded operator, the set of all elements of the form $g_F = P u_F$ is bounded in X^* and therefore there exists another constant $R > 0$ such that $\|g_F\| < R$ for any $F \in \Lambda$.

Now, for any subspace $F_0 \in \Lambda$ consider the set

$$U_{F_0} = \bigcup \{[u_F, g_F] \mid F \supseteq F_0\}.$$

We assert that the family $\{U_{F_0} \mid F_0 \in \Lambda\}$ has the finite intersection property. Indeed, consider a pair $F_0, G_0 \in \Lambda$. If $H_0 \in \Lambda$ is so that $F_0 \cup G_0 \subseteq H_0$, then

$$U_{F_0} \cap U_{G_0} \supseteq U_{H_0}.$$

On the other hand, for each $F_0 \in \Lambda$ we have

$$U_{F_0} \subseteq [B(0, r) \times B^*(0, R)] \cap G(T|_C),$$

where $B^*(0, R) \subset X^*$. Since X is reflexive, the balls $B(0, r)$ and $B^*(0, R)$ are both weakly compact and thus there exists a pair $[u, g]$ in $C \times X^*$ which belongs to the weak closure of each set U_{F_0} with $F_0 \in \Lambda$. If we fix now the subspace F_0, then, by Proposition I.1.4, there exists a sequence $\{[u_n, g_n]\} \subset U_{F_0}$ such that $[u_n, g_n] \to [u, g]$ in $X \times X^*$, i.e., $u_n \to u$ in X, $u \in C$ and $g_n \to g$ in X^*. Extend the monotone mapping T to a maximal monotone set of $X \times X^*$. As $n \to \infty$ in (1), we obtain

$$\limsup (P u_n, u_n) \leqslant (g, x) + (f, x - u) \quad (\forall)\, [x, f] \in G(T). \tag{2}$$

We show now that

$$(g, x) + (f, x - u) \leqslant (g, u).$$

Indeed, if this does not hold, we have

$$(f + g, x - u) > 0 \quad (\forall)\, [x, f] \in G(T), \tag{3}$$

and this implies that $[u, -g] \in G(T)$ because of the maximality of the extension of $G(T)$. In particular, setting $x = u$ and $f = -g$ in (3) we obtain the contradiction

$$0 = (g - g, u - u) > 0.$$

Therefore (2) yields $\lim \sup (Pu_n, u_n - u) \leqslant 0$ and because of the pseudo-monotonicity of P we have

$$\lim \inf (Pu_n, u_n - x) \geqslant (Pu, u - x) \quad (\forall) \ x \in C.$$

This inequality combined with (1) leads to

$$(f + Pu, x - u) \geqslant 0 \quad (\forall) \ [x, f] \in G(T),$$

which completes the proof.

According to Proposition I.1.4, the existence of the sequence $\{[u_n, g_n]\} \subset U_{F_0}$ which converges weakly in $X \times X^*$ to the element $[u, g]$ of the weak closure of U_{F_0} justifies the fact that ordinary sequences are sufficient to define pseudo-monotone mappings on reflexive Banach spaces.

REMARK If $[0, 0] \in G(T)$ we may weaken the assumptions on P replacing its boundedness by quasi-boundedness. Indeed, taking $x = 0$ and $f = 0$ in (1) we obtain $(P_F u_F, u_F) \leqslant 0$. Since the net $\{u_F\}$ is bounded by the quasi-boundedness of P, the boundedness of the net $\{g_F\}$, where $g_F = P_F u_F$, is obvious.

We observe that the condition $[0, 0] \in G(T)$ can be assumed without loss of generality because of the invariance of monotonicity under translations.

An important consequence of Debrunner-Flor's lemma is the following result on the inversion of functional equations.

THEOREM *Let C be a closed subset of a reflexive Banach space X. If $T : X \mapsto 2^{X^*}$ is a maximal monotone mapping with $0 \in D(T) \cap C$ and $P : C \mapsto X^*$ is a pseudo-monotone bounded operator which is coercive with respect to $h \in X$, then there exists at least one element $u \in X$ such that $h \in (T + P)u$.*

Proof: According to the previous proposition, there exists an element $u \in C$ with the property that

$$(f + Pu - h, x - u) \geqslant 0 \quad (\forall) \ [x, f] \in G(T).$$

Since T is maximal monotone, we have that $[u, h - Pu] \in G(T)$, i.e.,

$$h \in (T + P)u.$$

As coercivity implies coercivity with respect to each element of X, *under the hypotheses of the theorem, if P is coercive then the sum $T + P$ is surjective.*

COROLLARY *Let X be a reflexive Banach space, let $T : X \mapsto 2^{X^*}$ be a maximal monotone mapping and let $P : X \mapsto X^*$ be a hemicontinuous bounded coercive monotone operator with $D(P) = X$. Then $R(T + P) = X^*$.*

Proof: By Corollary 2.3, P is a maximal monotone operator and by Theorem 2.4 it is also pseudo-monotone.

Take $T : X \mapsto 2^{X^*}$ to be the mapping defined by $Tx = \{0\}$ for each $x \in X$. Consequently, by the above theorem *any pseudo-monotone bounded coercive operator is surjective*.

2.10. Using Corollary 2.9 and the properties of the duality map, we can prove the main surjectivity result on maximal monotone mappings.

THEOREM *If X is a (strictly convex) reflexive Banach space and $T: X \mapsto 2^{X^*}$ is a maximal monotone and coercive mapping, then T is surjective.*

Proof: As monotonicity is invariant under translations, it is sufficient to show that $0 \in R(T)$.

Consider a sequence of positive numbers $\{\varepsilon_n\}$ such that $\varepsilon_n \to 0$ and take the sequence of maps $\{T + \varepsilon_n J\}$. Since X is a strictly convex reflexive Banach space, Theorem 2.6 implies that J is a bounded demicontinuous and strictly monotone operator. It is also coercive since

$$(Jx, x) = \|x\|^2 \quad (\forall) \, x \in D(J) = X.$$

By Corollary 2.9 to each ε_n there corresponds an element $u_n \in X$ with the property that $0 \in (T + \varepsilon_n J)u_n$. This implies the existence of an element $f_n \in Tu_n$ such that $f_n + \varepsilon_n J u_n = 0$. Since T is coercive, we have that

$$0 = (f_n + \varepsilon_n J u_n, u_n) \geqslant c(\|u_n\|) \|u_n\| + \varepsilon_n \|u_n\|^2,$$

that is, $c(\|u_n\|) \leqslant 0$. Hence, there exists a constant $r > 0$ such that $\|u_n\| < r$, because $c(r) \to \infty$ as $r \to \infty$. Since X is reflexive, we can assume that the sequence $\{u_n\}$, or at least one of its subsequences converges weakly to u in X. We must show that $0 \in Tu$. Indeed, since $\varepsilon_n \to 0$ we have $\|f_n\| = \varepsilon_n \|J u_n\| = \varepsilon_n \|u_n\| \leqslant \varepsilon_n r \to 0$, i.e., $f_n \to 0$. By the monotonicity relation

$$(g - f_n, y - u_n) \geqslant 0 \quad (\forall) \, [y, g] \in G(T)$$

we have

$$(g, y - u) \geqslant 0 \quad (\forall) \, [u, g] \in G(T)$$

as $n \to \infty$. Hence $0 \in Tu$ by the maximality of T.

COROLLARY *If X is a (strictly convex) reflxive Banach space and $T : X \mapsto X^*$ with $D(T) = X$ is a coercive hemicontinuous monotone mapping, then T is surjective.*

Proof: Indeed, T is maximal monotone by Corollary 2.3 and it suffices to apply the theorem.

A maximality criterion

2.11. Using the results of previous sections we shall discuss presently the relation between surjectivity and maximality in reflexive Banach spaces. Subsequently, by virtue of Asplund's theorem, the reflexive Banach spaces X and X^* will be considered as strictly convex spaces. We first give some details concerning the duality structure of these spaces.

LEMMA *If* $\{x_n\}$ *is a sequence in the reflexive Banach space* X *such that*

$$(Jx_n - Jx, x_n - x) \to 0,$$

then $x_n \to x$ *in* X.

Proof: After some computation we obtain

$$(Jx - Jx, x_n - x) = (\|x_n\| - \|x\|)^2 + [\|x_n\| \|x\| - (Jx_n, x)] +$$
$$+ ([\|x_n\| \|x\| - (Jx, x_n)] \to 0,$$

where each term to the right is non-negative. Hence $\|x_n\| \to \|x\|$ and $(Jx, x_n) \to \|x\|^2$. Since X is a reflexive space, bounded sets in X are weakly sequentially compact and thus we can assume that $x_n \to y$ in X, whence $\|y\| \leqslant \lim \inf \|x_n\| = \|x\|$. Therefore for $n \to \infty$ we have $(Jx, x_n) \to (Jx, y) = \|x\|^2$ and $\|y\| > \|x\|$. Hence $\|x\| = \|y\|$ and $(Jx, x) = (Jx, y)$. By (ii) in Proposition I.1.5 the strict convexity of X implies that Jx attains its minimum in at most one point on the sphere $S(0, \|x\|)$, i.e., $x = y$ and $x_n \to x$ in X.

PROPOSITION *Let* X *be a reflexive Banach space and let* λ *be a positive number. If* $T : X \mapsto 2^{X^*}$ *is a maximal monotone mapping, then* $R(T + \lambda J) = X^*$ *and* $(T + \lambda J)^{-1} : X^* \mapsto X$ *is a demicontinuous single-valued maximal monotone map.*

Proof: Since $J : X \mapsto X^*$ is a coercive monotone demicontinuous and bounded operator with $D(T) = X$, Corollary 2.9 implies that $R(T + \lambda J) = X^*$ for any $\lambda > 0$. Thus we can define the inverse map $(T + \lambda J)^{-1}$ on X^*.

$(T + \lambda J)^{-1}$ *is single-valued.* Indeed, let $x, y \in (T + \lambda J)^{-1}(h)$ and put $f = h - \lambda Jx \in Tx$, $g = h - \lambda Jy \in Ty$. By the monotonicity of T we deduce that

$$0 = (h - h, x - y) = ((f + \lambda Jx) - (g + \lambda Jy), x - y) =$$
$$= (f - g, x - y) + \lambda (Jx - Jy, x - y) \geqslant \lambda (Jx - Jy, x - y),$$

that is, $x = y$ because of the strict monotonicity of J.

$(T + \lambda J)^{-1}$ *is demicontinuous.* Let $\{x_n\} \subset X$ be such that

$$(T + \lambda J) x_n \to (T + \lambda J)x \text{ in } X^*.$$

By Theorem 2.2, $\{x_n\}$ is bounded and

$$((T + \lambda J)x_n - (T + \lambda J)x, x_n - x) = (x_n - Tx, x_n - x) +$$
$$+ \lambda(Jx_n - Jx, x_n - x) \to 0.$$

Since both terms are non-negative, it follows that $(Jx_n - Jx, x_n - x) \to 0$. This, in virtue of the lemma, implies that $x_n \rightharpoonup x$ in X.

$(T + \lambda J)^{-1}$ *is a maximal operator* by Corollary 2.3.

We now deduce easily the following

THEOREM *Let X be a reflexive Banach space. If $T : X \mapsto 2^{X^*}$ is a monotone mapping, then a necessary and sufficient condition for the maximality of T is that*

$$R(T + \lambda J) = X^* \text{ for each real number } \lambda > 0.$$

Proof: Necessity follows by the proposition. The condition is also sufficient since, by Corollary 2.3, $T + \lambda J$ is a maximal monotone mapping for each $\lambda > 0$. The maximality of T follows from Theorem 2.5.

We can now establish a sufficient condition for the maximality of sums of monotone mappings.

COROLLARY *If $T : X \mapsto 2^{X^*}$ is a maximal monotone map and $P : X \mapsto X^*$ is a hemicontinuous bounded monotone operator with $D(P) = X$, then the sum $S = T + P$ is a maximal monotone mapping.*

Proof: For any $\lambda > 0$ the sum $P + \lambda J$ is a bounded hemicontinuous monotone and coercive operator. Thus Corollary 2.9 implies that $R(T + P + \lambda J) = X^*$ or $R(S + \lambda J) = X^*$, which in virtue of the theorem leads to the conclusion that $S = T + P$ is maximal monotone.

For example, let $X = X^* = \mathbb{R}^1$ and define on $\mathbb{R}_+$ the map $Tx = x^2$. Obviously T is monotone but it is not maximal monotone, because it admits the maximal monotone extension

$$\tilde{T}x = \begin{cases} x^2 & \text{if } x > 0 \\ (-\infty, 0) & \text{if } x = 0. \end{cases}$$

2.12. Using the previous results we can describe a maximal extension of a monotone mapping.

THEOREM *Let C be a closed convex subset of a reflexive Banach space X. If $T : X \mapsto 2^{X^*}$ is a monotone mapping, then there exists a maximal monotone extension $\tilde{T}$ of $T|_C$ such that $D(\tilde{T}) \subseteq C$.*

Proof: Zorn's lemma ensures the existence of a maximal monotone extension T' with $G(T') \subseteq X \times X^*$. Denote by $\tilde{T}$ the restriction of T' to C. Take $P = \lambda J$ with $\lambda > 0$ in Proposition 2.9. There exists an element $u \in C$ such that

$$(f + \lambda Ju - h, x - u) > 0 \qquad (\forall) \ [x, f] \in G(T').$$

As T' is maximal on X, we have $[u, h - \lambda Ju] \in G(T')$. Then $\tilde{T}: C \mapsto 2^{X*}$ is maximal monotone because $R(\tilde{T} + \lambda J) = X^*$.

Subdifferentials of convex functions

2.13. Let X be a reflexive Banach space endowed with a strictly convex norm.

PROPOSITION *Let $\varphi: X \mapsto \overline{\mathbb{R}}$ be a proper convex l.s.c. function. Then $\partial \varphi: X \mapsto 2^{X*}$ is a maximal monotone mapping.*

Proof: As it was shown in example 2 in 2.1, $\partial \varphi$ is monotone on $D(\partial \varphi)$. To prove the maximality of $\partial \varphi$ we apply theorem 2.11, that is, we must show that the equation $f \in (J + \partial \varphi)x$ has a solution $x_0 \in D(\partial \varphi)$ for each $f \in X^*$. Indeed, let $\Phi: X \mapsto \overline{\mathbb{R}}$ be the proper convex l.s.c. function defined by

$$\Phi(x) = \frac{1}{2} \|x\|^2 + \varphi(x) - (f, x).$$

Since $\Phi(x) \to \infty$ as $\|x\| \to \infty$, by Corollary I.2.4, there exists an $x_0 \in D(\partial \varphi)$ such that $\Phi(x_0) = \inf \{\Phi(x) \mid x \in X\}$, i.e., by Theorem I.2.6,

$$0 \in \partial \Phi(x_0) = (J + \partial \varphi) x_0 - f.$$

Hence $f \in (J + \partial \varphi) x_0$, i.e., $\partial \varphi$ is maximal monotone.

A mapping $T: X \to 2^{X*}$ is said to be *cyclically monotone* if for any cyclical set $\{x_0, \ldots, x_n, x_{n+1} = x_0\} \subset D(T)$ and $f_j \in Tx_j$, $0 \leqslant j \leqslant n$, we have

$$\sum_{j=0}^{n} (f_j, x_j - x_{j+1}) \geqslant 0. \tag{1}$$

T is called *cyclically maximal monotone* if it has no cyclically monotone extension. A cyclical map which is maximal monotone is also cyclically maximal monotone.

THEOREM *A map $T: X \mapsto 2^{X*}$ is cyclically maximal monotone if and only if T is the subdifferential of a proper convex l.s.c. function.*

Proof: The "if" part. Let $\varphi: X \mapsto \overline{\mathbb{R}}$ be a proper convex l.s.c. function such that $T = \partial \varphi$, let the set $\{x_0, x_1, \cdots, x_{n+1} = x_0\}$ be cyclical in $D(T)$ and let $f_j \in Tx_j$, $0 \leqslant j \leqslant n$. By the definition of subdifferential we can see that

$$(f_j, x_j - x_{j+1}) \geqslant \varphi(x_j) - \varphi(x_{j+1}), \quad 0 \leqslant j \leqslant n$$

and by addition we obtain (1).

The "only if "part. Take T cyclically maximal monotone and define for any natural number n the function

$$\varphi(x) = \sup \left\{ \sum_{j=0}^{n-1} (f_j, x_{j+1} - x_j) + (f_n, x - x_n) \mid [x_j \, f_j] \in G(T), 0 \leqslant j \leqslant n \right\}.$$

As an upper-bound of affine functions, this function is convex and l.s.c. It follows from (1) that $\varphi(x_0) \leqslant 0$, more precisely $\varphi(x_0) = 0$, that is, φ is a proper function. Moreover for any $[x, f] \in G(T)$ we have

$$(f_0, x_1 - x_0) + \ldots + (f_n, x - x_n) + (f, y - x) \leqslant \varphi(y,) \qquad (\forall) \, y \in X,$$

and this implies that $(f, y - x) \leqslant \varphi(y) - \varphi(x)$, i.e., $f \in \partial \varphi(x)$. Hence $T = \partial \varphi$, due to the maximality of T. The proof is complete.

Finally, we note an interesting property of maximal monotone mappings defined on real line:

Any monotone set in $\mathbb{R} \times \mathbb{R}$ is cyclically monotone.

Indeed, let $T : \mathbb{R} \mapsto 2^{\mathbb{R}}$ be a monotone map and let $\{x_0, \ldots, x_n, x_{n+1} = x_0\}$ be a cyclical set in $D(T)$. We may assume that $x_0 \leqslant x_1 \leqslant \cdots \leqslant x_n$, and this implies that for $f_j \in Tx_j$, $0 \leqslant j \leqslant n$, $f_{n+1} \leqslant f_1 \leqslant \ldots \leqslant f_n$. Therefore,

$$\sum_{j=0}^{n} f_j(x_j - x_{j+1}) = \sum_{j=1}^{n} (f_j - f_{n+1})(x_j - x_{j+1}) \geqslant 0.$$

If, moreover, T is maximal, then by the theorem there exists a convex l.s.c. function $\varphi : \mathbb{R} \mapsto \overline{\mathbb{R}}$ such that $T = \partial \varphi$.

Example. Let β be a nondecreasing real function and define a maximal monotone mapping by

$$Tx = [\beta(x - 0), \beta(x + 0)].$$

For any $x_0 \in D(T)$, we can construct φ as follows

$$\varphi(x) = \begin{cases} \varphi(x_0) + \displaystyle\int_{x_0}^{x} \beta(t) \, dt & \text{if } x \in D(T) \\ + \infty & \text{otherwise} \end{cases}$$

Clearly $T = \partial \varphi$.

2.14. We have used the normalized duality map in proving the maximality of maps. In a number of concrete examples it is easier to work with general duality maps which can be obtained as subdifferentials of a class of convex functions.

A real function $\varphi : \mathbb{R}_+ \mapsto \mathbb{R}_+$ is called the *gauge* if it is continuous, increasing, $\varphi(0) = 0$ and $\varphi(r) \to \infty$ as $r \to \infty$. The primitive Φ of a gauge function φ, given by

$$\Phi(t) = \int_0^t \varphi(r) \, dr, \qquad t \in [0, \infty), \tag{1}$$

is a strictly increasing, continuous, convex and positive function.

Let X be a reflexive Banach space. The subdifferential of $\Phi(\|x\|)$ denoted by $J_\varphi : X \mapsto 2^{X^*}$ is said to be the *duality map corresponding to the gauge* φ. The normalized duality map is the duality map which corresponds to the gauge $\varphi(r) = r$.

Due to the results of Section 2.13, we conclude that J_φ is a cyclically maximal monotone mapping.

The form of duality maps is described by the following

THEOREM *The map* $J_\varphi : X \mapsto 2^{X^*}$ *is a duality map corresponding to the gauge* φ, *if and only if*

$$J_\varphi(x) = \{f \in X^* \mid (f, x) = \|f\| \|x\|, \quad \|f\| = \varphi(\|x\|)\}. \tag{2}$$

Proof: By definition, $f \in J_\varphi(x)$ provided that

$$\Phi(\|y\|) - \Phi(\|x\|) \geqslant (f, y - x) \qquad (\forall) \, y \in X. \tag{3}$$

Choosing $\|y\| = \|x\|$ in (3) we can deduce that $(f, x) \geqslant \|f\| \|x\|$, that is, $(f, x) = \|f\| \|x\|$. Now take in (3) $y = tu$ and $x = su$, where $\|u\| = 1$ and $s, t > 0$. We have

$$\Phi(t) - \Phi(s) \geqslant \|f\| (t - s).$$

This relation may be put in the following double form :

$$\|f\| \leqslant \frac{1}{r} [\Phi(s + r) - \Phi(s)] \qquad \text{for } r = t - s$$

and

$$\|f\| \geqslant \frac{1}{r} [\Phi(s) - \Phi(s - r)] \qquad \text{for } r = s - t.$$

Therefore $\|f\| = \varphi(s) = \varphi(\|x\|)$.

Conversely, by (1) and (2) we can see that

$$\Phi(\|y\|) - \Phi(\|x\|) = \int_{\|x\|}^{\|y\|} \varphi(r) \, dr \geq \varphi(\|x\|) \, (\|y\| - \|x\|) =$$

$$= \varphi(\|x\|) \, \|y\| - (f, x) \geq (f, y - x),$$

which implies (3).

As in the case of the normalized duality map, J_φ is *defined on X.* In fact, by the Hahn-Banach theorem, to each $x \in S(0, 1)$ in X there corresponds uniquely an element $f \in X^*$ such that $\|f\| = 1$ and $(f, x) = 1$. Hence we can define

$$J_\varphi(\lambda x) = \varphi(\lambda) f \qquad \lambda \geq 0$$

and then $D(J_\varphi) = X$.

It is also easily seen that J_φ is a single-valued map whenever X^* is a strictly convex space.

We conclude this section with some examples of duality maps which often appear in applications.

Let Ω be a domain in $\mathbb{R}^N$ and let X be a reflexive Banach space with the norm $\| \cdot \|$. *The duality map in* $L^p(\Omega, X), 1 < p < \infty$, *corresponding to the gauge* $\varphi(r) = r^{p-1}$, *is*

$$J_\varphi(u) = \|u\|^{p-1}. \ \text{sign } u = \|u\|^{p-2} u.$$

Indeed, taking the Gâteaux differential of $\Phi(\|u\|_p)$, where $\Phi(t) = \dfrac{t^p}{p}$ we have

$$(J_\varphi(u), v) = \left[\frac{d}{dt} \Phi(\|u + tv\|_p) \right]_{t=0} = \frac{1}{p} \left[\frac{d}{dt} (\|u + tv\|_p^p) \right]_{t=0} =$$

$$= \frac{1}{p} \left[\frac{d}{dt} \int_\Omega \|u + tv\|^p \, dx \right]_{t=0} = \int_\Omega \|u\|^{p-1} (\text{sign } u) \, v \, dx = (\|u\|^{p-2} u, v).$$

For the Sobolev space $W_0^{1,p}(\Omega), 1 < p < \infty$, *the duality map corresponding to the gauge* $\varphi(r) = r^{p-1}$ *is*

$$J_\varphi(u) = - \sum_{i=1}^N \frac{\partial}{\partial x_i} \left(\left| \frac{\partial u}{\partial x_i} \right|^{p-2} \frac{\partial u}{\partial x_i} \right).$$

In fact, we have seen that the norm in $W_0^{1,p}(\Omega)$ is the norm of $L^p(\Omega, \mathbb{R}^N)$ applied to the function ∇u. Therefore

$$(J_\varphi(u), v) = \int_\Omega |\nabla u|^{p-2} \nabla u \cdot \nabla v \, dx.$$

Applying Green's formula on the right-hand side we obtain

$$(J_\varphi(u),\, v) = \int_\Omega \left[-\sum_{i=1}^N \frac{\partial}{\partial x_i} \left(\left| \frac{\partial u}{\partial x_i} \right|^{p-2} \frac{\partial u}{\partial x_i} \right) \right] v \, dx.$$

We observe that for $p = 2$,

$$J_\varphi(u) = -\Delta u.$$

Thus the duality map on $W_0^{1,p}(\Omega)$ turns out to be a nonlinear generalization of the Laplacian.

3. Perturbation of maximal monotone mappings

In this section we deal with the sum $T + P$ where T is a maximal monotone mapping, and P is a mapping with the generalized pseudo-monotone property. In order to deduce a general maximality criterion for $T + P$ we apply to such mappings an approximation procedure by considering the Yosida approximant T_λ.

The Yosida approximant

3.1. Let X be a reflexive Banach space and let X^* be its dual space (which are both assumed to be strictly convex spaces). Let $T: X \mapsto 2^{X^*}$ be a maximal monotone mapping with $[0, 0] \in G(T)$. By Corollary 2.9, the equation

$$0 \in J(x - u) + \lambda Tx$$

has a solution for any fixed $u \in X$ and $\lambda > 0$, i.e., there exists an $[x, f] \in G(T)$ such that

$$J(x - u) + \lambda f = 0. \tag{1}$$

Setting

$$f = T_\lambda u \quad \text{and} \quad x = \mathcal{J}_\lambda u, \tag{2}$$

we can define two new maps: the *Yosida approximant* $T_\lambda : X \mapsto X^*$ and the *resolvent* $\mathcal{J}_\lambda : X \mapsto D(T)$. Both these maps are operators because J is single-valued and $D(T_\lambda) = D(\mathcal{J}_\lambda) = X$.

We note that for all $u \in X$ we have the splitting

$$u = \mathscr{J}_\lambda u + \lambda J^{-1}(T_\lambda u) \tag{3}$$

because $J(u - \mathscr{J}_\lambda u) = \lambda T_\lambda u$.

Since $T_\lambda u \in T \mathscr{J}_\lambda u$, it is important to distinguish between the single-valued map T_λ and the multi-valued map $T \mathscr{J}_\lambda$. We can also write

$$T_\lambda u = \frac{1}{\lambda} J(u - x) = \frac{1}{\lambda} J(u - \mathscr{J}_\lambda u). \tag{4}$$

For any $[v, g] \in G(T)$, by the monotonicity of T we have

$$(g, \mathscr{J}_\lambda u - v) \leqslant (T_\lambda u, \mathscr{J}_\lambda u - v) = -\frac{1}{\lambda}(J(\mathscr{J}_\lambda u - u), \mathscr{J}_\lambda u - v) =$$

$$= -\frac{1}{\lambda}(J(\mathscr{J}_\lambda u - u), \mathscr{J}_\lambda u - u) - \frac{1}{\lambda}(J(\mathscr{J}_\lambda u - u), u - v)$$

whence

$$\|\mathscr{J}_\lambda u - u\|^2 \leqslant -\lambda(g, \mathscr{J}_\lambda u - v) - (J(\mathscr{J}_\lambda u - u), u - v). \tag{5}$$

This implies the boundedness of the net $\{\mathscr{J}_\lambda u - u\}$.

In particular, if $v = u$ then (5) yields

$$\|\mathscr{J}_\lambda u - u\| \leqslant \lambda\|g\|$$

and

$$\|\mathscr{J}_\lambda u\| \leqslant \lambda\|g\| + \|u\| \qquad (\forall) u \in D(T).$$

By (4) we have also

$$\|T_\lambda u\| = \frac{1}{\lambda}\|\mathscr{J}_\lambda u - u\| \leqslant \|g\| \qquad (\forall) u \in D(T).$$

Therefore, *the operators J_λ and T_λ are bounded on $D(T)$.*

PROPOSITION If $u \in \overline{\mathrm{conv}\, D(T)}$, then $\mathscr{J}_\lambda u \to u$ in X as $\lambda \to 0$.

Proof: Let $\{\lambda_n\}$ be a sequence such that $\lambda_n \to 0$ and $J(\mathscr{J}_{\lambda_n} u - u) \rightharpoonup h$ in X^*. By (5) we deduce that

$$\limsup \|\mathscr{J}_{\lambda_n} u - u\|^2 \leqslant (h, v - u) \qquad (\forall) v \in D(T),$$

and this holds also for $v \in \overline{\text{conv } D(T)}$. In particular, this inequality remains valid for $v = u$, whence the conclusion of the proposition.

Consequently $\overline{D(T)} = \overline{\text{conv } D(T)}$, that is, $\overline{D(T)}$ is a convex set of X. As T and T^{-1} are both maximal monotone mappings and $R(T) = D(T^{-1})$, the set $\overline{R(T)}$ is also convex.

For any set C of X we define

$$|C| = \begin{cases} \inf \{\|x\| \mid x \in C\} & \text{if } C \neq \emptyset, \\ + \infty & \text{otherwise.} \end{cases}$$

Obviously, in our case this definition is meaningful for $|Tx|$. Denote by $T^0 x$ the element of Tx for which $\|T^0 x\| = |Tx|$. In a strictly convex Banach space this element is either unique or it does not exist. If X is also reflexive, then $\|T^0 x\| = \inf \{\|f\| \mid f \in Tx\}$ is always attained (see I.3.3); hence $D(T^0) = D(T)$.

THEOREM *Let* $T: X \mapsto 2^{X^*}$ *be a maximal monotone mapping with* $[0, 0] \in G(T)$. *Then*

(i) T_λ *is a bounded maximal monotone operator and* $[0, 0] \in G(T_\lambda)$ *for each* $\lambda > 0$;

(ii) $T_\lambda u \to T^0 u$ *in* X *as* $\lambda \to 0$ *for all* $u \in D(T)$;

(iii) $\|T_\lambda u\| \to \infty$ *as* $\lambda \to 0$ *for all* $u \notin \overline{D(T)}$.

Proof: (i) Indeed, $f \in Tx$ and $\lambda f = J(u - x)$, or equivalently $x \in T^{-1}f$ and $u - x = J^{-1}(\lambda f) = \lambda J^{-1}f$. Hence $u \in (T^{-1} + \lambda J^{-1})f$ and $T_\lambda = (T^{-1} + \lambda J^{-1})^{-1}$. Since T is maximal monotone and J^{-1} is monotone bounded and coercive with $D(J^{-1}) = X$, Corollary 2.11 assures that the sum $T^{-1} + \lambda J^{-1}$ is maximal monotone and so is T_λ.

Now, since $T_\lambda u = (T^{-1} + \lambda J^{-1})^{-1}u$ for all $u \in X$ it follows that

$$u = (T^{-1} + \lambda J^{-1}) T_\lambda u = T^{-1}T_\lambda u + \lambda J^{-1}T_\lambda u.$$

Multiplying by $T_\lambda u$ we obtain

$$(T_\lambda u, u) = (T_\lambda u, T^{-1}T_\lambda u) + \lambda(T_\lambda u, J^{-1}T_\lambda u) \ .$$

Since $[0, 0] \in G(T)$, we get $(T_\lambda u, T^{-1}T_\lambda u) \geqslant 0$. Hence $\lambda \|T_\lambda u\| \leqslant \|u\|$, that is T_λ is bounded. Finally, we have

$$0 \in T^{-1}(0) = T^{-1}(0) + \lambda J^{-1}(0) = (T^{-1} + \lambda J^{-1}) (0),$$

and thus $[0, 0] \in G(T_\lambda)$.

(ii) Since $\|T_\lambda u\| \leqslant \|g\|$, for every $g \in Tu$ it follows that $\|T_\lambda u\| \leqslant |Tu| = \|T^0 u\|$. By the reflexivity of X, there exists a sequence $\{\lambda_n\}$ such that $T_{\lambda_n} u \to f$ in X^* as $\lambda_n \to 0$. Thus $\|f\| \leqslant \|T^0 u\|$. On the other hand, by the monotonicity of T we have

$$(h - T_{\lambda_n} u, y - \mathcal{J}_{\lambda_n} u) \geqslant 0 \qquad (\forall) \ [y, h] \in G(T)$$

and letting $\lambda_n \to 0$, we obtain $(h - f, y - u) \geqslant 0$. The maximality of T implies now that $f \in Tu$, and hence $\|f\| \geqslant \|T^0 u\|$. Consequently $\|f\| = \|T^0 u\|$, and thus $T_\lambda u \to T^0 u$ in X^*.

(iii) Suppose that, on the contrary, for at least one point $u \notin \overline{D(T)}$ there exists a constant $M > 0$ such that $\|T_\lambda u\| \leqslant M$. Then by (4) we have $\|J(u - \mathcal{J}_\lambda u)\| \leqslant M\lambda \to 0$ as $\lambda \to 0$. Hence $\mathcal{J}_\lambda u \to u$ in X, i.e., $u \in \overline{D(T)}$ which contradicts our assumption.

It is obvious that *the general properties of maximal monotone mappings assure the demicontinuity of T_λ on $D(T_\lambda) = X$* (see Remark 2.4).

In the special case of a Hilbert space H whose dual is identified with H, the duality map $J = I$ is the identity operator and equation (1) becomes

$$u \in (I + \lambda T)\, x.$$

Thus

$$\mathcal{J}_\lambda = (I + \lambda T)^{-1} \text{ and } T_\lambda = \frac{1}{\lambda}(I - \mathcal{J}_\lambda)$$

and we have the splitting

$$u = \mathcal{J}_\lambda u + \lambda T_\lambda u \qquad (\forall)\,\lambda > 0.$$

Put $\mathcal{J}_\lambda u = x$. Then by $(I + \lambda T)\, x \ni u$ it follows that

$$u = x + \lambda f \quad \text{with} \quad f \in Tx$$

and in virtue of Proposition 2.1 the monotonicity of T is equivalent to

$$\|\mathcal{J}_\lambda u - \mathcal{J}_\lambda v\| \leqslant \|u - v\| \qquad (\forall)\, u, v \in R(I + \lambda T).$$

When $T: H \mapsto 2^H$ is maximal monotone then, by Theorem 2.11, $R(I + \lambda T) = H$ and $\mathcal{J}_\lambda$ is a non-expansive map on H.

Moreover in this case T_λ is a Lipschitz continuous operator:

$$\|T_\lambda u - T_\lambda v\| \leqslant \frac{1}{\lambda}\|u - v\| \quad (\forall)\, u, v \in H.$$

Indeed, we can write

$$\|T_\lambda u - T_\lambda v\|\ \|u - v\| \geqslant (T_\lambda u - T_\lambda v, u - v) =$$

$$= (T_\lambda u - T_\lambda v, \mathcal{J}_\lambda u - \mathcal{J}_\lambda v) + \lambda(T_\lambda u - T_\lambda v, T_\lambda u - T_\lambda v) \geqslant \lambda\|T_\lambda u - T_\lambda v\|^2.$$

3.2. Let us discuss the properties of the Yosida approximant in the case of a subdifferential.

Let $T: X \mapsto 2^{X^*}$ be such that $T = \partial\varphi$, where φ is a proper convex l.s.c. function defined on a reflexive strictly convex Banach space X. By Theorem 2.13, the map T is cyclically maximal monotone.

We infer that in this case *the Yosida approximant $T_\lambda = \partial\varphi_\lambda$ is also cyclically maximal monotone.* Indeed, since $T_\lambda x \in T\mathcal{J}_\lambda x$, for any cyclical set $\{x_0, \ldots, x_n, x_{n+1} = x_0\} \subset X$, we have

$$\sum_{j=0}^n (T_\lambda x_j, x_j - x_{j+1}) = \sum_{j=0}^n (T_\lambda x_j, x_j - \mathcal{J}_\lambda x_j - x_{j+1} + \mathcal{J}_\lambda x_{j+1}) +$$

$$+ \sum_{j=0}^n (T_\lambda x_j, \mathcal{J}_\lambda x_j - \mathcal{J}_\lambda x_{j+1}) \geqslant \sum_{j=0}^n (T_\lambda x_j, T_\lambda x_j - T_\lambda x_{j+1}) =$$

$$= \lambda \left[\sum_{j=0}^n \|T_\lambda x_j\|^2 - \sum_{j=0}^n (T_\lambda x_j, T_\lambda x_{j+1}) \right] \geqslant 0.$$

The following theorem tells us about the form of the function φ_λ with the property that $\partial\varphi_\lambda = T_\lambda$.

THEOREM *Let $\varphi : X \mapsto \overline{\mathbb{R}}$ be a proper convex l.s.c. function on X and let $T = \partial\varphi$. Then the function*

$$\varphi_\lambda(x) = \inf \left\{ \frac{1}{2\lambda} \|x - y\|^2 + \varphi(y) \mid y \in X \right\}$$

is convex Gâteaux differentiable on X and $\partial\varphi_\lambda = T_\lambda$ for all $\lambda > 0$. Moreover, $\varphi_\lambda(x) \to \varphi(x)$ as $\lambda \to 0$.

Proof: It is obvious that φ_λ is convex and everywhere finite on X. Moreover, for any fixed $x \in X$ the function

$$\psi(y) = \frac{1}{2\lambda} \|x - y\|^2 + \varphi(y)$$

is proper convex l.s.c. and $\psi(y) \to \infty$ as $\|y\| \to \infty$. Hence, by Corollary I.2.4, there exists a minimizing point $y_0 \in X$ and therefore

$$0 \in \partial\psi(y_0) = \frac{1}{\lambda} J(y_0 - x) + \partial\varphi(y_0).$$

Then for any $\lambda > 0$ we have $y_0 = \mathcal{J}_\lambda x$ and

$$\varphi_\lambda(x) = \frac{1}{2\lambda} \|x - \mathcal{J}_\lambda x\|^2 + \varphi(\mathcal{J}_\lambda x) = \frac{\lambda}{2} \|T_\lambda x\|^2 + \varphi(\mathcal{J}_\lambda x).$$

Now we can see that

$$\frac{1}{2\lambda} (\|z - \mathcal{J}_\lambda z\|^2 - \|x - \mathcal{J}_\lambda x\|^2) \geqslant \frac{1}{\lambda} (J(x - \mathcal{J}_\lambda x), z - x + \mathcal{J}_\lambda x - \mathcal{J}_\lambda z) =$$

$$= (T_\lambda x, z - x) + (T_\lambda x, \mathcal{J}_\lambda x - \mathcal{J}_\lambda z),$$

and since $T_\lambda x \in T \mathcal{J}_\lambda x$, we have also

$$\varphi(\mathcal{J}_\lambda z) - \varphi(\mathcal{J}_\lambda x) \geqslant (T_\lambda x, \mathcal{J}_\lambda z - \mathcal{J}_\lambda x) \quad (\forall) z \in X.$$

Consequently

$$\varphi_\lambda(z) - \varphi_\lambda(x) - (T_\lambda x, z - x) \geqslant 0 \quad (\forall) z \in X,$$

that is, $T_\lambda x \in \partial \varphi_\lambda(x)$. Since T_λ is cyclically maximal monotone, we have that $T_\lambda = \partial \varphi_\lambda$. We take $z = x + tu$ and conclude that T_λ is G-differential of φ_λ. Now, by the definition,

$$\varphi(\mathcal{J}_\lambda x) \leqslant \varphi_\lambda(x) \leqslant \varphi(x)$$

and since $\mathcal{J}_\lambda x \to x$ in X, the lower semicontinuity of φ implies that

$$\varphi(x) \leqslant \lim_{\mathcal{J}_\lambda x \to x} \inf \varphi(\mathcal{J}_\lambda x) \leqslant \varphi(x).$$

The two relations above prove that $\varphi_\lambda(x) \to \varphi(x)$ as $\lambda \to 0$.

COROLLARY *Suppose that in the theorem $X = H$ is a Hilbert space. Then the function φ_λ is Fréchet differentiable on H and $\mathrm{d}\varphi_\lambda = T_\lambda$.*

Proof: Indeed, by the subgradient inequality we can see that

$$0 \leqslant \varphi_\lambda(y) - \varphi_\lambda(x) - (T_\lambda x, y - x) = \frac{1}{2\lambda}(\|y - \mathcal{J}_\lambda y\|^2 - \|x - \mathcal{J}_\lambda x\|^2) +$$

$$+ \varphi(\mathcal{J}_\lambda y) - \varphi(\mathcal{J}_\lambda x) - (T_\lambda x, y - x) \leqslant \frac{1}{\lambda}(y - \mathcal{J}_\lambda y, y - x + \mathcal{J}_\lambda x - \mathcal{J}_\lambda y)$$

$$+ (T_\lambda y, \mathcal{J}_\lambda y - \mathcal{J}_\lambda x) - (T_\lambda x, x - y) \leqslant (T_\lambda y, y - x + \mathcal{J}_\lambda x - \mathcal{J}_\lambda y) +$$

$$(T_\lambda y, \mathcal{J}_\lambda y - \mathcal{J}_\lambda x) - (T_\lambda x, x - y) \leqslant (T_\lambda y - T_\lambda x, y - x) \leqslant$$

$$\leqslant \frac{1}{\lambda}\|y - x\|^2,$$

which proves our assertion.

Example 1. Let C be a closed convex set of a reflexive Banach space X, and consider the *indicator function* of C

$$I_C x = \begin{cases} 0 & \text{if } x \in C \\ +\infty & \text{otherwise,} \end{cases}$$

which is proper convex and l.s.c. *Then*

$$\partial I_C x = \{f \in X^* \mid (f, x - y) \geqslant 0 \quad (\forall) y \in C\}$$

is a cone with 0 as vertex.

Obviously $D(\partial I_C) = C$ and $\partial I_C x = \{0\}$ for each $x \in \text{Int } C$. Moreover, if $x \in \partial C$, then $\partial I_C x$ coincides with the cone of normals to C at the point x. In particular, if C is a linear subspace of X, then

$$\partial I_C x = \begin{cases} C^\perp & \text{if } x \in C \\ 0 & \text{otherwise} \end{cases}$$

If $X = H$ is a Hilbert space, then the resolvent $(I + \lambda \partial I_C)^{-1}$ is the projection on C because $y = (I + \lambda \partial I_C)^{-1} x$ if and only if $x - y \in \partial I_C y$ and

$$(x - y, z - y) \leqslant 0 \qquad (\forall) \, z \in C,$$

i.e., $y = P_C x$. The Yosida approximant $(\partial I_C)_\lambda x = \dfrac{1}{\lambda}(x - P_C x)$ is a Fréchet

differential for the function $(I_C)_\lambda(x) = \dfrac{1}{2\lambda} \|x - P_C x\|^2$.

Example 2. (Convex integrands) Let $j: \mathbb{R} \mapsto \overline{\mathbb{R}}$ be a proper convex l.s.c. function. Then the function $\varphi: L^P(\Omega) \mapsto \overline{\mathbb{R}}$ defined by

$$\varphi(u) = \begin{cases} \displaystyle\int_\Omega j(u(x)) \, dx & \text{if } j(u) \in L^1(\Omega) \\ +\infty & \text{otherwise,} \end{cases}$$

where Ω is a bounded domain in $\mathbb{R}^N$, is l.s.c. and convex on $L^P(\Omega)$. Moreover, $f \in \partial \varphi(u)$ if and only if $f(x) \in \partial j(u(x))$ for almost every $x \in \Omega$.

Indeed, by the properties of j, φ is a well-defined proper convex function. To prove that φ is l.s.c. on $L^P(\Omega)$, we must show, in virtue of Proposition I.2.3, that the level set

$$\varphi^\leqslant(r) = \left\{ u \in L^P(\Omega) \,\middle|\, \int_\Omega j(u(x)) \, dx \leqslant r \right\}$$

is closed for each $r > 0$.

Let $\{u_n\} \subset \varphi^\leqslant(r)$ be such that $u_n \to u$ in $L^P(\Omega)$. Passing if necessary to a subsequence, we may assume that $u_n(x) \to u(x)$ for almost every $x \in \Omega$. Next, for $s \in D(\partial j)$ and $g \in \partial j(s)$ take $\tilde{j}(t) = j(t) - j(s) - (g, t - s)$ instead of $j(t)$ and assume that $j(t) \geqslant 0$. Since $j(u) \leqslant \lim \inf j(u_n)$, Fatou's lemma implies that

$$\int_\Omega j(u(x)) \, dx \leqslant \lim \inf \int_\Omega j(u_n(x)) \, dx \leqslant r,$$

i.e., $u \in \varphi^\leqslant(r)$ as required.

Now let $[u, f] \in G(\partial \varphi) \subset L^P(\Omega) \times L^{P'}(\Omega)$. By the definition of $\partial \varphi$ we have

$$\int_\Omega [j(v(x)) - j(u(x))] \, dx \geqslant \int_\Omega f(x) \, [v(x) - u(x)] \, dx \qquad (\forall) \, v \in L^P(\Omega)$$

For any measurable subset A of Ω, we define

$$w(x) = \begin{cases} v(x) & \text{if } x \in A, \\ u(x) & \text{if } x \in \Omega - A. \end{cases}$$

Let us write w istead of v in the above inequality. We have

$$\int_A \{j(u(x)) - j(v(x)) - f(x)\,[u(x) - v(x)]\}\,\mathrm{d}x \leqslant 0 \qquad (\forall)\,v \in L^p(\Omega).$$

Since A is arbitrarily chosen, we may conclude that

$$j(u(x)) - j(v(x)) \leqslant f(x)\,[u(x) - v(x)] \quad \text{a.a.} \ \ x \in \Omega,$$

for all $v \in L^p(\Omega)$. Hence $f(x) \in \partial j(u(x))$ for almost every $x \in \Omega$. Conversely, let $[u, f] \in L^p(\Omega) \times L^{p'}(\Omega)$ be such that for all $y \in \mathbb{R}$

$$j(y) - j(u\,(x)) \geqslant f(x)\,[y - v(x)] \quad \text{a.a.} \ \ x \in \Omega.$$

Then it is obvious that $f \in \partial \varphi$.

Example 3. Let Ω be a domain in $\mathbb{R}^N$ and let $\beta \colon \mathbb{R} \mapsto 2^{\mathbb{R}}$ be a maximal monotone mapping with $0 \in \beta(0)$. Then the *realization* $T \colon L^2(\Omega) \mapsto 2^{L^2(\Omega)}$ defined by $Tu = \beta(u(x))$ is maximal monotone.

The monotonicity is obvious. To show the maximality of T, by Theorem 2.11, it is sufficient to prove that $R(I + T) = L^2(\Omega)$. Let $f \in L^2(\Omega)$. As β is maximal monotone, there exists a $u(x)$ such that

$$f(x) \in u(x) + \beta(u(x)) \quad \text{a.a.} \ \ x \in \Omega,$$

i.e., $u(x) = \mathcal{J}_1 f(x)$, where $\mathcal{J}_\lambda = (I + \lambda\beta)^{-1}$ with $\lambda > 0$ denotes the resolven of β. Since $\mathcal{J}_\lambda \colon \mathbb{R} \mapsto \mathbb{R}$ is a non-expansive operator, it follows that $u \in L^2(\Omega)$

Surjectivity of perturbed mappings

3.3. Let X and X^* be (strictly convex) reflexive Banach spaces. We give first the following basic result.

PROPOSITION *Let $T \colon X \mapsto 2^{X^*}$ be a maximal monotone mapping with $[0, 0] \in G(T)$ and let $P \colon X \mapsto 2^{X^*}$ be a mapping with the generalized pseudo-monotone property. If $f \in X^*$ can be written in the form*

$$f = T_\lambda u_\lambda + g_\lambda \qquad (\forall)\,\lambda \in (0, \lambda_0),$$

where $g_\lambda \in Pu_\lambda$ with

$$\|T_\lambda u_\lambda\| \leqslant M \ \text{and} \ \|u_\lambda\| \leqslant M \qquad (\forall)\,\lambda \in (0, \lambda_0],$$

then $f \in R(T + P)$.

Proof: From $J(x_\lambda - u_\lambda) + \lambda T_\lambda u = 0$ with $x_\lambda \in D(T)$, we deduce that

$$\|x_\lambda - u_\lambda\| = \lambda \|T_\lambda u\| \leqslant \lambda M \leqslant \lambda_0 M \quad \text{and} \quad \|x_\lambda\| \leqslant (\lambda_0 + 1) M.$$

Since the closed balls in X and in X^* are weakly sequentially compact, given a sequence $\lambda_n \to 0$ we can find two sequences $\{x_n\}$, $x_n = x_{\lambda_n}$ and $\{u_n\}$, $u_n = u_{\lambda_n}$, which are both weakly convergent to the same limit u in X and such that $T_{\lambda_n} u_n = T_n u_n \rightharpoonup h$, $g_{\lambda_n} = g_n \rightharpoonup g$ in X^* with $h + g = f$. We must show that $h \in Tu$ and $g \subset Pu$.

Indeed, for a pair $[m, n]$ of natural numbers we can write

$$(T_n u_n - T_m u_m, u_n - u_m) + (g_n - g_m, \, u_n - u_m) = (f - f, \, u_n - u_m) = 0.$$

Since the first term is non-negative by the monotonicity of T, we deduce that

$$\limsup_{m,\, n \to \infty} (g_n - g_m, \, u_n - u_m) \leqslant 0.$$

According to Lemma 1.3, the generalized pseudo-monotone property of P implies that $g \in Pu$ and $\lim (g_n, u_n - u) = 0$. Moreover,

$$\lim (g_n, x_n - u) = \lim (g_n, u_n - u) + \lim (g_n, x_n - u_n) = 0.$$

Now, for each $n \in \mathbb{N}$,

$$(T_n u_n - h, \, x_n - u) + (g_n - g, x_n - u) = (f - f, x_n - u) = 0;$$

hence $\lim (T_n u_n, x_n - u) = 0$. Since $T_n u_n \in Tx_n$ and T is maximal monotone with the generalized pseudo-monotone property, $h \in Tu$. Therefore,

$$f = h + g \in (T + P) u$$

and the proof is complete.

LEMMA *Let P and T be two maximal monotone mappings from X into X^* such that $D(P) \cap D(T) \neq \emptyset$. Then for any $f \in X^*$ the set of solutions $\{u_\lambda\}$ of the equation $f \in (J + P + T_\lambda) u$ remains bounded as $\lambda \to 0$.*

Proof: Indeed, for any $u_0 \in D(P) \cap D(T)$, the elements $f_\lambda \in (J + P + T_\lambda) u_0$ are bounded in λ because of $\|T_\lambda u_0\| \leqslant |Tu_0|$. By the monotonicity of P and T_λ, we have

$$(f - f_\lambda, \, u_\lambda - u_0) \geqslant (\|u_\lambda\| - \|u_0\|)^2.$$

Thus

$$\|u_\lambda\|^2 \leqslant (\|f - f_\lambda\| + 2\|u_0\|) \, \|u_\lambda\| + (\|f - f_\lambda\| - \|u_0\|) \, \|u_0\|,$$

and $\{u_\lambda\}$ remains bounded as $\lambda \to 0$. The proof is complete.

We note that for any fixed $\lambda > 0$ and each $f \in X^*$ the equation

$$f \in (J + P + T_\lambda) u \tag{1}$$

has a unique solution $u_\lambda \in X$. Indeed if this is not so and $u_\lambda^1 \neq u_\lambda^2$ verify (1), then from the monotonicity of P and T_λ we have $(Ju_\lambda^1 - Ju_\lambda^2, u_\lambda^1 - u_\lambda^2) \leqslant 0$. This implies that $u_\lambda^1 = u_\lambda^2$ because of the strict convexity of X.

THEOREM *Let P and T be two maximal monotone maps with $D(P) \cap D(T) \neq \varnothing$, and let u_λ be the solution of equation (1). Then $f \in R(J + P + T)$ if and only if $\|T_\lambda u_\lambda\|$ is bounded as $\lambda \to 0$, for arbitrary f in X^*.*

Proof: We show first the "if" part. Suppose $f \in R(J + P + T)$. Then there exists a $u \in D(P) \cap D(T)$ such that

$$Ju + g + h = f, \quad g \in Pu, \quad h \in Tu.$$

Let u_λ be the solution of equation (1) i.e., $Ju_\lambda + g_\lambda + T_\lambda u_\lambda = f, \quad g_\lambda \in Pu_\lambda$. Then,

$$0 \leqslant (Ju_\lambda - Ju, u_\lambda - u) = (g - g_\lambda, u_\lambda - u) + (h - T_\lambda u_\lambda, u_\lambda - u) \leqslant$$
$$\leqslant (h - T_\lambda u_\lambda, u_\lambda - u),$$

because of the monotonicity of P. We use now the spliting $u_\lambda = \mathcal{J}_\lambda u_\lambda + \lambda J^{-1}(T_\lambda u_\lambda)$ and deduce that

$$0 \leqslant (h - T_\lambda u_\lambda, \mathcal{J}_\lambda u_\lambda - u) + (h - T_\lambda u_\lambda, \lambda J^{-1}(T_\lambda u_\lambda)) \leqslant$$
$$\leqslant (h - T_\lambda u_\lambda, \lambda J^{-1}(T_\lambda u_\lambda))$$

due to the monotonicity of T and $T_\lambda u_\lambda \in T\mathcal{J}_\lambda u_\lambda$. Hence

$$\|T_\lambda u_\lambda\|^2 \leqslant (h, J^{-1}(T_\lambda u_\lambda)) \quad \text{and} \quad \|T_\lambda u_\lambda\| \leqslant \|h\|,$$

and the necessity is proved.

The "only if" part follows by the proposition. Indeed, suppose that $\|T_\lambda u_\lambda\|$ remains bounded as $\lambda \to 0$. By the above lemma we can find an $M > 0$ such that both

$$\|T_\lambda u_\lambda\| \leqslant M \quad \text{and} \quad \|u_\lambda\| \leqslant M \quad (\forall) \ \lambda \in (0, \lambda_0]$$

hold. According to Corollary 2.11, the sum $J + P$ is maximal monotone and thus it has the generalized pseudo-monotone property. Hence, by the proposition, $f \in R(T + J + P)$.

COROLLARY *If $X = H$ is a Hilbert space, then we have also $u_\lambda \to u$ in X as $\lambda \to 0$.*

Proof: Indeed, let u_λ and u_μ be the solutions of (1) which correspond to λ and μ respectively. Then

$$\|u_\lambda - u_\mu\|^2 + (g_\lambda - g_\mu, u_\lambda - u_\mu) + (T_\lambda u_\lambda - T_\mu u_\mu, u_\lambda - u_\mu) = 0.$$

Since P is monotone we see that

$$\|u_\lambda - u_\mu\|^2 \leqslant (T_\lambda u_\lambda - T_\mu u_\mu, u_\mu - u_\lambda).$$

Using the representation (1) from 3.1, we have

$$u_\mu - u_\lambda = (\mathscr{I}_\mu u_\mu - \mathscr{I}_\lambda u_\lambda) + (\mu T_\mu u_\mu - \lambda T_\lambda u_\lambda),$$

and substituting this above, we get

$$\| u_\lambda - u_\mu \|^2 \leqslant \| T_\lambda u_\lambda - T_\mu u_\mu \| \; \| \mu T_\mu u_\mu - \lambda T_\lambda u_\lambda \|.$$

Therefore the net $\{u_\lambda\}$ is a (generalized) Cauchy sequence and thus $u_\lambda \to u$ as $\lambda \to 0$.

$\{T_\lambda u_\lambda\}$ remains bounded by hypotheses and the same is true for the net $\{g_\lambda\}$ with $g_\lambda = f - Ju_\lambda - T_\lambda u_\lambda$. Hence there exists a sequence $\lambda_n \to 0$ such that $T_{\lambda_n} u_{\lambda_n} \to h, g_{\lambda_n} \to g$ in X^* and $Ju + g + h = f$. Since $\mathscr{I}_{\lambda_n} u_{\lambda_n} \to u$, we have $h \in Tu$ and therefore $f \in R(J + P + T)$.

3.4. We consider the special case of the above result where we take a sum with a subdifferential is taken.

PROPOSITION *Let $T: X \mapsto 2^{X^*}$ be a maximal monotone mapping and let $\varphi: X \mapsto \bar{\mathbb{R}}$ be a proper convex l.s.c. function. If there exists a positive number C such that*

$$\varphi(\mathscr{I}_\lambda u) \leqslant \varphi(u) + C\lambda \qquad (\forall)\; \lambda > 0,\; u \in \mathrm{dom}\; \varphi, \tag{1}$$

then

(i) $T + \partial\varphi$ *is maximal monotone,*

(ii) $| Tu | \leqslant |(T + \partial\varphi)\, u | + \sqrt{C} \qquad (\forall)\; u \in D(T) \cap D(\partial\varphi).$

Proof: Apply Theorem 3.3 with $P = \partial\varphi$. If u_λ verifies the equation $f \in (J + \partial\varphi + T_\lambda)\, u_\lambda$ then

$$(f - Ju_\lambda - T_\lambda u_\lambda, v - u_\lambda) \leqslant \varphi(v) - \varphi(u_\lambda) \qquad (\forall)\; v \in \mathrm{dom}\; \varphi.$$

Taking $v = \mathscr{I}_\lambda u_\lambda$, we obtain $v - u_\lambda = \mathscr{I}_\lambda u_\lambda - u_\lambda = - \lambda J^{-1}(T_\lambda u_\lambda)$ and

$$(f - Ju_\lambda - T_\lambda u_\lambda, - \lambda J^{-1}(T_\lambda u_\lambda)) \leqslant C\lambda.$$

Hence $\| T_\lambda u_\lambda \|^2 \leqslant \| f - Ju_\lambda \| \; \| T_\lambda u_\lambda \| + C$ and thus

$$\| T_\lambda u_\lambda \| \leqslant \| f - Ju_\lambda \| + \sqrt{C}. \tag{2}$$

Since $\{u_\lambda\}$ is bounded as $\lambda \to 0$, (2) implies that $\{T_\lambda u_\lambda\}$ has the same property. By Theorem 3.3 and by Theorem 2.11, we conclude now easily that (i) holds.

For every $[u, g] \in G(\partial\varphi)$ we have $(g, \mathscr{I}_\lambda u_\lambda - u) \leqslant \varphi(\mathscr{I}_\lambda u) - \varphi(u)$ and then $(g, - \lambda J^{-1}(T_\lambda u)) \leqslant C\lambda$. Hence

$$(g, J^{-1}(T^0 u)) \geqslant - C \qquad (\forall)\; u \in D(T) \cap D(\partial\varphi),\; g \in \partial\varphi(u).$$

If $f = (T + \partial\varphi)^0 u$, then $f = h + g, h \in Tu, g \in \partial\varphi(u)$. Therefore

$$(f, J^{-1} (T^0 u)) = (h, J^{-1} (T^0 u)) + (g, J^{-1} (T^0 u)) \geqslant \| T^0 u \|^2 - C.$$

Thus, $\| T^0 u \| \leqslant \| f \| + \sqrt{C}$ as stated in (ii).

3.5. An operator $P: X \mapsto X^*$ is said to be *smooth* if it is bounded, coercive, maximal monotone and with $D(P) = X$. The duality map defined on a (strictly convex) reflexive Banach space is an example of smooth operator.

A mapping $T: X \mapsto 2^{X^*}$ with the generalized pseudo-monotone property is called *regular* if $R(T + P) = X^*$ for any smooth operator P. By Corollary 2.9 and Theorem 2.11, *any monotone mapping $T: X \mapsto 2^{X^*}$ with $0 \in D(T)$ is regular if and only if it is maximal.*

We now establish the following

THEOREM *Let $T: X \mapsto 2^{X^*}$ be a maximal monotone mapping with $0 \in D(T)$ and let $P: X \mapsto 2^{X^*}$ be a regular coercive mapping. Then the quasi-boundedness of either T or P implies that $R(T + P) = X^*$.*

Proof: Without loss of generality we may assume that $[0, 0] \in G(T)$. When the approximant T_λ is viewed as a smooth operator we have $R(T_\lambda + P) = X^*$ for any $\lambda > 0$. In order to prove that $f \in X^*$ belongs to $R(T + P)$ it is sufficient to verify the conditions of Proposition 3.3.

As $f \in R(T_\lambda + P)$ for any $\lambda > 0$, there exists an element $u_\lambda \in X$ such that

$$f = T_\lambda u_\lambda + g_\lambda \quad \text{with} \quad g_\lambda = P u_\lambda.$$

Since $[0, 0] \in G(T_\lambda)$ it follows that $(T_\lambda u, u) \geqslant 0$ for all $u \in X$. Hence the sum $T_\lambda + P$ is uniformly coercive for $\lambda > 0$, and

$$c(\|u_\lambda\|) \, \|u_\lambda\| \leqslant (T_\lambda u_\lambda, u_\lambda) + (g_\lambda, u_\lambda) = (f, u_\lambda) \leqslant \|f\| \, \|u_\lambda\|,$$

that is, the set $\{u_\lambda\}$ is bounded by a positive constant M.

Suppose that P is quasi-bounded. By the relations

$$(T_\lambda u_\lambda, u_\lambda) + (g_\lambda, u_\lambda) = (f, u_\lambda) \quad \text{and} \quad (g_\lambda, u_\lambda) \leqslant (f, u_\lambda) \leqslant \|f\| \, \|u_\lambda\|,$$

the nets $\{g_\lambda\}$ and $\{T_\lambda u_\lambda\}$ are both bounded.

Suppose now that T is quasi-bounded. Since $\|u_\lambda\| \leqslant M$, there exists a positive constant M_1 such that $|c(\|u_\lambda\|)| \leqslant M_1$ and thus the coercivity of P implies that

$$(g_\lambda, u_\lambda) \geqslant c(\|u_\lambda\|) \, \|u_\lambda\| \geqslant - M_1 \|u_\lambda\|.$$

Hence,

$$(T_\lambda u_\lambda, u_\lambda) = (f - g_\lambda, u) \leqslant (\|f\| + M_1) \, \|u_\lambda\| = M_2 \, \|u_\lambda\|.$$

On the other hand, by the definition of T_λ, we have

$$(\lambda T_\lambda u_\lambda, u_\lambda - x_\lambda) = \|u_\lambda - x_\lambda\|^2 \geqslant 0,$$

and thus $(T_\lambda u_\lambda, x_\lambda) \leqslant (T_\lambda u_\lambda, u_\lambda) \leqslant M_2 M$. As $[x_\lambda, T_\lambda u_\lambda] \in G(T)$, the net $\{T_\lambda u_\lambda\}$ is uniformly bounded for any $\lambda > 0$.

In both cases the conclusion of the theorem follows by Proposition 3.3.

As a special case we note the

COROLLARY *Let* $T: X \mapsto 2^{X^*}$ *be a maximal monotone mapping with* $0 \in D(T)$ *and let* $P_1: X \to 2^{X^*}$ *be a regular quasi-bounded mapping. If there exists a constant* $k > 0$ *such that*

$$(g, y) \geqslant -k \|y\| \quad (\forall) \ [y,g] \in G(P_1),$$

then the sum $T + P_1$ *is a regular mapping.*

Proof: Let $P_2: X \mapsto X^*$ be a smooth operator. Then, according to Theorem 1.3, the map $P = P_1 + P_2$ has the generalized pseudo-monotone property. Moreover, in our hypotheses P is quasi-bounded regular and coercive. Hence the theorem gives $R(T + P_1 + P_2) = X^*$, that is, $T + P_1$ is a regular mapping.

Maximality of sums

3.6. Now we shall give some sufficient conditions for the maximality of sums of maximal monotone mappings. These conditions are important since the sum of two maximal monotone mappings is not always maximal monotone. We shall give first an example of this kind.

Let $X = L^2(\mathbb{R}_+)$ and define the linear operators A and B by
$$Au = -u'', \quad D(A) = H^2(\mathbb{R}_+) \cap H^1_0(\mathbb{R}_+) \subset X,$$
$$Bu = -u'', \quad D(B) = \{u \in H^2(\mathbb{R}_+) \mid u'(0) = 0\} \subset X.$$
Then A and B are both monotone operators because

$$(Au, u) = (Bu, u) = -\int_0^\infty u''u \, dt = -- [u' \, u]_0^\infty + \int_0^\infty (u')^2 \, dt \geqslant 0.$$

To prove the maximality of these operators we shall apply Theorem 2.11, i.e., for each $\lambda > 0$ we show that there exists a $u \in D(A)$, $(u \in D(B))$ such that

$$-u'' + \lambda u = f, \quad f \in L^2(\mathbb{R}_+).$$

Indeed, by the method of the variation of constants we can easily see that the general solution of the above equation is

$$u(t) = e^{\mu s}\left[\alpha - \frac{1}{2\mu}\int_0^t e^{-\mu s} f(s) \, ds\right] + e^{-\mu t}\left[\beta + \frac{1}{2\mu}\int_0^t e^{\mu s} f(s) \, ds\right],$$

where $\mu^2 = \lambda$ and α and β are constants that will be determined from the boundary data. For the operator A we can see that due to $u(0) = 0$ the conditions $\alpha + \beta = 0$ and $\lim\limits_{t \to \infty} u(t) = 0$ imply that

$$\alpha = \frac{1}{2\mu} \int_0^\infty e^{-\mu s} f(s) \, ds \quad \text{and} \quad \beta = -\frac{1}{2\mu} \int_0^\infty e^{-\mu s} f(s) \, ds.$$

Therefore, applying the transformation

$$\text{sh } x = \frac{1}{2}(e^x - e^{-x}), \quad \text{ch } x = \frac{1}{2}(e^x + e^{-x}), \quad x \in \mathbb{R}, \tag{1}$$

we have

$$u(t) = \frac{e^{-\mu t}}{\mu} \int_0^t \text{sh } (\mu s) f(s) \, ds + \frac{\text{sh } (\mu t)}{\mu} \int_t^\infty e^{-\mu s} f(s) \, ds. \tag{2}$$

By similar arguments in the case of B we can see that $u'(0) = 0$ implies

$$\alpha = \beta = \frac{1}{2\mu} \int_0^\infty e^{-\mu s} f(s) \, ds,$$

and by (1), we obtain

$$u(t) = \frac{e^{-\mu t}}{\mu} \int_0^t \text{ch } (\mu s) f(s) \, ds + \frac{\text{ch } (\mu t)}{\mu} \int_t^\infty e^{-\mu s} f(s) \, ds. \tag{3}$$

Therefore both operators A and B are maximal monotone. Their sum $T = A + B$ is defined by

$$Tu = -2u'', \quad D(T) = D(A) \cap D(B) = \{u \in H^2(\mathbb{R}_+) \cap H_0^1(\mathbb{R}_+) \mid u'(0) = 0\}$$

If T is maximal, then necessarily

$$(T + \lambda J)(D(T)) = X^* \text{ for all } \lambda > 0.$$

Choose $g(t) = e^{-\lambda t}$. We can find a $\lambda_0 > 0$ such that g is orthogonal to $(T + \lambda_0 J)(D(T))$. In fact,

$$\int_0^\infty g(t) f(t) \, dt = \int_0^\infty e^{-\lambda t}[\lambda u(t) - 2u''(t)] \, dt = \lambda (1 - 2\lambda) \int_0^\infty e^{-\lambda t} u(t) \, dt.$$

Hence,

$$\int_0^\infty g(t) f(t) \, dt = 0 \text{ when } \lambda_0 = \frac{1}{2}.$$

Finally, let P be the linear operator in X defined by $Pu = u'$, $D(P) = H_0^1(\mathbb{R}_+)$. Then $(Tu, v) = (Pu, Pv)$ and obviously $D(P)$ is larger than $D(T)$. We remark that $D(A)$ and $D(B)$ are subpaces of X, whence

$$\text{Int } D(A) = \text{Int } D(B) = \varnothing.$$

Since maximal monotone mappings are regular, we may derive from the foregoing results various sufficient conditions for the maximality of the sum of maximal monotone mappings.

PROPOSITION *Let T_1 and T_2 be two maximal monotone mappings with $0 \in D(T_1) \cap D(T_2)$. If T_1 is quasi-bounded, then $T_1 + T_2$ is maximal monotone.*

Proof: Since $f \in T(0)$, we can see by the monotonicity of T that

$$(g, y) \geq - \|f\| \, \|y\| \text{ for any } [y, g] \in G(T_1).$$

Now we may apply Corollary 3.5.

In order to state the second sufficient condition we need the following

LEMMA *Any monotone mapping $T: X \to 2^{X^*}$ with $0 \in \mathrm{Int}\, D(T)$ is quasi-bounded.*

Proof: By hypothesis, there exists an $r > 0$ such that $B(0, r) \subseteq \mathrm{Int}\, D(T)$. As T is locally bounded at the origin (see Theorem 2.2), there exists a constant $M > 0$ such that $\|f\| \leq M$ for any $[x, f] \in G(T)$ with $x \in B(0, r)$. By the monotonicity of T we have

$$(g, y) \geq (g, x) + (f, y - x) \quad (\forall) \, [y, g] \in G(T).$$

Thus the inequality $|(f, y - x)| \leq M(\|y\| + r)$ implies that

$$(g, x) \leq (g, y) + M(\|y\| + r) \text{ for all } x \in B(0, r).$$

Since $\sup \{(g, x) \mid \|x\| \leq r\} = \|g\| \, r$, we obtain the estimate

$$\|g\| \leq \frac{1}{r} (g, y) + \frac{M}{r} \|y\| + M.$$

This implies easily the quasi-boundedness of T.

THEOREM *Let T_1 and T_2 be two maximal monotone mappings. If*

$$\mathrm{Int}\, D(T_1) \cap D(T_2) \neq \emptyset,$$

then $T_1 + T_2$ is maximal monotone.

Proof: Indeed, if $u \in \mathrm{Int}\, D(T_1) \cap D(T_2)$, then by using, if they are non-trivial, the translations $Ax = T_1(x + u) - T_1 u$ and $Bx = T_2(x + u) - T_2 u$, we can assume that $0 \in \mathrm{Int}\, D(T_1) \cap D(T_2)$ and $[0, 0] \in G(T_1) \cap G(T_2)$. According to the previous lemma, T_1 is quasi-bounded and we can apply the proposition.

3.7. In order to describe more exactly the structure of the class of regular mappings we prove the following

THEOREM *Let X be a reflexive Banach space, let $T: X \mapsto 2^{X^*}$ be a maximal monotone mapping with $0 \in D(T)$ and let $P: X \mapsto 2^{X^*}$ be a regular mapping such that $(f, x) \geqslant -k \|x\|, k > 0$ for all $[x, f] \in G(P)$. Suppose that*

(i) *$D(P) \subset D(T)$;*

(ii) *There exist two continuous functions c_1 and c_2 from $\mathbb{R}_+$ to $\mathbb{R}_+$ with $c_1(r) < 1$ for all $r > 0$ such that*

$$|Tx| \leqslant c_1(\|x\|) |Px| + c_2(\|x\|) \quad (\forall)\ x \in D(P).$$

Then $R(T + P + Q) = X^*$ for all smooth operators $Q: X \mapsto X^*$.

Proof: We may assume without loss of generality that $[0, 0] \in G(T)$. Let Q be a smooth operator. Since for each $\lambda > 0$ the sum $T_\lambda + Q$ is again smooth, the regularity of P implies that $R(P + T_\lambda + Q) = X^*$. Let $S = P + Q$. As $R(T_\lambda + S) = X^*$ any $f \in X$ may be put in the form

$$f = T_\lambda u_\lambda + g_\lambda \text{ with } g_\lambda \in S u_\lambda \tag{1}$$

for some $u_\lambda \in X$ and each $\lambda > 0$. By Proposition 3.3, $R(T + S) = X^*$ whenever $\{u_\lambda\}$ and $\{T_\lambda u_\lambda\}$ are both uniformly bounded for $0 < \lambda < \lambda_0$.

First we note that S is coercive on $D(P)$,

$$\frac{(Sx, x)}{\|x\|} \geqslant -k + \frac{(Qx, x)}{\|x\|} \to \infty \text{ as } \|x\| \to \infty$$

and thus $T_\lambda + S$ is uniformly coercive for $\lambda > 0$ since

$$\frac{((T_\lambda + S)x, x)}{\|x\|} = \frac{(T_\lambda x, x)}{\|x\|} + \frac{(Sx, x)}{\|x\|} \geqslant -|T(0)| + \frac{(Sx, x)}{\|x\|}.$$

Thus $\{u_\lambda\}$ is uniformly bounded, say $\|u_\lambda\| \leqslant M$, because by (1) we have

$$c(\|u_\lambda\|) \|u_\lambda\| \leqslant ((T_\lambda + S)u_\lambda, u_\lambda) = (f, u_\lambda) \leqslant \|f\| \|u_\lambda\|,$$

where $c(r) \to \infty$ as $r \to \infty$. By definition,

$$\lambda T_\lambda u_\lambda + J(x_\lambda - u_\lambda) = 0 \text{ with } x_\lambda \in D(T).$$

Since $g_\lambda \in S u_\lambda$, u_λ lies in $D(P)$ and thus in $D(T)$ by (i). Multiplying by $x_\lambda - u_\lambda$, we obtain

$$\|x_\lambda - u_\lambda\|^2 = \lambda(T_\lambda u_\lambda, u_\lambda - x_\lambda) \leqslant \lambda(h, u_\lambda - x_\lambda),$$

whence $\|x_\lambda - u_\lambda\| \leqslant \lambda \|h\|$ for every $h \in Tu$. Hence $\|T_\lambda u_\lambda\| \leqslant |Tu|$ for each $\lambda > 0$. In virtue of (ii) we have

$$\|T_\lambda u_\lambda\| \leqslant |Tu_\lambda| \leqslant c_1(\|u_\lambda\|) |Pu_\lambda| + c_2(\|u_\lambda\|), \tag{2}$$

and since $\|u_\lambda\| \leqslant M$, there exist positive constants $c_1 < 1$ and $c_2 > 0$ such that $c_1(\|u_\lambda\|) \leqslant c_1$ and $c_2(\|u_\lambda\|) \leqslant c_2$ for any $\lambda \in (0, \lambda_0)$. As $|Pu_\lambda| \leqslant \|g_\lambda\| \leqslant \leqslant \|f\| + \|T_\lambda u_\lambda\|$, by (2) it follows that

$$\|T_\lambda u_\lambda\| \leqslant c_1(\|f\| + \|T_\lambda u_\lambda\|) + c_2,$$

i.e.,

$$\|T_\lambda u_\lambda\| \leqslant \frac{c_1\|f\| + c_2}{1 - c_1} \qquad (\forall) \, \lambda > 0, \quad \lambda \in (0, \lambda_0)$$

According to Proposition 3.3, $R(T + S)$ is all of X^*.

Under the assumptions of the theorem the sum $T + P$ has the generalized pseudo-monotone property due to Theorem 1.3. Thus $T + P$ is regular.

If we take $Q = \lambda J$ as smooth operator, then Theorem 2.11 implies the following

COROLLARY *Let T_1 and T_2 be two maximal monotone mappings from a reflexive Banach space X into 2^{X^*}. If the conditions (i) and (ii) are satisfied, then sum $T_1 + T_2$ is also maximal monotone.*

3.8. In the case when $X = H$ is a Hilbert space, we use also the following criterion:

THEOREM *Let T and P be two maximal monotone mappings from H into H, such that*

$$(Pu, \, T_\lambda u) \geqslant 0 \qquad (\forall) \, u \in D(P), \, \lambda > 0. \tag{1}$$

Then the sum $T + P$ is also maximal monotone.

Proof: Given $f \in H$, denote by u_λ the solution of the equation

$$f \in (I + P + T_\lambda) u. \tag{2}$$

By Corollary 3.4, it is sufficient to prove that $\|u_\lambda\|$ and $\|T_\lambda u_\lambda\|$ remain bounded as $\lambda \to 0$, because in this case $u_\lambda \to u$ and $f \in (I + P + T) u$.

Multiplying (2) by $T_\lambda u_\lambda$ and making use of (1), we get

$$(u_\lambda, \, T_\lambda u_\lambda) + \|T_\lambda u_\lambda\|^2 \leqslant \|f\| \, \|T_\lambda u_\lambda\|. \tag{3}$$

Take v in $D(P)$. We have $(Pu_\lambda - Pv, \, u_\lambda - v) \geqslant 0$, and by (2) we obtain

$$(f - T_\lambda u_\lambda - u_\lambda - Pv, \, u_\lambda - v) \geqslant 0.$$

Hence

$$\|u_\lambda\|^2 + (T_\lambda u_\lambda, \, u_\lambda) \leqslant (Pv - f, \, v) + (f - Pv + v, \, u_\lambda) + (T_\lambda u_\lambda, \, v) \leqslant$$

$$\leqslant c_1 + c_2 \|u_\lambda\| + c_3 \|T_\lambda u_\lambda\|.$$

For any $w \in D(T)$, we have $(T_\lambda u_\lambda - T_\lambda w, u_\lambda - w) \geqslant 0$ and, since $\|T_\lambda w\| \leqslant |Tw|$, we get

$$(T_\lambda u_\lambda, u_\lambda) \geqslant - |Tw| \, \|w\| - |Tw| \, \|u_\lambda\| - \|T_\lambda u_\lambda\| \, \|w\| =$$

$$= - c_4 - c_5 \|u_\lambda\| - c_6 \|T_\lambda u_\lambda\|.$$

Consequently,

$$\|u_\lambda\|^2 \leqslant - (T_\lambda u_\lambda, u_\lambda) + c_1 + c_2 \|u_\lambda\| + c_3 \|T_\lambda u_\lambda\| \leqslant$$

$$\leqslant (c_1 + c_4) + (c_2 + c_5) \|u_\lambda\| + (c_3 + c_6) \|T_\lambda u_\lambda\|$$

and by (3) we deduce that

$$\|T_\lambda u_\lambda\|^2 \leqslant - (u_\lambda, T_\lambda u_\lambda) + \|f\| \, \|T_\lambda u_\lambda\| \leqslant c_4 + c_5 \|u_\lambda\| + (c_6 + \|f\|) \, \|T_\lambda u_\lambda\|.$$

This proves the boundedness of $\|T_\lambda u_\lambda\|$ as $\lambda \to 0$. The proof is complete.

4. Noncoercive operators

Operators of type (S)

4.1. Let C be a closed convex set of a Banach space X. An operator $T: C \mapsto X^*$ is *of type* (S) if any sequence $\{x_n\} \subset C$ converging weakly to x in X, for which

$$\lim \, (Tx_n, x_n - x) = 0,$$

is in fact strongly convergent in X. The operator T is said to be *of type* (S_+) provided that whenever $x_n \rightharpoonup x$ in X and

$$\lim \sup \, (Tx_n, x_n - x) \leqslant 0, \tag{1}$$

then $x_n \to x$ in X. We remark that condition (1) can be written as

$$\lim \sup \, (Tx_n - Tx, x_n - x) \leqslant 0. \tag{2}$$

Clearly, the class of all operators of type (S_+) is a convex subclass of operators of type (S).

Example 1. If X is a locally uniform convex Banach space, then the duality map $J: X \mapsto X^*$ is of type (S_+). Indeed, let $\{x_n\}$ be such that $x_n \rightharpoonup x$ in X and

$$(\|x_n\| - \|x\|)^2 \leqslant (Jx_n - Jx, x_n - x).$$

Condition (2) and Proposition I.1.6 imply that $x_n \to x$ in X.

Example 2. Let $T: H_0^1(\Omega) \mapsto H_0^1(\Omega)$ be defined by the linear Dirichlet form

$$(Tu, v) = \int_\Omega (u\,v + \nabla u \,\nabla v) \,\mathrm{d}x,$$

where Ω is a domain in $\mathbb{R}^N$. Then T is of type (S) because of

$$(Tu_n - Tu, u_n - u) = \int_\Omega [|u_n - u|^2 + |\nabla (u_n - u)|^2] \,\mathrm{d}x = \|u_n - u\|_{1,2}^2.$$

Here T is the variational operator associated to the Dirichlet problem

$$\begin{cases} - \triangle u + u = 0 \text{ in } \Omega \\ \qquad\quad u = 0 \text{ on } \partial\Omega. \end{cases}$$

Any demicontinuous operator $T: C \mapsto X^*$ *of type* (S_+) *is pseudo-monotone.* Indeed, let $x_n \rightharpoonup x$ be such that $\lim \sup (Tx_n, x_n - x) \leqslant 0$. Then $x_n \to x$, by condition (S_+), and for any $y \in X$ it follows that

$$\lim (Tx_n, x_n - y) \geqslant (Tx, x - y).$$

Surjectivity of noncoercive mappings

4.2. Let X be a reflexive Banach space and let $T : X \mapsto 2^{X^*}$ be a monotone mapping. We have seen in Theorem 2.2 that T is locally bounded on Int $D(T)$. By means of the local boundedness of T^{-1} we establish a more general result on the surjectivity of maximal monotone mappings.

The local boundedness of T^{-1} means that to any $f \in X^*$ there corresponds an $r > 0$ such that the set $\{x \in X|\ B(f, r) \cap Tx \neq \emptyset\}$ is bounded. In a Banach space T^{-1} is locally bounded if and only if any sequence $\{x_n\} \subset X$, with $\{[x_n, f_n]\} \subset G(T)$ and $f_n \to f$ in X^*, remains bounded.

We remark that any coercive mapping has this property. In fact, from

$$(f_n, x_n) \geqslant c(\|x_n\|)\,\|x_n\|$$

and $f_n \to f$ in X^*, it follows that $c(\|x_n\|) < \infty$. As $c(r) \to \infty$ for $r \to \infty$, we obtain the boundedness of $\{x_n\}$. However, the set of all maps which have locally bounded inverses is wider than the set of coercive mappings. The following example gives a noncoercive map with a locally bounded inverse.

Let $T : X \mapsto X^*$ be a linear operator of type (S) such that $0 \in \rho(T)$ (the resolvent set) and let $P : X \mapsto X^*$ be a completely continuous nonlinear operator, which is asymptotically zero $\left(\text{i.e. } \dfrac{Px}{\|x\|} \to 0 \text{ as } \|x\| \to \infty\right)$. Then the sum $T + P$ may be non-coercive. However its inverse is locally bounded. Indeed, if this is not the case, consider $\{x_n\} \subset X$ with $\|x_n\| \to \infty$ and $(T+P)x_n \to f$ in X^*. Let $y_n = \dfrac{x_n}{\|x_n\|}$. Then from $Ty_n + \dfrac{Px_n}{\|x_n\|} \to 0$ it follows that $Ty_n \to 0$. By the reflexivity of X, we can assume that $y_n \rightharpoonup y$ in X, which together with the type (S) of T imply that $v_n \to y$ in X. Hence $Ty = 0$ with $y \neq 0$, which contradicts the assumption that $0 \in \rho(T)$.

We can state now the following generalization of Theorem 2.9:

THEOREM *Let X be a reflexive Banach space and let $T: X \mapsto 2^{X^*}$ be maximal monotone. Then the mapping T is surjective if and only if T^{-1} is locally bounded on X^*.*

Proof: Since T is maximal monotone, T^{-1} is also maximal monotone. If $R(T) = X^*$, then by Theorem 2.2, T^{-1} is locally bounded on X^*.

The converse is obtained by proving that $R(T)$ is both an open and a closed set of X^*.

$R(T)$ *is closed:* Let $\{f_n\}$ be a sequence in X^* such that $f_n \in Tx_n$ and $f_n \to f$ in X^*. Since T^{-1} is locally bounded, $\{x_n\}$ is a bounded sequence. By the reflexivity of X, there is a subsequence which we denote again by $\{x_n\}$, such that $x_n \rightharpoonup x$ in X. Hence $[x, f] \in G(T)$ because of the demiclosedness of T.

$R(T)$ *is open:* Let $[x, f] \in G(T)$ and let $r > 0$ be such that T^{-1} is bounded on $B(f, r)$. For $g \in B\left(f, \dfrac{r}{2}\right)$ we show that $g \in R(T)$. Due to Corollary 2.9, there exists for any $\lambda > 0$ a solution $x_\lambda \in D(T)$ of the equation

$$\lambda J(x_\lambda - x) + g_\lambda = g, \quad g_\lambda \in Tx_\lambda.$$

By the monotonicity of T in x_λ and x, we have

$$(g - \lambda J(x_\lambda - x) - f, \quad x_\lambda - x) \geqslant 0,$$

and this implies that $\lambda \|x_\lambda - x\| \leqslant \|g - f\| < \dfrac{r}{2}$. Then

$$\|g - g_\lambda\| = \lambda \|x_\lambda - x\| < \|g - f\| < \dfrac{r}{2}$$

and $\|g_\lambda - f\| < r$. Since T^{-1} is bounded on $B(f, r)$, the set of solutions $\{x_\lambda\}$ remains bounded and

$$\|g - g_\lambda\| = \lambda \|x_\lambda - x\| \to 0 \quad \text{as } \lambda \to 0.$$

Thus $g \in R(T)$ because of the closedness of $R(T)$. The proof is complete.

In particular, *if T is maximal monotone and $D(T)$ is bounded, then T is surjective.*

COROLLARY *If $T: X \mapsto 2^{X^*}$ is maximal monotone and weakly coercive (i.e., $\lim\limits_{\|x\| \to \infty} \|T^0 x\| = +\infty$, $x \in D(T)$), then $R(T) = X^*$.*

Proof: Indeed, the weak coerciveness implies the boundedness of T^{-1} and we can apply the theorem.

An operator $T : X \mapsto X^*$ is called *strongly monotone* if there exists a continuous increasing function $c : \mathbb{R}_+ \mapsto \mathbb{R}_+$ with $c(0) = 0$ such that

$$(Tx - Ty, x - y) \geq c(\|x - y\|) \|x - y\| \qquad (\forall)x, y \in D(T).$$

The above definition implies that $c(\|x - y\|) \leq \|Tx - Ty\|$, that is, T is one-to-one from $D(T)$ into $R(T)$.

Moreover, *any strongly monotone operator is of type (S).*

PROPOSITION *Let $T : X \longrightarrow X^*$ be a hemicontinuous strongly monotone operator with $D(T) = X$. Then $R(T) = X^*$.*

Proof: The function $c(r)$ has a continuous increasing inverse $c^{-1} : \mathbb{R}_+ \mapsto \mathbb{R}_+$ with $c^{-1}(0) = 0$. Putting $Tx = f$ and $Ty = g$, we can write $\|T^{-1}f - T^{-1}g\| \leq c^{-1}(\|f - g\|)$ which implies the continuity of T^{-1} on $R(T)$. Now, for all $f \in R(T)$ such that $\|f\| < r, r > 0$ we have $\|T^{-1}f\| \leq c^{-1}(r) + \|T^{-1}(0)\|$, that is, T^{-1} maps bounded sets of X^* into bounded sets of X. As T is also maximal monotone, it follows by the theorem that $R(T) = X^*$.

4.3. We extend the surjectivity results about noncoercive mappings to regular ones.

LEMMA 1 *Let C be a bounded and weakly closed subset of a reflexive Banach space X. If $T : X \to 2^{X^*}$ is a mapping with the generalized pseudo-monotone property, then $T(C) = \{f \in X^* | f \in Tx, x \in C\}$ is strongly closed in X^*.*

Proof: Let $\{f_n\} \subset T(C)$ be such that $f_n \to f$ in X^*. For each $n \in \mathbb{N}$ there exists an $\{x_n\} \subset C$ such that $f_n \in Tx_n$. As C is bounded and weakly closed, we can assume that $x_n \rightharpoonup x$ in X and $x \in C$. Hence $\lim (f_n, x_n - x) = 0$. The generalized pseudo-monotone property of T implies that $f \in Tx$, that is, $f \in T(C)$.

PROPOSITION *Let X be a reflexive Banach space and let $T : X \mapsto 2^{X^*}$ be a regular mapping. Suppose that*
  (i) *T^{-1} is bounded;*
  (ii) *There exists a constant $k > 0$ such that*

$$(f, x) \geq - k\|x\| \qquad (\forall) [x, f] \in G(T).$$

Then $R(T) = X^*$.

Proof: Let $f \in X^*$ be given. Since the duality map $J : X \mapsto X^*$ is a smooth operator, the regularity of T implies that the equation

$$f \in (T + \lambda J)x$$

has a solution $x_\lambda \in D(T)$ for each $\lambda > 0$. Since $f_\lambda = f - \lambda \, Jx_\lambda \in Tx_\lambda$, by (ii), we have

$$-k \, \|x_\lambda\| \leqslant (f_\lambda, x_\lambda) = (f, x_\lambda) - \lambda \|x_\lambda\|^2,$$

and thus

$$\lambda \|x_\lambda\|^2 \leqslant k\|x_\lambda\| + \|f\| \, \|x_\lambda\|.$$

Hence $\lambda \|x_\lambda\| \leqslant k + \|f\|$ and therefore

$$\|f_\lambda\| \leqslant \|f\| + \lambda \|x_\lambda\| \leqslant k + \|f\|,$$

that is, the family $\{f_\lambda | \lambda > 0\}$ is contained in a fixed bounded set of X^*. If we apply (i), it follows that there exists an $r > 0$, independent of λ, with $\|x_\lambda\| \leqslant r$. Therefore $\|f - f_\lambda\| = \lambda \|Jx_\lambda\| = \lambda \|x_\lambda\| \to 0$ as $\lambda \to 0$. Thus f lies in the strong closure of $T(\bar{B}(0, r))$ in X^*. Since $\bar{B}(0, r) \subset X$ is convex and closed, by the previous lemma $T(\bar{B}(0, r))$ is strongly closed in X^* and hence it contains f. Consequently, $f \in R(T)$, and this completes the proof.

A mapping $T : X \mapsto 2^{X^*}$ is called *coercive in* $x_0 \in X$ if there exists a function $c : \mathbb{R}_+ \mapsto \mathbb{R}$ with $c(r) \to \infty$ as $r \to \infty$ such that

$$(f, x - x_0) \geqslant c(\|x\|) \, \|x\| \qquad (\forall)[x, f] \in G(T).$$

Obviously, a coercive mapping is coercive in $0 \in X$.

COROLLARY *Let* $T: X \mapsto 2^{X^*}$ *be a regular mapping which is coercive in* $x_0 \in X$ *and which satisfies* (ii). *Then* T *is surjective.*

Proof: It is sufficient to show that (i) is fulfilled. By the coercivity of T in x_0 it follows that

$$\frac{(Tx, x - x_0)}{\|x\|} \to \infty \quad \text{as} \quad \|x\| \to \infty.$$

But generally we have

$$\frac{(Tx, x - x_0)}{\|x\|} \leqslant \|T^0 x\| \left(1 + \frac{\|x_0\|}{\|x\|}\right),$$

which implies that $\|T^0 x\| \to \infty$ as $\|x\| \to \infty$. Hence T^{-1} maps bounded sets of X^* into bounded sets of X.

So far we have often used an approximation of a mapping T by $T + \lambda J$ with $\lambda > 0$ and then we took $\lambda \to 0$. We shall discuss now another method of approximation: the *truncation* of monotone mappings.

Let us denote the indicator function of a closed ball $\overline{B}(0, R)$ by I_R. We note first that the subdifferential of I_R has the form

$$\partial I_R = \begin{cases} 0 & \text{if } \|x\| < R, \\ \lambda Jx & \text{if } \|x\| = R, \quad \text{for all } \lambda \geqslant 0, \\ \varnothing & \text{otherwise.} \end{cases}$$

Indeed, let $x \in \partial B(0, R)$. By the definition of ∂I_R we have

$$(f, y - x) \geqslant 0 \qquad (\forall) \, y \in B(0, R), \quad f \in \partial I_R \, y.$$

The subgradient inequality yields

$$(\lambda Jx, y - x) \leqslant \frac{\lambda}{2} \, [\|y\|^2 - \|x\|^2] \leqslant 0 \qquad (\forall) \, \lambda > 0.$$

Thus $(f - \lambda Jx, y - x) \geqslant 0$. The maximality of ∂I_R implies that

$$\lambda Jx \in \partial I_R x \text{ for all } \lambda > 0.$$

THEOREM *Let X be a reflexive Banach space and let $T : X \mapsto 2^{X^*}$ be a regular mapping such that $(f, x) \geqslant - k\|x\|$ for some constant $k > 0$ and all $[x, f] \in G(T)$. Suppose that there exists an $R > 0$ such that*

$$0 \notin (T + \lambda J)x \quad \text{for all } \lambda > 0 \quad \text{and} \quad x \in \partial B(0, R). \tag{1}$$

Then $0 \in Tx_0$ for some $x_0 \in D(T)$ with $\|x_0\| < R$.

Proof: The map ∂I_R is maximal monotone. Since $0 \in D(\partial I_R)$, by Lemma 3.6, ∂I_R is regular and quasi-bounded. Thus by Corollary 3.5, the sum $T + \partial I_R$ is regular too. Since $D(\partial I_R)$ is bounded, $(T + \partial I_R)^{-1}$ is bounded. Hence, by the proposition, there exists an $x_0 \in B(0, R)$ such that $0 \in (T + \partial I_R)x_0$. We have $\|x_0\| < R$ because $\|x_0\| = R$ contradicts (1). The proof is, complete.

It is evident that a map T, which is coercive with respect to the origin, i.e.

$$(f, x) > 0 \quad \text{for} \quad \|x\| \geqslant R, \, f \in Tx,$$

satisfies condition (1).

5. Mappings of type (M)

The mappings of type (M) extend the class of the pseudo-monotone operators and they have found recent applications in nonlinear strongly elliptic and Hammerstein equations. The original definition of a single-valued map of type (M) between two locally convex vector spaces dual to each other has been given by Brézis in [1]. In particular, an operator T from a Banach space X to its dual X^* is of type (M) provided that

(m$_1$) If $\{x_\alpha\}$ is a bounded net in X such that $x_\alpha \rightharpoonup x$ in X, $Tx_\alpha \rightharpoonup f$ in X^* and

$$\limsup (Tx_\alpha, x_\alpha - x) \leqslant 0,$$

then $Tx = f$;

(m$_2$) The restriction of T to each finite-dimensional subspace of X is continuous in the weak* topology of X^*.

In the case of reflexive Banach spaces, Proposition I.1.4 shows that we may replace the nets by ordinary sequences. In this way Kemnochi [1] defined the multivalued mappings of type (M).

Extent of the class

5.1. Let X be a reflexive Banach space, let X^* be its dual space and let $T : X \mapsto 2^{X^*}$ be a nonlinear multivalued map. The map T is said to be of *type* (M) provided that

(M$_1$) The set Tx is nonempty, bounded, convex and closed, for each $x \in X$;

(M$_2$) If $\{[x_n, f_n]\}$ is a sequence in $G(T)$ such that $[x_n, f_n] \rightharpoonup [x, f]$ in $X \times X^*$ and

$$\limsup (f_n, x_n - x) \leqslant 0,$$

then $[x, f] \in G(T)$;

(M$_3$) T is continuous from the finite-dimensional subspaces of X to X^* in the weak topology.

We remark that (M$_1$) implies $D(T) = X$.

In particular, an operator $T : X \mapsto X^*$ is of type (M) provided that $D(T) = X$ and the conditions (m$_1$) $-$ (m$_2$) are satisfied for ordinary sequences.

Condition (M$_2$) is weaker than the one in the Definition 1.3 of the mappings with the generalized pseudo-monotone property. We may remark that condition (M$_2$) for nonlinear mappings corresponds in a certain sense to the condition in the weakly closed graph theorem for linear maps.

PROPOSITION *Let* $L : X \mapsto 2^{X^*}$ *be a linear mapping. Then* L *is bounded if and only if* L *satisfies condition* (M$_2$).

Proof: If L is linear bounded, then $G(L)$ is closed and convex. Hence $G(L)$ is weakly closed. Therefore $x_n \rightharpoonup x$ in X and $f_n \rightharpoonup f$ in X^* with $f_n \in Lx_n$ imply $f \in Lx$ and condition (M$_2$) is obviously fulfilled.

Suppose now that L satisfies condition (M$_2$) and let $\{[x_n, f_n]\}$ be a sequence in $G(L)$ such that $[x_n, f_n] \to [x, f]$. It is immediate that $(f_n, x_n) \to (f, x)$ and thus by (M$_2$) we have $f \in Lx$. Then the graph of L is closed and hence the closed graph theorem (e.g. Yosida [1, p. 79]) implies that L is bounded.

In particular, the above proposition shows that a linear mapping L from X to X^* is of type (M) if and only if it is bounded.

5.2. To give some examples of mappings of type (M) we first prove the following

PROPOSITION *Any pseudo-monotone mapping $T : X \mapsto 2^{X^*}$ is of type (M).*

Proof: Condition (M_2) follows from the generalized pseudo-monotone property of pseudo-monotone mappings.

In particular, by Theorem 2.4, *any maximal monotone operator T with $D(T) = X$ is of type (M).*

The sum of two mappings of type (M) is not necessarily of type (M). The following counterexample has been given by Brézis [1]:

Let us consider a Hilbert space H with an orthonormal basis $\{e_n\}$, the identity operator I and the projection T on the unit sphere. Obviously, the map $-I$ if of type (M) and T is a continuous monotone operator with $D(T) = X$. We show that the sum $S = T - I$ is not of type (M). Indeed, for $x_n = e_1 + e_n$ we have $Sx_n = \left(\dfrac{1}{\sqrt{2}} - 1 \right) x_n$. Thus $x_n \to e_1$ and $Sx_n \to \left(\dfrac{1}{\sqrt{2}} - 1 \right) e_1$ in H. Evidently, for $y = \displaystyle\sum_{k=1}^{\infty} y_k e_k$ with $y_k \to 0$ as $k \to \infty$, it follows that

$$(x_n, y) = (e_1, y) + \sum_{k=1}^{\infty} y_k(e_n, e_k) = (e_1, y) + y_n.$$

As $n \to \infty$, we obtain $(x_n, y) \to (e_1, y)$. On the other hand,

$$\lim \sup (Sx_n, x_n - x) = 2\left(\frac{1}{\sqrt{2}} - 1 \right) - \left(\frac{1}{\sqrt{2}} - 1 \right) = \frac{1}{\sqrt{2}} - 1 < 0$$

and $Se_1 \to 0$. Since $\left(\dfrac{1}{\sqrt{2}} - 1 \right) e_1 \neq 0$, condition (M_2) is not fulfilled and thus S is not of type (M).

5.3. Recall that an operator $C : X \mapsto X^*$ is *compact* if it is continuous and it maps bounded sets of X into relatively compact sets of X^*. A compact operator need not be of type (M). Here is a counterexample (Petryshyn and Fitzpatrick [1]):

In the Hilbert space l^2, take the compact operator

$$Cx = (2 - \|x\|, 0, 0, \ldots).$$

Let $x_n = (x_1^n, x_2^n, \ldots)$, where $x_j^n = \delta_j^n$ (the Knonecker symbol). For each $n \in \mathbb{N}$ we have $Cx_n = (1, 0, 0, \ldots)$ Obviously $x_n \to 0$, $Cx_n \to g = (1, 0, 0, \ldots)$ in l^2 and $\lim \sup (Cx_n, x_n) = 0 = (g, 0)$. The map C is not of type (M) because $C(0) = (2, 0, 0, \ldots)$.

Since the zero map is of type (M), we see from the above counterexample that the sum of a map of type (M) and a compact one is not necessarily of type (M).

Every demicontinuous operator of type (S_+) is also of type (M). The converse implication is not true because the operator $-I$ in a Hilbert space is of type (M) but it does not verify condition (S_+).

PROPOSITION *Let* $T : X \mapsto X^*$ *be a demicontinuous operator of type* (S_+) *and let* $C : X \mapsto X^*$ *be a compact map. Then the sum* $T + C$ *is of type* (M).

Proof: Let $\{x_n\} \subset X$ be a sequence such that $x_n \rightharpoonup x$ in X, $(T + C) x_n \rightharpoonup f$ in X^* and $\lim \sup ((T + C) x_n, x_n) \leqslant (f, x)$. Since C is a compact operator, there exist a $g \in X^*$ and a subsequence, also denoted by $\{x_n\}$, such that $Cx_n \to g$ in X^*. Hence $Tx_n \rightharpoonup f - g$ in X^*. Since the map T is demicontinuous and of type (S_+), $\lim \sup (Tx_n, x_n - x) \leqslant 0$ implies that $x_n \to x$ and $Tx_n \rightharpoonup Tx$. By the continuity of C, we have $Cx_n \to Cx$ and $f = (T + C)x$. As $(T + C)$ is obviously demicontinuous, it is of type (M) and the proof is complete.

COROLLARY *If H is a Hilbert space and $C : H \mapsto H$ a compact map, then the sum* $(I + C)$ *is of type* (M).

Proof: It is sufficient to show that I satisfies condition (S_+). Indeed, if $x_n \rightharpoonup x$ in H and $\lim \sup (x_n, x_n - x) \leqslant 0$, then the weakly lower semicontinuity of the norm implies that

$$\|x\|^2 \leqslant \lim \inf \|x_n\|^2 \leqslant \lim \sup \|x_n\|^2 \leqslant \|x\|^2.$$

Since $\|x_n\| \to \|x\|$, we get $x_n \to x$ in H.

5.4. We can easily see that the function $f(x) = (Qx, x)$ is weakly lower-semi-continuous whenever the map $Q : X \mapsto X^*$ is completely continuous (i.e., $x_n \rightharpoonup x$ in X implies that $Qx_n \to Qx$ in X^*).

PROPOSITION *Let* $T : X \mapsto X^*$ *be a map of type* (M) *and let* $P : X \mapsto X^*$ *be a weakly continuous operator such that the function* $f(x) = (Px, x)$ *is weakly lower semicontinuous. Then the sum* $(T + P)$ *is of type* (M).

Proof: Let $\{x_n\} \subset X$ be such that $x_n \rightharpoonup x$ in X, $(T + P) x_n \rightharpoonup g$ in X^* and $\lim \sup ((T + P) x_n, x_n) \leqslant (g, x)$. Since P is weakly continuous, we have $Px_n \rightharpoonup Px$ and $Tx_n \rightharpoonup g - Px$. Accordingly,

$$\lim \sup (Tx_n, x_n) = \lim \sup ((T + P)x_n, x_n) - \lim \inf (Px_n, x_n) \leqslant (g - Px, x)$$

and T being of type (M), we conclude that $Tx = g - Px$, i.e. $(T+P)x = g$ as required.

We complete this result by proving the following

LEMMA *If $T : X \mapsto X^*$ is monotone and weakly continuous, then $f(x) = (Tx, x)$ is a weak lower-semicontinuous function.*

Proof: If $x_n \rightharpoonup x$ in X, then by the monotonicity of T

$$(Tx_n, x_n - x) \geqslant (Tx, x_n - x) \qquad (\forall)\ n \in \mathbb{N},$$

and it follows that

$$\lim \inf (Tx_n, x_n - x) \geqslant \lim \inf (Tx, x_n - x) = 0.$$

By the weak continuity of T,

$$0 \leqslant \lim \inf (Tx_n, x_n - x) = \lim \inf (Tx_n, x_n) - (Tx, x),$$

and thus

$$f(x) \leqslant \lim \inf f(x_n).$$

This proves the lemma.

Now let $T : X \mapsto X^*$ be a map of type (M), let $P : X \mapsto X^*$ be a monotone weakly continuous mapping and let $Q : X \mapsto X^*$ be a completely continuous operator. In virtue of the above results the sum $(T + P + Q)$ is of type (M).

Surjectivity of mappings of type (M)

In various applications one needs mappings of type (M) which are densely defined in a reflexive Banach space X. We shall investigate presently the range of these maps.

THEOREM *Let X_0 be a dense subspace of X, let $T: X \mapsto 2^{X^*}$ be a maximal monotone mapping with $[0, 0] \in G(T)$ where $G(T)$ is weakly closed in $X \times X^*$. Let $P : X \mapsto 2^{X^*}$ be a quasi-bounded mapping of type (M) with $X_0 \subset D(P)$. If there exists an $r > 0$ such that $(Px, x) > 0$ for $\|x\| \geqslant r$, then there exists a solution $x \in D(T) \cap B(0, r)$ of the equation $0 \in (T + P)x$. Furthermore, the set $\{x \in X \mid 0 \in (T + P)x\}$ is weakly compact.*

Proof: For the sake of simplicity only, we suppose X separable. Since X is also reflexive, we may assume that X and X^* are strictly convex.

Let $J : X \mapsto X^*$ be the duality map. Yosida's approximant $T_\lambda = (T^{-1} + \lambda J^{-1})^{-1}$ with $\lambda > 0$ has all the properties stated in Theorem 3.1. The separability assumption on X implies the existence of an increasing sequence of subspaces $X_1 \subset X_2 \subset \ldots \subset X_n \subset \ldots \subset X_0$ with $\dim X_n = n$, such that $\bigcup\{X_n \mid n \in \mathbb{N}\} = X_0$. For $n \in \mathbb{N}$, let $j_n : X_n \mapsto X$ denote the inclusion map and $j_n^* : X^* \mapsto X_n^*$ the corresponding projection map. Then the mapping

$$S_{n\lambda} = j_n^*(T_\lambda + P) j_n : X_n \mapsto X_n^*$$

is continuous and $(S_{n\lambda}x, x) > 0$ for $x \in X_n$, with $\|x\| \geqslant r$. In fact, the condition $[0, 0] \in G(T)$ implies that $(T_\lambda x, x) \geqslant 0$ and

$$(S_{n\lambda}x, x) = (T_\lambda x, x) + (Px, x) \geqslant (Px, x) > 0 \qquad (\forall)\ x \in X_n - \bar{B}(0, r) \cap X_n.$$

By Proposition 2.8, there exists an $x_{n\lambda} \in X_n \cap B(0, r)$ such that $S_{n\lambda}x_{n\lambda} = 0$, or equivalently, $(T_\lambda + P)x_{n\lambda} = h_{n\lambda} \in X_n^\perp$ (the annihilator of X_n in X^*). Again, from $(T_\lambda x_{n\lambda}, x_{n\lambda}) \geqslant 0$ it follows that $(Px_{n\lambda}, x_{n\lambda}) \leqslant 0$ for all $n \in \mathbb{N}$. By the quasi-boundedness of P, there exists an $R > 0$ (independent of n and λ) such that $\|Px_{n\lambda}\| \leqslant R$ for all $n \in \mathbb{N}$ and $\lambda > 0$.

Since T_λ is bounded, $(h_{n\lambda}, y) \to 0$ for each $y \in X_0$, i.e., $h_{n\lambda} \to 0$ in X^*. The boundedness of the sequences $\{x_{n\lambda}\}$ and $\{Px_{n\lambda}\}$ and the reflexivity of X allow us to select subsequences such that $x_{n\lambda} \rightharpoonup x_\lambda$ in X and $Px_{n\lambda} \rightharpoonup g_\lambda$ in X^*.

We assert that $[x_\lambda, g_\lambda] \in G(P)$. Indeed, using the monotonicity of T_λ and the fact that $(h_{n\lambda}, y) \to 0$ for each $y \in X_0$, we see from

$$0 \leqslant (T_\lambda x_{n\lambda} - T_\lambda x_\lambda, x_{n\lambda} - x_\lambda) = (h_{n\lambda} - Px_{n\lambda} - T_\lambda x_\lambda, x_{n\lambda} - x_\lambda)$$

and $x_{n\lambda}, x_\lambda \in X_0$ that $\limsup (Px_{n\lambda}, x_{n\lambda}) \leqslant (g_\lambda, x_\lambda)$. Hence $g_\lambda = Px_\lambda$, because P is of type (M).

Now we have

$$x_{n\lambda} \in T^{-1}(h_{n\lambda} - Px_{n\lambda}) + \lambda J^{-1}(h_{n\lambda} - Px_{n\lambda}),$$

and thus there exists a $z_{n\lambda} \in T^{-1}(h_{n\lambda} - Px_{n\lambda})$ such that

$$x_{n\lambda} = z_{n\lambda} + \lambda J^{-1}(h_{n\lambda} - Px_{n\lambda}).$$

Using the fact that $0 \in T^{-1}(0)$, we get

$$(h_{n\lambda} - Px_{n\lambda}, x_{n\lambda}) \geqslant \lambda \|h_{n\lambda} - Px_{n\lambda}\|^2.$$

Since $(h_{n\lambda}, x_{n\lambda}) = 0$ and $\|x_{n\lambda}\| \leqslant r$, $\|Px_{n\lambda}\| \leqslant R$, there exists a constant $C > 0$ independent of n and λ, such that $\lambda \|h_{n\lambda} - Px_{n\lambda}\|^2 \leqslant C^2$ and therefore

$$\|x_{n\lambda} - z_{n\lambda}\| = \lambda \|h_{n\lambda} - Px_{n\lambda}\| \leqslant \sqrt{\lambda}\ C. \tag{1}$$

Passing, if necessary, to another subsequence, we may assume the existence of $z_\lambda \in X$ and $w_\lambda \in X^*$ such that $z_{n\lambda} \rightharpoonup z_\lambda$ in X and $h_{n\lambda} - Px_{n\lambda} \to w_\lambda$ in X^*. Since $G(T)$ is weakly closed in $X \times X^*$, we can see that $[z_\lambda, w_\lambda] \in G(T)$. As $h_{\lambda n} \rightharpoonup 0$ in X^*, we have $w_\lambda = -g_\lambda = -Px_\lambda$.

Now, using the weakly lower semicontinuity of the norm in X and taking $\varepsilon_\lambda = x_\lambda - z_\lambda$, we conclude by (1) that $\varepsilon_\lambda \to 0$ in X as $\lambda \to 0$. Let $\{\lambda_k\}$ be a sequence of positive numbers such that $\lambda_k \to 0$ as $k \to \infty$. Then we may assume that there exist $x \in X$ and $g \in X^*$ such that both $x_{\lambda_k} \rightharpoonup x$ in X and $g_{\lambda_k} = Px_{\lambda_k} \rightharpoonup g$ in X^*. Next $z_{\lambda_k} = x_{\lambda_k} - \varepsilon_{\lambda_k} \rightharpoonup x$ in X and $Tz_{\lambda_k} = w_{\lambda_k} = -Px_{\lambda_k} \rightharpoonup -g$ in X^* imply that $[x, -g] \in G(T)$, because $G(T)$ is weakly closed in $X \times X^*$.

To complete the proof of the theorem it is sufficient to show that $g = Px$ such that we may conclude that $0 \in (T + P)x$. Using the monotonicity of T and the fact that $-g \in Tx$, we get from

$$0 \leqslant (w_{\lambda_k} + g, z_{\lambda_k} - x) = (-g_{\lambda_k} + g, x_{\lambda_k} - \varepsilon_{\lambda_k} - x)$$

that $\limsup (g_{\lambda_k}, x_{\lambda_k}) \leqslant (g, x)$. This implies that $x \in D(P)$ and $g \in Px$, since P is of type (M).

Finally, the set $\{x \in X \mid 0 \in (T + P)x\}$ is bounded because $[0, 0] \in G(T)$ and P is coercive. From our assumptions it is easy to deduce that this set is weakly sequentially closed. Since X is reflexive, by Proposition I.1.4, this set is weakly compact.

COROLLARY *Under the assumptions of the theorem, if P is a coercive mapping, then $R(T + P) = X^*$.*

We remark that in the case of a linear maximal monotone operator L, $G(L)$ is always weakly closed.

Applying now the theorem with $Tx = \{0\}$ for all $x \in X$, we can state the following fact: *If $P : X \mapsto 2^{X^*}$ is a quasi-bounded coercive map of type (M), then $R(P) = X^*$.*

The analogue of the theorem when T is a general maximal monotone mapping still remains a conjecture. Let us remark that the condition $[0, 0] \in G(T)$ can be replaced by the condition $D(T) \cap X_0 \neq \emptyset$.

6. Related topics and exercises

Throughout this section X is a reflexive Banach space with its dual space X^*, while H is a Hillbert space.

6.1. Let $T : X \mapsto 2^{X^*}$ satisfy (P_1), (P_2) and the condition

(P_4) For each $x_0 \in X$ and each bounded subset H of X, there exists a constant $c(B, x_0)$ such that

$$(f, x - x_0) \geqslant c(B, x_0) \qquad (\forall) \; [x, f] \in G(T) \text{ with } x \in B.$$

Then T satisfies (P_3) (Kemnochi [1]). The definition of a pseudo-monotone mapping by conditions (P_1), (P_2) and (P_4) generalizes the definition due to Brézis [1].

6.2. Let $T : X \mapsto 2^{X^*}$ satisfy (P_1), (P_2) and the condition

(P_5) T is locally bounded on each finite-dimensional subspace F of X. Then T is pseudo-monotone (Browder and Hess [1]).

6.3. Let $L : X \mapsto X^*$ be a linear monotone operator. Then L is maximal monotone if and only if $G(L)$ is maximal among all linear monotone graphs.

Proof: If L is a maximal monotone mapping, then obviously L is maximal in the set of all linear monotone operators. Conversely, if $[x, f] \in G(L)$ and $(g - f, y - x) \geqslant 0$ for all $[y, g] \in G(L)$, then the linear space $G(\tilde{L})$ spanned by $G(L) \cup [x, f]$ is also monotone, since for $[y, g] \in G(L)$ and $t \neq 0$, we have

$$(tf + g, tx + y) = t^2 \left(f + \frac{g}{t}, x + \frac{y}{t} \right) \geqslant 0.$$

6.4. Let L be linear and single-valued. Prove that L is maximal monotone if and only if $\overline{D(L)} = X$ and L is maximal among all linear single-valued monotone operators.

HINT If L is maximal monotone and single-valued, then $L(0) = \{0\} = L^*(0) = D(L)^{\perp}$ and hence $\overline{D(L)} = X$.

6.5. Let C be a closed convex set in X, let $T: C \mapsto X^*$ be a maximal monotone operator and let $P: C \mapsto X^*$ be a map with the generalized pseudo-monotone property. Prove that $T + P$ has the generalized pseudo-monotone property.

HINT See the proof of Proposition 2.5.

6.6 Let L_0 and L_1 be two linear monotone mappings from X to 2^{X^*} such that $L_0 \subseteq L_1^*$. Then there exists a maximal monotone linear mapping L with the property $L_0 \subseteq L \subseteq L_1^*$ (Brézis and Browder [5]).

6.7. Let $T: \mathbb{R}^N \mapsto \mathbb{R}^N$ be a surjective monotone mapping. Prove that

$$\lim_{\|x\| \to \infty} \|Tx\| = \infty.$$

SOLUTION Suppose there exist $f \in \mathbb{R}^N$ and $\{x_n\} \subset \mathbb{R}^N$ such that $\|x_n\| \to \infty$, $\dfrac{x_n}{\|x_n\|} \to z$ and $Tx \to f$. Passing to the limit in

$$\left(Tx_n - Ty, \frac{x_n}{\|x_n\|} - \frac{y}{\|x_n\|} \right) \geqslant 0,$$

we obtain $(f, z) \geqslant (Ty, z)$ for all $y \in X$. Since T is surjective, to each $n \in \mathbb{N}$ there corresponds a $y_n \in X$ such that $Ty_n = nz$ and hence $(f, z) \geqslant n(z, z)$. Thus $z = 0$, and this contradicts $\|z\| = 1$.

6.8. Suppose that $T: \mathbb{R}^N \to \mathbb{R}^N$ is monotone on $B(0, r + \varepsilon) \subset \mathbb{R}^N$. Prove that $T(\bar{B}(0, r))$ is a bounded set of $\mathbb{R}^N$.

SOLUTION Assume that there exists a sequence $\{x_n\} \subset B(0, r)$ such that $x_n \to x_0$ and $\|Tx_n\| \to \infty$. Then there exist a subsequence $\{x_j\}$ of $\{x_n\}$ and a $z \in \partial B(0, 1)$ such that $\dfrac{Tx_j}{\|Tx_j\|} \to z$. We have

$$\left(\frac{Tx_j}{\|Tx_j\|} - \frac{Ty}{\|Tx_j\|}, x_j - y \right) \geqslant 0 \quad (\forall) \; y \in B(0, r + \varepsilon).$$

Letting $j \to \infty$, we get $(z, x_0) \geqslant (z, y)$. Taking $y = x_0 + \dfrac{\varepsilon}{2} z$, we obtain $z=0$ which

contradicts $\|z\| = 1$.

6.9. There exists a mapping $T : \mathbb{R}^N \mapsto \mathbb{R}^N$, $N \geqslant 2$, monotone on $\bar{B}(0, \ 1) \subset \mathbb{R}^N$, such that $T(\bar{B}(0, 1))$ is unbounded.

HINT Let $\{x_n\} \subset \partial B(0, 1)$ be such that $x_n \neq x_m$ for $m \neq n$. Then the mapping

$$Tx = \begin{cases} x, & \text{for } x \in \bar{B}(0, 1), \quad x \neq x_n; \\ (n + 1) \, x_n, & \text{for } x = x_n, \end{cases}$$

has the required properties.

6.10. Prove that the normalized duality map is linear if and only if X is a Hilbert space.

SOLUTION Suppose that J is linear and let $f \in Jx$ and $g \in Jy$ for $x, y \in X$. Then $f + g \in J(x + y), f - g \in J(x - y)$ and $\|x + y\|^2 = (f + g, x + y) = \|x\|^2 + (f, y) + (g, x) + \|y\|^2$, $\|x - y\|^2 \ (f - g, x - y) = \|x\|^2 - (f, y) - (g, x) + \|y\|^2$. Consequently $\|x + y\|^2 + \|x - y\|^2 = 2(\|x\|^2 + \|y\|^2)$, which characterizes a Hilbert space.

6.11. Let $T : H \mapsto 2^H$ be a maximal monotone mapping. Prove that for all $\lambda, \mu > 0$

(i) $\quad \mathcal{I}_\lambda = -\dfrac{1}{\lambda} R\left(-\dfrac{1}{\lambda}, T \right)$;

(ii) $\quad \mathcal{I}_\lambda x = \mathcal{I}_\mu \left(\dfrac{\mu}{\lambda} x + \dfrac{\lambda - \mu}{\lambda} \mathcal{I}_\lambda x \right)$, (the resolvent identity);

(iii) $\quad (T_\lambda)_\mu = T_{\lambda + \mu}$.

6.12. Let $T : X \to 2^{X^*}$ be a maximal monotone mapping and define $\varphi(x) = \inf \{ \|f\| \mid f \in Tx \}$. Prove that φ is l.s.c.

SOLUTION Let $x_n \to x$ be such that $\varphi(x) > \liminf \varphi(x_n)$. Then, without loss of generality, we can choose $f_n \in Tx_n$ and an $a > 0$ such that $\|f_n\| < \varphi(x) - a$. By the local boundedness of T, there exists a subsequence $\{f_j\}$ such that $f_j \to f$ in X^*. By the maximality of T, it follows that $f \in Tx$. Since the norm in X^* is weakly lower-semicontinuous, we have $\|f\| \leqslant \liminf \|f_n\| \leqslant \varphi(x) - a$. This is a contradiction.

6.13. Let Ω be a bounded domain in $\mathbb{R}^N$ for which the Sobolev imbedding theorem holds. Suppose that $j : \mathbb{R} \mapsto \overline{\mathbb{R}}$ is a convex l.s.c. function and let $\beta = \partial j$. For any $u \in L^2(\Omega)$ define the proper convex l.s.c. function

$$\varphi(u) = \begin{cases} \dfrac{1}{2} \displaystyle\int_\Omega |\nabla u|^2 \, dx + \displaystyle\int_{\partial\Omega} j(u) \, d\sigma, & \text{if } u \in H^1(\Omega) \text{ and } j \in L^1(\partial\Omega), \\[2mm] + \infty & , \text{ otherwise.} \end{cases}$$

Then $\partial \varphi = - \triangle$ with $D(\partial \varphi) = \left\{ u \in H^2(\Omega) \left| -\dfrac{\partial u}{\partial n} \in \beta(u) \text{ a.e. on } \partial\Omega \right. \right\}$. Moreover, there exist positive constants c_1 and c_2 such that

$$\|u\|_{2,2} \leqslant c_1 \|u - \triangle u\|_2 + c_2 \qquad (\forall)\ u \in D(\partial \varphi)$$

(see Barbu [2, pp. 63–67] and Brézis [3]).

6.14. Let $L: X \mapsto X^*$ be a linear monotone operator such that $R(L)$ is closed. Prove that $R(L) = X^*$ if and only if L is one-to-one.

SOLUTION By Theorem 2.2, the linear operator L is locally bounded and therefore continuous. When L is one-to-one, the closed graph theorem implies that L^{-1} is closed. Then L^{-1} is also locally bounded and $R(L) = X^*$, by Theorem 4.2.

Conversely, the condition $R(L) = X^*$ and Theorem 4.2 ensure that L^{-1} is locally bounded. Thus L is one-to-one since in the contrary case there exists a sequence $\{x_n\} \subset N(L)$ with $\|x_n\| \to \infty$ and $Lx_n = 0$. This would contradict the local boundedness of L^{-1}.

6.15. Let $\varphi: X \mapsto \overline{\mathbb{R}}$ be a proper convex l.s.c. function. Then the subdifferential $T = \partial \varphi$ is surjective if and only if

$$\varphi(x) - (f, x) \to \infty \quad \text{as} \quad \|x\| \to \infty$$

for all $f \in X^*$.

Proof: For any $f \in X^*$, the function $\Phi(x) = \varphi(x) - (f, x)$ is proper l.s.c. convex and $\Phi(x) \to \infty$ as $\|x\| \to \infty$. By Corollary I.2.4, there exists an x_0 depending on f, which is the minimizing point for Φ on X, i.e., we have

$$\varphi(x) - (f, x_n) \geqslant \varphi(x_0) - (f, x_0) \qquad (\forall)\ x \in X.$$

Thus $f \in \partial \varphi(x_0)$, that is, $R(T) = X^*$.

Conversely, suppose that T is surjective and $\{x_n\} \subset X$ is a sequence with $\|x_n\| \to \infty$ such that $\Phi(x_n)$ remains bounded. Clearly, we can determine a $g \in X^*$ such that $(g, x_n) \to \infty$ as $\|x_n\| \to \infty$. By the surjectivity of T, there exists a $y \in D(T)$ such that $f + g \in Ty$. By the subgradient inequality we obtain

$$(g, x_n) \leqslant \varphi(x_n) - (f, x_n) - \varphi(y) + (f + g, y),$$

i.e., (g, x_n) remains bounded. This contradicts the choice of g.

6.16. Let φ be a proper convex l.s.c. function and let $T = \partial \varphi$. Prove that the following statements are equivalent:

(i) For any $x_0 \in \text{dom } \varphi$, $\displaystyle\lim_{\substack{\|x\| \to \infty \\ f \in Tx}} \dfrac{(f, x - x_0)}{\|x\|} = \infty$;

(ii) There exists an $x_0 \in X$ such that $\displaystyle\lim_{\substack{\|x\| \to \infty \\ x \in D(T)}} \dfrac{(T^0 x, x - x_0)}{\|x\|} = \infty$;

159

(iii) $\displaystyle\lim_{\substack{\|x\|\to\infty \\ x\in D(T)}} \|T^0 x\| = \infty$;

(iv) $R(T) = X^*$ and T^{-1} is bounded;

(v) $\displaystyle\lim_{\|x\|\to\infty} \frac{\varphi(x)}{\|x\|} = \infty$.

SOLUTION It is obvious that (i) $\Rightarrow$ (ii). Since

$$\frac{(T^0 x,\ x - x_0)}{\|x\|} \leqslant \|T^0 x\| \left(1 + \frac{x_0}{\|x\|}\right),$$

it follows that (ii) $\Rightarrow$ (iii). Corollary 4.2 implies that (iii) $\Rightarrow$ (iv). To prove that (iv) $\Rightarrow$ (v), we can suppose (adding, if need be, a positive constant) that $\varphi \geqslant 0$. Let $r > 0$ be arbitrarily chosen. Then to each $f \in B(0, r) \subset X^*$ there corresponds, by (iv), a $y \in D(T)$ such that $f \in Ty$ and $\|y\| < M$. By the subgradient inequality $\varphi(x) - \varphi(y) \geqslant (f, x - y)$ we can see that

$$(f, x) \leqslant \varphi(x) + Mr \qquad (\forall)\ x \in \mathrm{dom}\ \varphi.$$

Consequently $r\|x\| \leqslant \varphi(x) + Mr$ and $\dfrac{\varphi(x)}{\|x\|} \geqslant r - \dfrac{Mr}{\|x\|}$. Hence $\displaystyle\liminf_{\|x\|\to\infty} \frac{\varphi(x)}{\|x\|} \geqslant r$. Finally (v) $\Rightarrow$ (i), since for $f \in Tx$ we have $(f, x - x_0) \geqslant \varphi(x) - \varphi(x_0)$ and

$$\frac{(f, x - x_0)}{\|x\|} \geqslant \frac{\varphi(x) - \varphi(x_0)}{\|x\|} \to \infty \quad \text{as} \quad \|x\| \to \infty.$$

6.17. Let $T: X \mapsto 2^{X^*}$ be a maximal monotone mapping and let P be a pseudo-monotone bounded operator from $D(T)$ into X^*. If $(T + P)^{-1}$ is bounded and there exists an $x_0 \in X$ and two real numbers k, r such that

$$(f + Px, x - x_0) \geqslant -k\|x\| \quad (\forall)\ [x, f] \in G(T),\ \|x\| \geqslant r, \tag{1}$$

then $R(T + P) = X^*$.

Proof: Since we may use translations it is sufficient to show that $0 \in R(T + P)$. Let $\varphi(r) = \sup \{\|Px\| + 1 \mid \|x\| \leqslant r\}$ and $Gx = \varphi(\|x\|) Jx$. By Proposition 2.4, for all $\lambda > 0$, the sum $P + \lambda G$ is pseudo-monotone and bounded. Moreover, $P + \lambda G$ is also coercive, i.e.,

$$\frac{((P + \lambda G) x, x)}{\|x\|} \geqslant -\varphi(\|x\|) + \varphi(\|x\|) \|x\|, \quad x \in \overline{D(T)}.$$

Now Theorem 2.8 implies that $R(T + P + \lambda G) = X^*$. In particular, for each $\lambda > 0$, there exists an $x_\lambda \in D(T)$ such that $0 \in (T + P + \lambda G) x_\lambda$. Assuming $\|x_\lambda\| \geqslant r$, we have by (1) that

$$\lambda\varphi(\|x_\lambda\|)\,\|x_\lambda\| \leqslant \lambda\varphi(\|x_0\|)\,\|x_\lambda\| + k.$$

Consequently $\lambda\varphi(\|x_\lambda\|)\|x_\lambda\|$ is bounded as $\lambda\to 0$ and so is λGx. Since $(T+P)^{-1}$ is bounded, the nets $\{x_\lambda\}$ and $\{Px_\lambda\}$ remain also bounded. Let us choose a sequence $\lambda_n\to 0$ such that $x_n=x_{\lambda_n}\rightharpoonup x$ in X and $Px_n\to h$ in X^*. By the monotonicity of T, we get

$$(-Px_n-\lambda_n Gx_n-g,\,x_n-y)\geq 0 \qquad (\forall)\ [y,g]\in G(T), \tag{2}$$

and thus

$$\limsup (Px_n,\,x_n-x)\leq (h+g,\,y-x) \qquad (\forall)\ [y,g]\in G(T).$$

As in the latter part of the proof of Theorem 2.8, we have

$$\inf\{(h+g,\,y-x)\mid [y,g]\in G(T)\}\leq 0.$$

The pseudo-monotonicity of P implies that

$$(Px,\,x-y)\leq \liminf (Px_n,\,x_n-y) \qquad (\forall)\ y\in G(T).$$

We deduce from (2), that $(g+Px,\,y-x)\geq 0$ for all $[y,g]\in G(T)$. Hence by the maximality of T we have that $0\in (T+P)x$.

6.18. Let T be a maximal monotone mapping and let P be a bounded pseudo-monotone operator from $D(T)$ into X^*. If P is coercive in $x_0\in D(T)$, prove that $T+P$ is surjective.

HINT Let $(Px,\,x-x_0)\geq c(\|x\|)\|x\|$ with $c(r)\to\infty$ as $r\to\infty$. Let $r_0>0$ be such that $\|x\|\geq r_0$ implies that $\|x-x_0\|\leq 2\|x\|$. From the monotonicity of T we deduce that

$$(f+Px,\,x-x_0)\geq -\|f_0\|\|x-x_0\|+c(\|x\|)\|x\|\geq -k\|x\|,$$

where $f\in Tx$, $f_0\in Tx_0$ and $k=2\|f_0\|+c(0)$. The boundedness of $(T+P)^{-1}$ follows from the inequality

$$|(T+P)x|\,\|x-x_0\|\geq (T^0x+Px,\,x-x_0).$$

Next we apply 6.17.

6.19. A *semigroup of (nonlinear) contractions* is a function $S:(0,+\infty)\times H\to H$ satisfying the following conditions:

i) $S(t+s)x=S(t)S(s)x$ $(\forall)\ x\in X,\ t,s\geq 0$;

ii) $S(0)x=x$ $(\forall)\ x\in X$;

iii) $S(t)x$ is continuous in $t\to 0$, for every $x\in X$;

iv) $\|S(t)x-S(t)y\|\leq \|x-y\|$ $(\forall)\ t\geq 0,\ x,y\in X$.

For every $t>0$ set $T_t x=\dfrac{1}{t}[S(t)x-x]$. Let $D(T)$ be the set of all $x\in X$ for which $Tx=\lim\limits_{t\to 0} T_t x$ exists. The operator $T:H\to H$ is called the *generator* of the semigroup S.

We note that $-T$ *is a monotone operator.* Indeed,

$$- (T_t x - T_t y, x - y) = \frac{1}{t} \, [(x - y, x - y) - (S(t) \, x - S(t) y, x - y)] \geqslant$$

$$\geqslant \frac{1}{t} \, [\| x - y \|^2 - \| x - y \|^2] = 0.$$

This fact is crucial for the connection between nonlinear semigroups and monotone mappings.

Bibliographical comments

As we have mentioned at the beginning of the chapter, the pseudo-monotone operators were introduced by Brézis [1] and Browder [10]. The mappings of the variational calculus type have been investigated by Leray — Lions [1]. Our general Definition 1.2 of pseudo-monotonicity for multivalued mappings is due to Kemnochi [1]. The discussion of the structure of the classes of pseudo-monotone mappings and of those with generalized pseudo-monotone property follows the work of Browder — Hess [1].

The first monotonicity assumption for operators in a Hilbert space was used by Golomb [1] in 1935 to show uniqueness of the solution of a Hammerstein nonlinear integral equation. In 1959 Vainberg [1] replaced the Lipschitz condition by monotonicity for the contraction argument. In 1960, Kachurovskii [1] noticed that the differential of a convex function is a monotone operator and invented the term of monotone mapping. In 1962 Minty [1] succeeded in proving by the methods of convex analysis existence theorems for nonlinear functional equations in Hilbert spaces under monotonicity assumptions. From 1962 onwards there took place an intensive development of monotonicity as a powerful tool for proving the existence of solutions of nonlinear problems. This was due especially to the substantial contributions of Browder, who defined the multivalued monotone mappings from a Banach space to the parts of its dual space (see Browder [3]).

Many proofs of local boundedness of monotone mappings are known. Among these, we note those due to Rockafellar [2] and Pazy [1]. In this respect Lemma 2.2 due to Fitzpatrick [1] provides a more natural proof of the local boundedness of monotone mappings. The interdependence between the continuity and the local boundedness of monotone mappings was investigated by Kato [1], Browder [2], [10] and Rockafellar [2].

The pseudo-monotonicity of a single-valued hemicontinuous monotone map (Theorem 2.5) was first discussed by Brézis [1]. The proof of this fact for multivalued maximal monotone mappings (Theorem 2.4) is due to Browder — Hess [1]. Various properties of linear monotone operators and the criterion of their maximality were given by Brézis [6].

Minty [1] and Browder [1] obtained the first results on the surjectivity of coercive monotone mappings. In considering the surjectivity of a sum of a maximal monotone mapping and a pseudo-monotone operator, it becomes necessary to extend to infinite-dimensional spaces Debrunner and Flor's lemma. For this purpose, some preliminary results of Browder [6] and a proof method due to Brézis [2] have been used.

Both the criterion of maximality 2.10 and the maximal monotonicity of the subdifferential of a convex l.s.c. function were established by Rockafellar in [1] and [4], respectively. Pascali [4] succeeded in giving more direct proof of the latter property. We also mention that Calvert [1] has proved that criterion 2.10 fails in a non-reflexive space. The characterization of duality maps as subdifferentials of certain convex functions in 2.14 is due to Asplund [1]. In non-reflexive Banach spaces, the duality maps have been studied by Gossez [1].

The properties of the resolvent and the Yosida approximant for a maximal monotone mapping have been stated by Brézis — Crandall — Pazy [1]. In Brézis' book [5] these properties are treated largely in Hilbert spaces. In these works much attention is given to the generalized pseudo-monotone property of maximal monotone mappings. Browder — Hess [1] show that the class of nonlinear mappings with the generalized pseudo-monotone property may be adequately used in the study of perturbations of maximal monotone operators. Thus Lemma 3.3 and the use of smooth operators permits us to give direct proofs of sufficient condition on maximality for sums of monotone operators (Rockafellar [4]). Zini [1] has extended some of these results to general spaces. A general result about the continuous perturbations of a maximal monotone operators is due to Barbu [1]. The interested reader should consult the expository lectures by Rockafellar [5] which contain many results on convex integrands.

We also mention the lectures of Biroli [1], Showalter [1] and Zeidler [1].

The surjectivity of non-coercive maximal monotone mappings with locally bounded inverses has been first proved by Browder [6] in the case of Banach spaces with uniformly convex duals. A direct proof of Theorem 4.2 in reflexive spaces was given by Brézis [6]. This theorem has recently been extended by Brézis — Browder [3] to the case of general Banach spaces. By using a suitable truncation method, De Figueiredo [2] succeeded in eliminating the coercivity condition in the proof of surjectivity for pseudo-monotone operators (Theorem 4.3).

A recent direction in the study of monotone operators concerns the fact that the multivalued monotone mappings are almost everywhere single-valued (see Fitzpatrick [2], Kenderov [1] and Zarantonello [2]).

The mappings of type (M) form a class larger than the pseudo-monotone ones and our approach to the structure of this class follows the works of Petryshyn — Fitzpatrick [1] and Kemnochi [2]. Theorem 5.4 concerning the surjectivity of a maximal monotone operator perturbed by a map of type (M) in a reflexive Banach space has been obtained by Gupta [4].

The extension to a Banach space X of the connection between semigroups of contractions and monotone operators requires the introduction of nonlinear accretive (dissipative) operators from X into itself. There are two ways in defining these mappings. First, by means of the duality map, as it was done by Browder — De Figueiredo [1]. Second, by the subdifferential of the norm, which was used by Da Prato [3], Martin Jr. [1] and Sinestrari [1].

Chapter IV

Hammerstein equations

An integral equation (generally nonlinear) of Hammerstein type has the form

$$u(x) + \int_\Omega k(x, y) f(y, u(y)) \, dy = w(x), \tag{1}$$

where Ω is a domain of σ-finite measure dy in $\mathbb{R}^N$. The kernel $k : \Omega \times \Omega \to \mathbb{R}$ and the given function $f : \Omega \times \mathbb{R} \to \mathbb{R}$ are both measurable. Here the unknown function u and the inhomogeneous term w lie in a given Banach space X of measurable real-valued functions on Ω.

If we introduce the linear integral operator

$$Av = \int_\Omega k(., y) \, v(y) \, dy$$

and the *Nemitskyi* or *superposition operator*

$$Fu = f(., u(.)),$$

then we can formally rewrite the Hammerstein equation (1) in operator terms as

$$u + AFu = w. \tag{2}$$

As a rule, every elliptic boundary-value problem whose linear part posesses a Green's function can be transformed into a Hammerstein equation. Among these, we mention the problem of the forced oscillations of finite amplitude of a pendulum. The amplitude of oscillation $v(t)$ is a solution of the problem

$$\begin{cases} \dfrac{d^2 v}{dt^2} + a^2 \sin v(t) = z(t), & t \in [0,1], \\ v(0) = v(1) = 0, \end{cases} \tag{3}$$

where the driving function $z(t)$ is periodical and odd. The constant $a \neq 0$ depends on the length of the pendulum and on gravity.

Since Green's function for the problem

$$v''(t) = 0, \qquad v(0) = v(1) = 0,$$

is the triangular function

$$k(t, x) = \begin{cases} t(1 - x) & \text{if } 0 \leqslant t \leqslant x, \\ x(1 - t) & \text{if } x \leqslant r \leqslant 1, \end{cases}$$

problem (3) is equivalent to the nonlinear integral equation

$$v(t) = - \int_0^1 k(t, x) \, [z(x) - a^2 \sin v(x)] \, dx. \tag{4}$$

If

$$\int_0^1 k(t, x) \, z(x) \, dx = g(t) \quad \text{and} \quad v(t) + g(t) = u(t),$$

then (4) can be written as the Hammerstein equation

$$u(t) + \int_0^1 k(t, x) f(x, u(x)) \, dx = 0,$$

where $f(x, u(x)) = a^2 \sin [u(x) - g(x)]$.

Since its inception in the work of Hammerstein [1], the study of the integral equation (1) has mainly benefited by the results from the variational calculus and the theory of fixed points for compact mappings.

The breaking-up of the operator AF from a Banach space X into itself into its constituent parts $F : X \mapsto X^*$ and $A : X^* \mapsto X$ with $R(F) \subseteq D(A)$, where X^* is the dual space of X, has revealed the interdependence between the theory of nonlinear integral equations and that of monotone mappings.

We shall investigate some classes of nonlinear mappings related to the solvability of the abstract Hammerstein equation (2). Our purpose in this chapter is to give a presentation of the most recent methods in the theory of Hammerstein equations.

1. The Nemitskyi operator

We proceed from the study of the Nemitskyi operator, whose properties will be indispensable in the following chapters. According to the kind of Banach space, which is clearly distinguished in applications, we have two cases.

The case $L^p(\Omega)$ with $1 < p < \infty$

1.1. Let Ω be a domain of σ-finite measure in $\mathbb{R}^N$, let $M > 0$ be an integer and let $f : \Omega \times \mathbb{R}^M \to \mathbb{R}$ be a function satisfying the Caratheodory conditions:
  (i) $f(.,r) : \Omega \mapsto \mathbb{R}$ is measurable for all fixed $r \in \mathbb{R}^M$;
  (ii) $f(x,.) : \mathbb{R}^M \mapsto \mathbb{R}$ is continuous for almost all $x \in \Omega$.
  To such a function we associate the *superposition* or *Nemitskyi operator*

$$Fu(x) = f(x, u(x))$$

defined on classes of functions $u : \Omega \mapsto \mathbb{R}^M$, $(u(x) = (u_1(x), \ldots, u_M(x))$. The structure of this operator determines the general form of coefficients in the nonlinear equations that will be considered. Some additional assumptions about the function f will cause that the Nemitskyi operator belong to various classes of nonlinear mappings.
  First, observe that *F carries a measurable function into a measurable function.* Indeed, we may assume without loss of generality that $f(x, r)$ is continuous in r for each $x \in \Omega$. If $u(x) = \lim u_n(x)$, where $u_n(x)$ is a simple function for any $n \in \mathbb{N}$, then $Fu(x) = \lim Fu_n(x)$ for each $x \in \Omega$. As Fu_n is the sum of a countable family of functions of the form $f(x, c_j)\chi_j(x)$, where χ_j are characteristic functions of measurable subsets in Ω and c_j are constants, each function Fu_n is measurable and so is Fu.
  Now consider the case when F maps $L^p(\Omega)$ into $L^q(\Omega)$, where $1 < p, q < \infty$.

THEOREM *Suppose that f verifies the conditions* (i) $-$ (ii) *and an inequality of the form*

$$|f(x, u)| \leqslant g_1(x) + c(x) \sum_{i=1}^{M} |u_j|^{\frac{p}{q}},$$ (1)

where $g_1 \in L^q(\Omega)$ and c is any non-negative $L^\infty(\Omega)$-function. Then F is a well-defined bounded continuous operator from $[L^p(\Omega)]^M$ into $L^q(\Omega)$.

Proof: For the sake of simplicity, we take $M = 1$. By (3) we have

$$\| Fu \|_q \leqslant \left[\int_\Omega (|g_1(x)| + |c(x)| \, |u(x)|^{\frac{p}{q}})^q \, dx \right]^{\frac{1}{q}} \leqslant \| g_1 \|_q + \| c \|_\infty \| u \|_p^{\frac{p}{q}},$$

so that F is a bounded map of $L^p(\Omega)$ into $L^q(\Omega)$.
  To prove the continuity of F, given a sequence $\{u_n\}$ which converges in L^p to u, we prove the existence of a subsequence $\{u_k\} \subseteq \{u_n\}$ such that $Fu_k \to Fu$ in $L^q(\Omega)$. We proceed by selecting a subsequence $\{u_k\} \subseteq \{u_n\}$ so that $u_k(x) \to u(x)$, a.a. $x \in \Omega$. As $f(x, u)$ is continuous in u, the sequence $Fu_k(x) = f(x, u_k(x))$ converges almost everywhere to $f(x, u(x)) = Fu(x)$.

Now, we apply Vitali's theorem of convergence. For every $\varepsilon > 0$ there exists a subdomain $\Omega_\varepsilon \subset \Omega$ with $\mu(\Omega_\varepsilon) < \infty$ and a real number $\delta > 0$ such that for any $A \subset \Omega$ with $\mu(A) < \delta$ we have

$$\int_A |u_k(x)|^p \, dx \leqslant \varepsilon \text{ and } \int_{\Omega - \Omega_\varepsilon} |u_k(x)|^p \, dx < \varepsilon \text{ uniformly in } k \in \mathbb{N}.$$

We deduce

$$\left(\int_A |Fu_k(x)|^\nu \, dx \right)^{\frac{1}{q}} \leqslant \left[\int_A (g_1(x) + |c(x)| \, |u_k|^{\frac{p}{q}})^q \, dx \right]^{\frac{1}{q}} \leqslant$$

$$\leqslant \left(\int_A |g_1(x)|^q \, dx \right)^{\frac{1}{q}} + \|c\|_\infty \varepsilon^{\frac{1}{q}}.$$

A similar inequality holds for the integral over Ω_ε. The sequence $\{Fu_k\}$ satisfies the conditions of Vitali's theorem and thus $Fu_k \to Fu$ in $L^q(\Omega)$. The theorem is proved.

In particular, when $q = p'$, $\dfrac{1}{p} + \dfrac{1}{p'} = 1$, we deduce the following: *If f verifies the conditions* (i) $-$ (ii) *and an inequality of the form*

$$|f(x, u)| \leqslant g_1(x) + c \sum_{j=1}^{M} |u_j|^{p-1},$$

where $g_1 \in L^{p'}(\Omega)$ and $c > 0$, then F is a well-defined bounded continuous operator from $[L^p(\Omega)]^M$ into $L^{p'}(\Omega)$.

COROLLARY *If, moreover, $f(x, r)$ is nondecreasing in r for each fixed $x \in \Omega$, then F is a monotone operator from $L^p(\Omega)$ into $L^{p'}(\Omega)$*

Proof: For any $u, v \in L^p(\Omega)$ we have

$$(Fu - Fv, u - v) = \int_\Omega [f(x, u(x)) - f(x, v(x))] \, [u(x) - v(x)] \, dx \geqslant 0 \quad (2)$$

since both factors of the integrand bear the same sign.

Next suppose that f verifies the assumptions of the previous theorem and that it is strictly monotone in r for fixed $x \in \Omega$, i.e.,

$$f(x, r) > f(x, s) \text{ for } r > s.$$

PROPOSITION *Assume further that f satisfies the inequality*

$$f(x, u)u \geqslant k|u|^p - g_2(x), \text{ for } x \in \Omega, \quad (3)$$

where $g_2 \in L^1(\Omega)$ and $k > 0$. Then $F : L^p(\Omega) \mapsto L^{p'}(\Omega)$ is of type (S_+).

Proof: We shall prove that if $\{u_n\} \subset L^p(\Omega)$ converges weakly to u and

$$\lim \sup (Fu_n - Fu, u_n - u) \leqslant 0$$

then $\{u_n\}$ is strongly convergent in $L^p(\Omega)$. By the monotonicity of F, we have

$$(Fu_n - Fu, u_n - u) \to 0 \text{ as } n \to \infty$$

and by (2), there exists a subsequence $\{u_j\} \subseteq \{u_n\}$ such that

$$[f(x, u_j(x)) - f(x, u(x))] [u_j(x) - u(x)] \to 0 \text{ as } j \to \infty$$

for almost every $x \in \Omega$. The above convergence holds outside subset $\Omega_0 \subset \Omega$ of null measure and $u(x)$ is finite on $\Omega - \Omega_0$. By our hypotheses, for any $x \in \Omega - \Omega_0$ we have

$$[f(x, u_j(x)) - f(x, u(x))] [u_j(x) - u(x)] \geqslant k(|u_j(x)|^p + |u(x)|^p) - 2g_2(x) -$$

$$- |c(x)|(|u_j(x)|^{p-1}|u(x)| + |u(x)|^{p-1}|u_j(x)|) - g_1(x) (|u(x)| + |u_j(x)|), \qquad (4)$$

that is, $|u_j(x)|$ remains bounded in j for each $x \in \Omega - \Omega_0$. Therefore for each $x \in \Omega - \Omega_0$ we can select a subsequence of $\{u_j(x)\}$ converging to a limit u_0. But

$$[f(x, u_0) - f(x, u_j(x))] [u_0 - u(x)] = 0$$

implies that $u_0 = u(x)$. Moreover, there exists a subsequence $\{u_k\} \subset \{u_j\}$ such that $u_k(x) \to u(x)$ for all $x \in \Omega - \Omega_0$. Indeed, in the contrary case there would exist an $\varepsilon > 0$, an element $x \in \Omega - \Omega_0$ and a sequence of integers $k_1 < k_2 < \ldots < k_n < \ldots$ such that

$$|u_{k_n}(x) - u(x)| > \varepsilon \qquad (\forall) \ k_n \in \mathbb{N},$$

which contradicts the pointwise convergence on $\Omega - \Omega_0$.

To avoid further proliferation of symbols we shall write u_n instead of u_{k_n}. Using again Vitali's theorem we prove now that $u_n \to u$ in $L^p(\Omega)$. For this purpose we rewrite (4) in the form

$$k|u_n(x)|^p \leqslant [f(x, u_n(x)) - f(x, u(x))] [u_n(x) - u(x)] - k|u(x)|^p + 2g_2(x) +$$

$$+ (g_1(x) + |c(x)| |u_n(x)|^{p-1}) |u(x)| + (g_1(x) + |c(x)| |u(x)|^{p-1}) |u_n(x)|.$$

Denote by E either a subset of small enough measure or the complement of a subset of finite measure in Ω. We shall prove the uniform equi-integrability of each term at right-hand side of the last inequality. For the first three terms this is obvious. Next, we have

$$\int_E (g_1(x) + |c(x)| \, |u_n(x)|^{p-1}) \, |u(x)| \, dx \leqslant$$

$$\leqslant \left\{ \int_E [g_1(x) + |c(x)| \, |u_n(x)|^{p-1}]^{p'} \, dx \right\}^{\frac{1}{p'}} \left\{ \int_E |u(x)|^p \, dx \right\}^{\frac{1}{p}} \leqslant$$

$$\leqslant \left\{ \int_\Omega [g_1(x) + |c(x)| \, |u_n(x)|^{p-1}]^{p'} \, dx \right\}^{\frac{1}{p'}} \left\{ \int_E |u(x)|^p \, dx \right\}^{\frac{1}{p}} \leqslant$$

$$\leqslant (\|g_1\|_{p'} + \|c\|_\infty \|u_n\|_p^{p-1}) \left\{ \int_E |u(x)|^p \, dx \right\}^{\frac{1}{p}},$$

where the last integral can be made small uniformly in $n \in \mathbb{N}$. Similar arguments can be used for the last term because

$$\int_E (g_1(x) + |c(x)| \, |u(x)|^{p-1}) \, |u_n(x)| \, dx \leqslant (\|g_1\|_{p'} + \|c\|_\infty \|u\|_p^{p-1}) \left\{ \int_E |u_n(x)|^p \, dx \right\}^{\frac{1}{p}}.$$

The condition of Vitali's theorem being verified, it follows that $u_n \to u$ in $L^p(\Omega)$ and hence the map F is of type (S_+). The proof is complete.

In the particular case of domains for which the Sobolev imbedding theorem holds, the condition

$$|f(x, u, \ldots, D^{m-1}u)| \leqslant g_1(x) + \sum_{|\alpha| \leqslant m-1} |D^\alpha u|, \quad m \geqslant 1,$$

assures the compactness of F from $W^{m,2}(\Omega)$ into $L^2(\Omega)$ as a composite of the compact imbedding $W^{m,2}(\Omega) \hookrightarrow W^{m-1,2}(\Omega)$ by the continuous injection $W^{m-1,2}(\Omega) \mapsto L^2(\Omega)$.

1.2. We shall give some results about the differentiability of the Nemitskyi operator $F : L^p(\Omega) \mapsto L^{p'}(\Omega)$ with $\dfrac{1}{p} + \dfrac{1}{p'} = 1$ and $p \geqslant 2$. Here we have to distinguish between the two cases $p = 2$ and $p > 2$.

PROPOSITION *Let $p = 2$. Assume that both $f(x, r)$ and its partial derivative $f_r'(x, r)$ verify the condition $(i) - (ii)$. If there exists a constant $C > 0$ such that*

$$|f_r'(x, r)| \leqslant C \qquad (\forall) \, x \in \Omega, \, r \in \mathbb{R},$$

then $F : L^2(\Omega) \to L^2(\Omega)$ is Gâteaux differentiable and

$$DF(u, v) = f_r'(x, u(x))v((x), \qquad (\forall) \, u, \, v \in L^2(\Omega).$$

Proof: Let us show the existence of

$$\lim_{t \to 0} \frac{1}{t} [F(u + tv) - F(u)] = DF(u, v).$$

By the mean value theorem, there exists an $s \in (0, 1)$ such that

$$\frac{1}{t} [F(u + tv) - F(u)] - DF(u, v) = [f_r'(.,u + stv) - f_r'(.,u)]v.$$

If $t \to 0$ then $stv(x) \to 0$ and thus

$$f_r'(x, u(x) + stv(x)) - f_r'(x, u(x)) \to 0 \text{ a.a. } x \in \Omega.$$

Hence

$$\lim_{t \to 0} \frac{1}{t} [F(u + tv) - F(u)] = DF(u, v).$$

The linear map $DF(u,.)$ is bounded because $|DF(u, v)| \leqslant C\|v\|$.

THEOREM *Let $p > 2$ and let $f(x, r)$ be a function satisfying the conditions $(i) - (ii)$. Assume that $f(x, r)$ is partially differentiable in r and f_r' verifies an inequality of the form*

$$|f_r'(x, u)| \leqslant g(x) + b|u|^{p-2}, \tag{1}$$

where $g \in L^{\frac{p}{p-2}}(\Omega)$ and $b > 0$. Then $F : L^p(\Omega) \mapsto L^{p'}(\Omega)$ is Fréchet differentiable and

$$dF(u, v) = f_r'(., u(.))v(.).$$

Proof: By the integration of (1), we obtain

$$|f(x, u)| \leqslant g(x) |u| + b_1 |u|^{p-1},$$

which implies that $f(., u(.)) \in L^{p'}(\Omega)$ for all $u \in L^p(\Omega)$. For any $u, v \in L^p(\Omega)$, consider

$$H(u, v) = \| F(u + v) - F(u) - f_r'(., u)v \|_{p'} =$$

$$= \left\{ \int_\Omega |f(x, u(x) + v(x)) - f(x, u(x)) - f_r'(x, u(x))v(x)|^{p'} \, dx \right\}^{\frac{1}{p'}}.$$

As in the above proposition

$$f(x, u + v) - f(x, u) = f_r'(x, u + sv) v$$

for some $s \in (0, 1)$ and thus

$$H(u, v) = \left(\int_\Omega |v(x) \, w(x)|^{p'} \, dx \right)^{\frac{1}{p}},$$

where $w(x) = f_r'(x, u(x) + sv(x)) - f_r'(x, u(x))$. By the Hölder inequality we deduce that

$$H(u, v) \leqslant \|v\|_p \|w\|_q \quad \text{with} \quad q = \frac{p}{p-2}, \quad \left(\frac{1}{p} + \frac{1}{q} = \frac{1}{p'} \right).$$

By Theorem 1.1, the Nemitskyi operator which corresponds to f_r' is continuous from $L^p(\Omega)$ into $L^q(\Omega)$ and thus

$$\|w\|_q = \|f_r'(.,u + sv) - f_r'(.,u)\|_q \to 0 \text{ as } v \to 0 \text{ in } L^p(\Omega).$$

Therefore,

$$\frac{H(u, v)}{\|v\|_p} \leqslant \|w\|_q \to 0 \text{ as } \|v\|_p \to 0,$$

which proves the theorem.

We complete these results by a condition that makes the Nemitskyi operator of potential type (see I.1.9.).

COROLLARY *Let* $1 < p < \infty$ *and let* $f(x, r)$ *be a function which verifies the conditions of Theorem 1.1 with* $q = p'$. *Then the function* $\varphi : L^p(\Omega) \mapsto \mathbb{R}$ *given by*

$$\varphi(u) = \int_\Omega \Phi(x, u(x)) \, dx,$$

where

$$\Phi(x, z) = \int_0^z f(x, r) \, dr,$$

satisfies $\varphi'(u) = Fu$.

Proof: We shall show that

$$\lim_{t \to 0} \frac{1}{t} [\varphi(u + tv) - \varphi(u)] = (Fu, v).$$

To calculate

$$\varphi(u + tv) - \varphi(u) - (Fu, v) = \int_\Omega dx \int_{u(x)}^{u(x)+tv(x)} f(x, r) \, dr - \int_\Omega f(x, u(x)) \, v(x) \, dx,$$

we take $r = u + stv$ and we then have

$$\int_\Omega dx \int_0^1 [f(x, u(x) + stv(x)) - f(x, u(x))] \, v(x) \, ds =$$

$$= \int_0^1 ds \int_\Omega [f(x, u(x) + stv(x)) - f(x, u(x))] \, v(x) \, dx.$$

Then, by Hölder's inequality

$$|\varphi(u + tv) - \varphi(u) - (Fu, v)| \leqslant \int_0^1 \|v\|_p \|F(u + stv) - F(u)\|_{p'} \, ds \to 0$$

as $t \to 0$, due to the continuity of F (see Theorem 1.1.). This implies the corollary.

The case $L^1(\Omega)$

1.3. Let us consider now the case $p = 1$ and $q = \infty$. Assume that $f(x, r)$ satisfies (i), (ii) and

(iii) $f(x, r)$ is monotone non-decreasing in r for a fixed $x \in \Omega$;

(iv) If we set $f_s(x) = f(x, s)$ for each constant s, then $f_s \in L^1(\Omega)$.

Under these assumptions the Nemitskyi operator F maps $L^\infty(\Omega)$ into $L^1(G)$. As we have seen in Section 1.1, (i) and (ii) imply that F carries measurable functions into measurable functions. Moreover, if

$$s_1 \leqslant v(x) \leqslant s_2 \qquad \text{a.e. on } \Omega,$$

then, by (iii), we have

$$f_{s_1} \leqslant Fv \leqslant f_{s_2}$$

and hence $Fv \in L^1(\Omega)$ for all v in $L^\infty(\Omega)$ due to (iv).

We show now that *F is continuous and bounded.* Indeed, any sequence $\{v_n\} \subset L^\infty(\Omega)$ strongly convergent to v in $L^\infty(\Omega)$ is bounded, i.e., $\|v_n\|_\infty \leqslant s$, for a certain $s > 0$ and, by (iv), $\|Fv_n\|_1 \leqslant f_s$. To prove the convergence of $\{Fv_n\}$ in $L^1(\Omega)$ we apply Lebesgue's theorem of dominated convergence. This argument shows also that F is bounded.

We have also that *F carries bounded sets of $L^\infty(\Omega)$ into weakly compact sets of $L^1(\Omega)$*, since for every $v \in [L^1(\Omega)]^* = L^\infty(\Omega)$, we have

$$|(v, Fu)| = \int_\Omega v(x) \, Fu(x) \, dx \leqslant \|v\|_\infty \int_\Omega Fu(x) \, dx \leqslant \text{const.},$$

and thus the set $\{(v, Fu)| \, u \in L^1(\Omega)\}$ contains a weakly convergent sequence.

Condition (iii) implies the monotonicity of F in the same way as in Corollary 1.1. To be more exact we set

$$\varphi(x, r) = \int_0^r f(x, s) \, ds.$$

This function is convex in r for almost all x in Ω. Thus for any real numbers r and s we have

$$\varphi(x, r) - \varphi(x, s) \geqslant (r - s) f(x, s).$$

Define the function $j : L^\infty(\Omega) \to \mathbb{R}$ by setting

$$j(u) = \int_\Omega \varphi(x, u(x)) \, dx.$$

If we integrate over Ω the inequality

$$\varphi(x, u(x)) - \varphi(x, v(x)) \geqslant (u(x) - v(x)) f(x, v(x))$$

for given u and v in $L^\infty(\Omega)$, then we obtain

$$j(u) - j(v) \geqslant \int_\Omega (u(x) - v(x)) f(x, v(x)) \, dx = (u - v, Fv).$$

By cyclic addition of this inequality written for three elements u, v, w in $L^\infty(\Omega)$ we obtain

$$(Fv, u - v) + (Fw, v - w) + (Fu, w - u) \leqslant 0,$$

which means the *trimonotonicity* or *3-cyclic monotonicity* of F. We observe that the set of all trimonotone mappings contains the gradients of convex l.s.c. functions.

We may rewrite the last inequality also in the form

$$(Fu - Fw, w - v) \leqslant (Fu - Fv, u - v),$$

which will lead to the introduction of a very useful concept.

2. Abstract Hammerstein equations

Let X be a Banach space with a strictly convex dual space X^*. Consider on X the Hammerstein equation

$$(I + AF) u \ni w,$$

where $F: X \mapsto 2^{X^*}$ is a nonlinear mapping and $A: X^* \mapsto 2^X$ is a linear map, such that $R(F) \subseteq D(A)$. Denote $P = AF: X \mapsto 2^X$ and write the above equation as $(I + P) u \ni w$.

Equations with linear compact mappings

2.1. When A is a compact map, and certain weak assumptions on the continuity of F are made, then P is compact.

PROPOSITION *If A is linear compact and F is bounded demicontinuous, then $P = AF$ is a compact mapping.*

Proof: Indeed, F carries bounded sets of X into bounded sets of X^* while A maps bounded sets of X^* into relatively compact sets of X.

The continuity of P follows from the demicontinuity of F and the complete continuity of A. Hence $P = AF$ is compact.

Within this framework we have the following existence results.

THEOREM *Let $A: X^* \mapsto X$ be a compact monotone linear map and let $F: X \mapsto X^*$ be a bounded demicontinuous operator. Suppose further that there exist a bounded open set $\Omega \subset X$ and an element $w \in \Omega$ such that*

$$(Fu, u - w) \geqslant 0 \ \text{ for all } \ u \in \partial\Omega.$$

Then the Hammerstein equation $(I + AF)u = w$ has at least one solution in $\bar{\Omega}$.

Proof: Let $J: X \mapsto X^*$ be the duality map. For each $\varepsilon > 0$, we consider the perturbation $F_\varepsilon: X \mapsto X^*$ defined by

$$F_\varepsilon u = Fu + \varepsilon J (u - w). \qquad (\forall) \ u \in D(F).$$

This map has the same properties as F and, by the proposition, $P_\varepsilon = AF_\varepsilon: X \to X$ is compact.

First, we prove the existence of a solution $u_\varepsilon \in \Omega$ of the equation

$$u_\varepsilon + P_\varepsilon u_\varepsilon = w. \tag{1}$$

In virtue of the Leray-Schauder principle, we must prove that the equation $(I + tP_\varepsilon) u_\varepsilon = w$ has no solutions on $\partial\Omega$ for all $t \in [0, 1]$. Suppose, on the contrary, that for some t in $[0,1]$ there exists a solution $u \in \partial\Omega$ of this equation. Then

$$0 = (F_\varepsilon u, u + tP_\varepsilon u - w) = (F_\varepsilon u, u - w) + t(F_\varepsilon u, AF_\varepsilon u).$$

Since A is linear monotone, $(F_\varepsilon u, AF_\varepsilon u) \geqslant 0$ and by the assumptions we have

$$0 \geqslant (F_\varepsilon u, u - w) = (Fu, u - w) + \varepsilon(J(u - w), u - w) \geqslant \varepsilon\|u - w\|^2.$$

Therefore $u = w$, which implies that $w \in \partial\Omega$, a contradiction.

Thus there exists a solution $u_\varepsilon \in \Omega$ of the equation (1) for arbitrary $\varepsilon > 0$. Moreover,

$$(I + P)u_\varepsilon = w - \varepsilon AJ(u_\varepsilon - w) = w_\varepsilon.$$

From the last equality it follows that $w_\varepsilon \to w$ in X as $\varepsilon \to 0$, that is, w lies in the closure of $(I + P)(\Omega)$. Since $P = AF$ is compact, we have $w \in (I + P)(\bar\Omega)$. This proves the assertion.

2.2. Consider in $L^1(\Omega)$ the Hammerstein equation

$$(I + AF)u = w \tag{1}$$

where $A: L^1(\Omega) \mapsto L^\infty(\Omega)$ and $F: L^\infty(\Omega) \mapsto L^1(\Omega)$ are defined by

$$(Au)(x) = \int_\Omega k(x, y)\, u(y)\, dx \quad \text{and} \quad (Fu)(x) = f(x, u(x))$$

on a σ-finite domain Ω in $\mathbb{R}^N$. As usually, we assume that $f(x, r)$ satisfies the Caratheodory conditions (i) — (ii) from section 1.

Let us assume also that

(iii') There exist $R > 0$ and φ in $L^1(\Omega)$ such that

$$[f(x, r) - \varphi(x)]\, r \geqslant 0 \quad \text{for all } |r| \geqslant R;$$

(iv') For each $S > 0$, there exists a g_S in $L^1(\Omega)$ such that

$$|f(x, s)| \leqslant g_S(x) \text{ for all } x \in \Omega \text{ and all } |s| \leqslant S;$$

(v) A is a linear compact operator from $L^1(\Omega)$ to $L^\infty(\Omega)$;

(vi) $(Au, u) \geqslant 0$ for all $u \in L^1(\Omega)$.

Observe that hypotheses (iii) and (iv) in 1.3 imply assumptions (iii') and (iv'). In fact, if we set $\varphi(x) = f(x, 0)$, then

$$[f(x, r) - \varphi(x)]\, r \geqslant 0.$$

By assumption (iv), φ lies in $L^1(\Omega)$. Thus (iii) and (iv) imply (iii'). Similarly (iii) and (iv) imply (iv').

On the other hand, (iv') implies that F maps $L^\infty(\Omega)$ into $L^1(\Omega)$.

LEMMA *Under assumptions* (iii'), (iv') *and* (vi) *there exists a function* $\rho(r)$ *for* $r > 0$, *which depends on* R, $\|\varphi\|_1$, $\|A\|$ *and* $\|g_S\|$, *such that if* u *is a solution of* (1), *then*

$$\|u\|_\infty \leqslant \rho(\|w\|_\infty).$$

Proof: By (1), we can write

$$u + A(Fu - \varphi) = w - A\varphi.$$

We set $R_1 = \max \{R, \|w - A\varphi\|_\infty + 1\}$ and apply $Fu - \varphi$ to both sides of the above equation. We obtain

$$(u, Fu - \varphi) + (A(Fu - \varphi), \ Fu - \varphi) = (w - A\varphi, Fu - \varphi).$$

By (vi), $(Au, u) \geq 0$ for all $u \in L^1(\Omega)$; hence

$$(u, Fu - \varphi) \leq \|w - A\varphi\|_\infty \|Fu - \varphi\|_1.$$

On the other hand,

$$(u, Fu - \varphi) = \int_\Omega u(y)\, [f(y, u(y)) - \varphi(y)]\, dy = I_1 + I_2,$$

where

$$I_1 = \int_{|u(y)| < R_1} u(y)\, [f(y, u(y)) - \varphi(y)]\, dy \quad \text{and}$$

$$I_2 = \int_{|u(y)| \geq R_1} u(y)\, [f(y, u(y)) - \varphi(y)]\, dy.$$

For I_1 we have the estimate

$$|I_1| \leq R_1 \int_\Omega (|g_{R_1}(y)| + |\varphi(y)|)\, dy = R_1(\|g_{R_1}\|_1 + \|\varphi\|_1).$$

For I_2 we have

$$I_2 = \int_{|u(y)| \geq R_1} |u(y)\,[f(y, u(y)) - \varphi(y)]|\, dy \geq R_1 \int_{|u(y)| \geq R_1} |f(y, u(y)) - \varphi(y)|\, dy =$$

$$= R_1 \|Fu - \varphi\|_1 - R_1 \int_{|u(y)| < R_1} |f(y, u(y)) - \varphi(y)|\, dy \geq R_1 \|Fu - \varphi\|_1 -$$

$$- R_1(\|g_{R_1}\|_1 + \|\varphi\|_1).$$

Hence,

$$R_1 \|Fu - \varphi\|_1 - 2R_1(\|g_{R_1}\|_1 + \|\varphi\|_1) \leq \|w - A\varphi\|_\infty \|Fu - \varphi\|_1 \leq$$

$$\leq (R_1 - 1)\, \|Fu - \varphi\|_1.$$

Thus

$$\| Fu - \varphi \|_1 \leqslant 2R_1(\| g_{R_1} \|_1 + \| \varphi \|_1),$$

and finally we obtain

$$\| u \|_\infty \leqslant \| w - A\varphi \|_\infty + \| A \| \, \| Fu - \varphi \|_1 \leqslant R_1(1 + 2\| A \| \, \| g_{R_1} \|_1 + 2\| A \| \, \| \varphi \|_1).$$

THEOREM *Under hypotheses* (i) − (v), *equation* (1) *has at least one solution* $u \in L^\infty(\Omega)$ *for each* $w \in L^\infty(\Omega)$.

Proof: As A is compact and F is demicontinuous, the product $P = {=} AF : L^\infty(\Omega) \rightarrow L^\infty(\Omega)$ is compact. To apply proposition II.3.7, we must prove the existence of a constant $M > 0$ such that

$$\| u_t \|_\infty \leqslant M \quad \text{where} \quad (I + tAF)\, u_t = w, \quad (\forall) \; t \in [0, 1].$$

Since $\| tA \| \leqslant \| A \|$, the lemma implies

$$\| u_t \|_\infty \leqslant \rho \, (\| w \|_\infty) = M.$$

This completes the proof.

Finally we remark that in dealing with the compactness of linear integral operators, possibly of potential type, we can use the results of both II.1.2 and II.1.3.

Pseudo-monotonicity methods

2.3. Pseudo-monotonicity methods have found an important application in the theory of Hammerstein equations. We shall give here some results on the existence and uniqueness of solutions for Hammerstein equations, most of them being direct consequences of the results presented in Chapter III.

Consider a reflexive Banach space X, its dual space X^* and the Hammerstein equation

$$(I + AF)u \ni w, \tag{1}$$

where $F : X \mapsto 2^{X^*}$ and $A : X^* \mapsto 2^X$ are in general nonlinear mappings and w is a given element in X.

THEOREM *Suppose that* $A : X^* \mapsto 2^X$ *is maximal monotone and that* $F : X \mapsto X^*$ *is a bounded coercive pseudo-monotone operator. Then for each* $w \in X$ *there exists a solution* $u \in X$ *of equation* (1).

Proof: By (1), $u \in X$ is a solution provided that

$$0 \in Fu - A^{-1}(w - u).$$

Define a new mapping $A_w : A \mapsto 2^{X^*}$ by setting

$$A_w u = -A^{-1}(w - u). \tag{2}$$

Since $G(A_w) = -G(A^{-1}) + [w, 0]$, it follows that A_w is maximal monotone. Hence, by Theorem III.2.9, there exists at least one element $u \in X$ such that $0 \in (F + A_w)\, u$, that is, $w \in (I + AF)\, u$.

Combining Theorem III.2.7 and Corollary III.2.9, one deduces the following

COROLLARY *Let* $A : X^* \mapsto 2^X$ *be a closed linear monotone map such that its adjoint map* A^* *is monotone. If* $F : X \mapsto X^*$ *is a coercive hemicontinuous monotone operator with* $D(F) = X$, *then equation* (1) *has a solution.*

2.4. The extension of the above results to Hammerstein equations in reflexive Banach spaces with noncoercive mappings is based on the results in III.4.2.

THEOREM *Let* $A : X^* \mapsto X$ *be a hemicontinuous monotone operator (not necessarily linear) with* $D(A) = X^*$, *and let* $F : X \mapsto 2^{X^*}$ *be a bounded maximal monotone mapping. If* $(I + AF)^{-1} : X \mapsto 2^X$ *is a locally bounded map, then for each* $w \in X$ *there exists a* $u \in X$ *such that* $(I + AF)\, u = w$.

Proof: Introduce the new unknown $v = u - w$ and consider the map $F_w v = = F(v + w)$ Then the Hammerstein equation can be written in the homogeneous form

$$v + AF_w v = 0.$$

Since maximality is invariant under translations, F_w is a maximal monotone mapping. The map $(I + AF_w)^{-1}$ remains locally bounded. Indeed, if $y_n \in (I + AF_w)\, u_n$ with $y_n \to y$ in X, then

$$y_n + w \in (u_n + w) + AF(u_n + w) \text{ and } y_n + w \to y + w \text{ in } X.$$

By the local boundedness of $(I + AF)^{-1}$ it follows that $\{u_n + w\}$ is bounded in X and so is $\{u_n\}$. Hence, we may take $w = 0$ without loss of generality. Suppose that $u \in X$ is a solution of the equation

$$u + AFu = 0 \tag{1}$$

and let $g \in Fu$. Then $u \in F^{-1} g$ and (1) is equivalent to

$$0 \in (F^{-1} + A)\, g. \tag{2}$$

Since F^{-1} is maximal monotone and $D(A)=X^*$, we conclude by Theorem III.3.6 that $F^{-1} + A$ is maximal monotone.

On the other hand, the map $(F^{-1} + A)^{-1}: X \mapsto 2^{X^*}$ is locally bounded. In fact, let $\{g_n\} \subset X^*$ and $\{x_n\} \subset X$ be two sequences such that $x_n \in (F^{-1}+A)g_n$ for each $n \in \mathbb{N}$ and $x_n \to x$ in X. We must show that $\{g_n\}$ is bounded. Consider $u_n \in F^{-1}g_n$ such that

$$x_n = u_n + Ag_n = u_n + AFu_n = (I + AF)u_n.$$

By hypothesis, $(I + AF)^{-1}$ is locally bounded and thus $\{u_n\}$ is bounded. As F is a bounded map, the sequence $\{g_n\}$ with $g_n \in Fu_n$ is also bounded.

Now, by Theorem III.4.2, the map $F^{-1} + A$ is surjective and so there exists a $g \in X^*$ which is a solution of (2).

In the case of Hammerstein equations with A linear, we can leave out the boundedness of F and we get the

PROPOSITION *Let $A:X^* \mapsto X$ be a linear monotone operator with $D(A) = X^*$ and let F be a monotone hemicontinuous mapping of $D(F) = X$ into X^*. Suppose that the mapping $(I + AF)^{-1}: X \mapsto 2^X$ is locally bounded. Then for each $w \in X$ there exists a $u \in X$ such that $(I + AF)u = w$.*

Proof: We have seen in the previous proof that it is sufficient to show that $0 \in R(I + AF)$, or equivalently, that there exists a $u \in X$ such that

$$0 \in (A^{-1} + F)u.$$

As A is a linear monotone operator with $D(A) = X^*$, Proposition III.2.2 implies that it is continuous. Hence, by Corollary III.2.3, both operators A and F are maximal. By Theorem III.3.6, the sum $S = A^{-1} + F$ is also maximal monotone. In order to apply Theorem III.4.2 we shall prove that S^{-1} is locally bounded that is, for any given $\{u_n\} \subset X$ the conditions $g_n \in Pu_n$ and $g_n \to g$ in X^* imply the boundedness of $\{u_n\}$.

In fact, since $g_n = f_n + Fu_n$ with $f_n \in A^{-1}u_n$, we have

$$Ag_n = Af_n + AFu_n = (I + AF)u_n$$

and by the continuity of A it follows that $Ag_n \to Ag$ in X. Thus the local boundedness of $(I + AF)^{-1}$ implies that $\{u_n\}$ is bounded. This completes the proof.

2.5. We shall present the properties of the Hammerstein operator $(I + AF)^{-1}$ in terms of pseudo-monotonicity. Assume, as previously, that a Hammerstein equation is given on a reflexive Banach space X.

PROPOSITION *Suppose that A is a linear maximal monotone operator of X^* into X and F is a bounded operator of type (M) of X into X^*. Then*
  (i) *For any sequence $\{u_n\} \subset X$ such that $u_n \rightharpoonup u$ in X and $(I + AF)u_n \to w$ in X, it follows that $(I + AF)u = w$.*

(ii) *For any closed bounded convex subset C of X, the image $(I + AF)(C)$ is closed in X.*

Proof: (i) Denote $g_n = Fu_n$ and $(I + AF) u_n = w_n$. By hypothesis, $w_n \to w$ in X. Due to the boundedness of F and the reflexivity of X^*, by taking a subsequence, if necessary, we can assume that $g_n \to g$ in X^*. Set $w_n = u_n + Ag_n$. Since A is linear maximal monotone, A is bounded and weakly continuous. Hence $Ag_n \to Ag$ in X and thus $w = u + Ag$. Therefore it is sufficient to prove that $g = Fu$.

For each n we have

$$(Fu_n, u_n - u) = (Fu_n, w_n - Ag_n - u) = (Fu_n, w_n - u) - (g_n, Ag_n).$$

Since $w_n - u \to w - u = Ag$ on X and $Fu_n \to g$ on X^* it follows that $(Fu_n, w_n - u) \to (g, Ag)$. Hence

$$\lim \sup (Fu_n, u_n - u) \leqslant (g, Ag) - \lim \inf (g_n, Ag_n).$$

Since A is monotone and weakly continuous, Lemma III.5.4 implies that $\varphi(g) = (g, Ag)$ is a weakly l.s.c. function on X^* and

$$(g, Ag) - \lim \inf (g_n, Ag_n) \leqslant 0.$$

As F is of type (M), we conclude that $Fu = g$.

(ii) If w belongs to the closure of the image $(I + AF)(C)$ then there exists a sequence $\{u_n\} \subset C$ such that $(I + AF) u_n \to w$ in X. Since C is bounded and X is reflexive we can assume that $u_n \to u$ in X. But C is closed convex, hence it is weakly closed and thus $u \in C$. As $(I + AF)u_n \to w$, by (i), we can deduce that $(I + AF) u = w$, and therefore $w \in (I + AF)(C)$.

A special case of the proposition is given by the following

COROLLARY *Suppose that A is a linear monotone operator of X^* into X and F is a bounded demicontinuous operator of type (S_+) from X into X^*. Then $(I + AF)$ is a proper mapping from bounded closed subsets of X into X in the following sense: for each compact subset K of X and each closed ball B of X the set $(I + AF)^{-1}(K) \cap B$ is compact.*

Proof: Let $\{u_n\}$ be a sequence in X such that $(I + AF) u_n = w_n$ lies in a compact set K. Without loss of generality we may assume that $w_n \to w$ in X with $w \in K$. By the boundedness of F, by possibly taking subsequences, we may assume that $u_n \to u$ in X and $Fu_n \to g$ in X^*. Applying the argument from the above proof we obtain

$$w - u = Ag \quad \text{and} \quad \lim \sup (Fu_n, u_n - u) \leqslant 0.$$

Since F is demicontinuous and satisfies condition (S_+), it follows that $u_n \to u$ in X and $Fu_n \to Fu$ in X^*. Hence $g = Fu$ and we have $(I + AF) u = w$. This implies that $u \in (I + AF)^{-1}(K)$. Thus $(I + AF)^{-1}(K) \cap B$ is compact for each closed ball B in X, as claimed.

Finally, we collect the above facts in the following

THEOREM *Let X be a reflexive Banach space and let $A: X^* \to X$ and $F: X \to X^*$ be hemicontinuous monotone operators defined on X^* and X respectively. Then the Hammerstein operator $T = (I + AF)^{-1}: X \to 2^X$ has the following properties:*

(j) *For each $w \in X$, the set Tw is closed and convex in X;*

(jj) *If F is strictly monotone, then the set Tw contains at most one single point;*

(jjj) *If T and F are bounded, then for each $w \in X$ the set Tw is non-empty, bounded, closed and convex in X;*

(jv) *If A is linear and the mapping T is bounded, then T is continuous from the strong topology of X to the weak topology of X;*

(v) *If A is linear, T is bounded and if F is bounded demicontinuous and satisfies condition (S_+), then T is continuous from the strong topology of X to the strong topology of X. In addition, if F is strictly monotone, then T is a continuous operator.*

Proof: (j) As in Theorem 2.3, the set Tw consists of the elements $u \in X$ for which $0 \in (A_w + F) u$. The sum $A_w + F$ is maximal monotone and by the properties of maximal monotone mappings the set $(A_w + F)^{-1} (0)$ is closed and convex in X.

(jj) If F is strictly monotone, then the sum $(A_w + F)$ is strictly monotone too. Hence $(A_w + F)^{-1} (0)$ consists of at most one single point.

(jjj) By Theorem 2.4, it follows under our hypotheses that $R (I + AF) = X$.

(jv) Since T is bounded we can consider, by possibly taking subsequences, that $w_n \to w$ and $u_n \rightharpoonup u$ with $u_n \in Tw_n$. Using the same argument as in the proof of the previous proposition we have $(I+AF) u = w$, i.e., the continuity of T from the strong topology of X to the weak topology of X.

(v) By the corollary, the mapping T is proper, that is, $w_n \in T^{-1}u_n$ and $w_n \to w$ in X imply that the sequence $\{u_n\}$ is contained in a compact set of X. Passing, if need be, to a subsequence, we have $u_n \to u$ in X and $u \in Tw$, i.e., $T: X \mapsto 2^X$ is a continuous mapping. For proving the last assertion we use condition (jj).

In the above discussion the assumptions made upon the mapping A and F could be weakened because the Banach spaces in which they act are assumed to be reflexive.

Hilbertean case

2.6. In Hilbert spaces the conditions in the existence theorems for Hammerstein equations involving unbounded linear maps can be refined.

Let H be a real Hilbert space and, as usually, identify H with its dual space. Consider the mappings A and F from H into itself. The domain $D(A)$ being generally different from H, the linear map A may be unbounded.

THEOREM *Let A be a linear maximal monotone operator and let F be a bounded mapping of type (M). Suppose that F is coercive with respect to the element $w \in H$, i.e.,*

$$\lim_{\|u\| \to \infty} \inf \, (F(u+w), u) > 0. \tag{1}$$

Then the equation $u + AFu = w$ is solvable in H.

Proof: Without loss of generality we may assume that $w = 0$ and consider only the homogeneous equation

$$(I + AF)u = 0.$$

By (1), there exists an $r > 0$ such that $(Fu, u) > 0$ for all $\|u\| \geqslant r$. By Theorem III.2.11, the contractive mapping $\mathcal{J}_\lambda = (I + \lambda A)^{-1}$ is linear monotone bounded and defined on H. Hence, in view of Proposition III.5.4, the sum $\mathcal{J}_\lambda + F$ is a bounded mapping of type (M) and

$$(\mathcal{J}_\lambda u + Fu, u) = (\mathcal{J}_\lambda u, u) + (Fu, u) > 0 \text{ for } \|u\| > r.$$

Therefore, by Corollary III.5.4, for each $\lambda > 0$, there exists a $u_\lambda \in H$ with $\|u_\lambda\| \leqslant r$ and $(\mathcal{J}_\lambda + F)u_\lambda = 0$. Since F is bounded, we may select a sequence $\{\lambda_n\}$ with $\lambda_n \to 0$ as $n \to \infty$ such that $u_{\lambda_n} \rightharpoonup u$ and $Fu_{\lambda_n} \rightharpoonup v$ in H.

We assert that $v \in D(A)$ and $Av = -u$. Indeed, since the graph of A is weakly closed in $H \times H$, it follows by

$$u_{\lambda_n} \rightharpoonup u, \quad Fu_{\lambda_n} \rightharpoonup v \text{ and } u_{\lambda_n} + AFu_{\lambda_n} + \lambda_n Fu_{\lambda_n} = 0$$

that $AFu_{\lambda_n} \rightharpoonup -u$. Hence $v \in D(A)$ and $Av = -u$.

Let us prove now that $Fu = v$. By the monotonicity of A we have that

$$0 \leqslant (Av - AFu_{\lambda_n}, v - Fu_{\lambda_n}) = (Av, v - Fu_{\lambda_n}) + (u_{\lambda_n}, v - Fu_{\lambda_n}) +$$

$$\lambda_n(Fu_{\lambda_n}, v - Fu_{\lambda_n}).$$

This implies that

$$(u_{\lambda_n}, Fu_n) \leqslant (Av, v - Fu_{\lambda_n}) + (v, u_{\lambda_n}) + \lambda_n(Fu_{\lambda_n}, v - Fu_{\lambda_n})$$

and thus $\lim \sup (u_{\lambda_n}, Fu_{\lambda_n}) \leqslant (u, v)$. Since F is of type (M), we deduce that $Fu = v$ and $AFu = -u$, which proves the theorem.

We note that *the solution of the equation $(I + AF)u = w$ is unique provided that F fulfils the additional condition:*

$$(Fu - Fv, u - v) \leqslant 0 \text{ implies that } Fu = Fv.$$

In fact, if $u, v \in H$ are such that $(I + AF) u = (I + AF) v = w$, then

$$0 = (u - v, Fu - Fv) + (AFu - AFv, Fu - Fv) \geqslant (u - v, Fu - Fv)$$

and $Fu = Fv$. It follows that $u = v$ because $u + AFu = v + AFv$ and $AFu = AFv$.

COROLLARY *Let A be a linear maximal monotone operator, let F be a bounded hemicontinuous and let $P : H \mapsto H$ be a compact operator such that*

$$(Fu - Fv, u - v) \geqslant (Pu - Pv, u - v) \qquad (\forall) \ u, v \in H.$$

Suppose further that

$$\lim_{||u|| \to \infty} \frac{(F(u + w), u)}{||u||} = \infty \qquad (\forall) \ w \in H.$$

Then the equation $u + AFu = w$ is solvable in H, for each given $w \in H$.

Proof: By the theorem it is sufficient to show that F is of type (M) which follows from Proposition III.5.4, because the two hemicontinuous monotone operators $F = (F - P) + P$ and $F - P$ are of type (M).

A remarkable property of Hammerstein operator is revealed by the following

PROPOSITION *Let A be a linear maximal monotone operator and let F be a bounded hemicontinuous strongly monotone mapping with the constant $c > 0$, i.e.,*

$$(Fu - Fv, u - v) \geqslant c \, ||u - v||^2 \qquad (\forall) \ u, v \in D(F).$$

Then the Hammerstein mapping $T = (I + AF)^{-1}$ exists as a bounded continuous operator from H into H.

Proof: First, we remark that $R \, (I + AF) = H$ and that the mapping $I + AF$ is one-to-one, because of the strong monotonicity of F and due to the above remark on the uniqueness of solution of Hammerstein equation.

We shall show that T is bounded and continuous on every ball of radius $r > 0$ centred at the origin of H. Let $w, z \in H$ be such that $||w|| \leqslant r$ and $||z|| \leqslant r$. Then there exist the solutions $u, v \in H$ for the equations $(I + AF) u = w$ and $(I + AF) v = z$, Thus,

$$(u - v, Fu - Fv) + (AFu - AFv, Fu - Fv) = (w - z, Fu - Fv)$$

which yields

$$c \, ||u - v||^2 \leqslant (w - z, Fu - Fv) \leqslant ||w - z|| \, ||Fu - Fv||. \tag{2}$$

On the other hand, since $(u-w,Fu)+(AFu,Fu)=0$ and thus $(u-w,Fu)\leqslant 0$, it follows that

$$c\,\|u - v\|^2 \leqslant (u - w,Fu - Fw) \leqslant - (u - w,Fw) \leqslant \|u - w\|\,\|Fw\|.$$

Hence $c\,\|u - w\| \leqslant \|Fw\|$ and similarly $c\,\|v - z\| \leqslant \|Fz\|$. As F is bounded and $\|w\| \leqslant r$, there exists a constant $M(r) > 0$ such that $\|Fw\| \leqslant M(r)$ for all $w \in H$ with $\|w\| \leqslant r$. Hence

$$\|u\| \leqslant r + \frac{1}{c}\, M(r) \quad\text{and}\quad \|v\|\leqslant r+\frac{1}{c}\, M(r)$$

and by (2) it follows that

$$\|u - v\|^2 \leqslant \frac{2}{c}\, M(r)\,\|w - z\|,$$

i.e., we have shown the boundedness and the continuity of T on every ball of radius r and center $0 \in H$. This proves the proposition.

Finally, the following example illustrates the conditions under which the Nemitskyi operator is strongly monotone. Let Ω be a domain of σ-finite measure in $\mathbb{R}^N$, $H = L^2(\Omega)$ and $f: \Omega \times \mathbb{R} \mapsto \mathbb{R}$ a function satisfying Caratheodory's conditions. Suppose that f satisfies further the following conditions:

(j) there exists a $g_1 \in L^2(\Omega)$ and a constant $b \geqslant 0$ such that

$$|f(x, r| \leqslant g_1(x) + b\,|r| \qquad (\forall)\; x \in \Omega,\; r \in \mathbb{R};$$

(jj) there exists a constant $c > 0$ such that

$$f(x, r_1) - f(x, r_2) \geqslant c\,(r_1 - r_2) \qquad (\forall)\; x \in \Omega,\; r_1,\; r_2 \in \mathbb{R}.$$

Then the associated Nemytsky operator $F: H \mapsto H$, defined by $Fu = f(.,u(.))$, is bounded continuous and strongly monotone with the constant c.

2.7. We shall investigate the case of an indefinite linear operator A. Let H be a Hilbert space, let H_1 be a closed subspace of H and let H_2 be its orthogonal complement. Assume that A can be broken up into two constituent parts $A_i: D(A_i) \to H_i$ with $D(A_i) \subset H_i$ for $i = 1, 2$.

THEOREM *Suppose that $A_i: H_i \mapsto H_i$, $i = 1,2$, are linear maximal monotone operators and $F:H\mapsto H$ is a continuous strongly monotone mapping with the constant $c > 0$. If there is a constant $m > 0$ such that $cm > 1$ and*

$$(A_2 u, u) \geqslant m\,\|u\|^2 \qquad (\forall)\; u \in D(A_2),$$

then the equation

$$u + (A_1 P_1 - A_2 P_2)Fu = w \tag{1}$$

has a unique solution in H for any given $w \in H$ (here P_i denotes the orthogonal projection of H onto H_i for $i = 1, 2$).

Proof: It is sufficient to prove that the homogeneous equation

$$u + (A_1P_1 - A_2P_2)Fu = 0 \tag{2}$$

has a unique solution. Indeed, if $w \neq 0$, we take $v = u - w$ and observe that the mapping $F_w: H \mapsto H$ defined by $F_w v = F(v + w)$ is continuous, strongly monotone with the same constant and that it reduces equation (1) to an equation (2) in v. If $u = u_1 + u_2$ with $u_1 \in H_1$ and $u_2 \in H_2$, we can write (2) as the following equivalent system

$$u_1 + A_1P_1F(u_1 + u_2) = 0, \tag{3}$$

$$u_2 - A_2P_2F(u_1 + u_2) = 0. \tag{4}$$

We first assert that for each fixed $u_1 \in H_1$ there exists exactly one $u_2 \in H_2$ which solves equation (4). In fact, the map $S_{u_1}(u_2) = P_2F(u_1 + u_2)$ from H_2 into H_2 is continuous and strongly monotone with the constant c. So, by Proposition III.4.2, this mapping is surjective on H_2 and it has a Lipschitz continuous single-valued inverse, i.e.,

$$\| S_{u_1}^{-1}x - S_{u_1}^{-1}y \| \leq \frac{1}{c} \|x - y)\| \qquad (\forall)x, y \in H_2.$$

Since A_2 is maximal and strongly monotone, it is surjective on H_2 having a Lipschitz continuous inverse with the constant $\frac{1}{m}$. Hence (4) is equivalent to the equation

$$u_2 = S_{u_1}^{-1}A_2^{-1}u_2. \tag{5}$$

As $cm > 1$, we can see that (5) has a unique solution in H_2 by the Banach contraction mapping principle. Thus, we may define a new map $R: H_1 \rightarrow H_2$, which associates to any $u_1 \in H_1$ the solution $u_2 \in H_2$ of (5).

Then equation (4) can be written in the form

$$Ru_1 - A_2P_2F(u_1 + Ru_1) = 0. \tag{6}$$

We assert that R is continuous under the conditions of the theorem. By the strong monotonicity of F it follows that

$$c\,\| Ru_1 - Rv_1 \|^2 \leq (F(u_1 + Ru_1) - F(v_1 + Rv_1), Ru_1 - Rv_1) +$$

$$+ (F(v_1 + Rv_1) - F(u_1 + Rv_1), Ru_1 - Rv_1), \quad u_1, v_1 \in H_1.$$

185

Using (6), we conclude that the first term of the right-hand side of the above inequality is equal to

$$(A_2^{-1}Ru_1 - A_2^{-1}Rv_1, Ru_1 - Rv_1) \leqslant \frac{1}{m}\|Ru_1 - Rv_1\|^2. \tag{7}$$

Hence

$$\left(c - \frac{1}{m}\right)\|Ru_1 - Rv_1\| \leqslant \|F(v_1 + Rv_1) - R(u_1 + Rv_1)\|$$

and thus the continuity of R in $v_1 \in H_1$ follows by the continuity of F in $v_1 + Rv_1 \in H$.

To complete the proof, we have only to show the existence and the uniqueness of the solution in H_1 of equation

$$u_1 + A_1P_1F(u_1 + Ru_1) = 0, \tag{8}$$

which is equivalent to (3). We remark that the mapping $Q = P_1F(I + R)$ is continuous because F and R are continuous. Next, using the strong monotonicity of F and the orthogonality of H_1 and H_2, we get

$$c\|u_1 - v_1\|^2 + c\|Ru_1 - Rv_1\|^2 \leqslant (F(u_1 + Ru_1) - F(v_1 + Rv_1), u_1 - v_1) +$$

$$+ (F(u_1 + Ru_1) - F(v_1 + Rv_1), Ru_1 - Rv_1)$$

and by (7), we obtain

$$c\|u_1 - v_1\|^2 + \left(c - \frac{1}{m}\right)\|Ru_1 - Rv_1\|^2 \leqslant$$

$$\leqslant (F(u_1 + Ru_1) - F(v_1 + Rv_1), u_1 - v_1),$$

that is, the strong monotonicity of Q:

$$c\|u_1 - v_1\|^2 \leqslant (Qu_1 - Qv_1, u_1 - v_1) \qquad (\forall) \; u_1, v_1 \in H_1.$$

Equation (8) can be written as

$$u_1 + A_1Qu_1 = 0.$$

By Proposition 2.6, it admits at least one solution in H_1.

To prove the uniqueness, assume that there exist two solutions $u_1, v_1 \in H_1$ of the equation (8). Then, by the strong monotonicity of Q and the monotonicity of A_1, we have

$$c\|u_1 - v_1\|^2 \leqslant (u_1 - v_1 + A_1Qu_1 - A_1Qv_1, Qu_1 - Qv_1) = 0.$$

This implies that $u_1 = v_1$. Thus the theorem is completely proved.

We shall now construct a class of Hammerstein equations in $L^2(0,1)$ to which the above theorem applies.

Consider in the Hilbert space $L^2 = L^2(0,1)$ a sequence $\{e_n\}$ of ortho-normal functions with pairwise disjoint supports. In order to obtain such a sequence we consider a strictly increasing sequence $\{a_n\} \subset [0,1]$ and define for every $n \in \mathbb{N}$ the function

$$e_n(x) = \begin{cases} \dfrac{1}{(a_{2n} - a_{2n-1})^{\frac{1}{2}}} & \text{if } x \in [a_{2n-1}, a_{2n}], \\ 0 & \text{otherwise.} \end{cases}$$

Denote by H the closed subspace of L^2 spanned by the sequence $\{e_n\}$ and let $H^\perp$ be its orthogonal complement. For any sequence of real numbers $\{\lambda_n\}$ define the kernel $k: [0,1] \times [0,1] \mapsto \mathbb{R}$ as follows

$$k(x, y) = \sum_{n=1}^{\infty} \lambda_n e_n(x) e_n(y).$$

For each pair $[x,y] \in [0,1] \times [0,1]$, at most one term of the above series is different from zero. Let A be the linear integral operator in L^2 with the kernel $k(x,y)$. Thus. for $v \in L^2$ we have

$$(Av)(x) = \int_0^1 k(x, y)v(y)\,\mathrm{d}y = \sum_{n=1}^{\infty} \lambda_n(v, e_n)\, e_n(x),$$

where (v,e_n) denote the Fourier coefficients of v with respect to the sequence $\{e_n\}$. Since

$$D(A) = \left\{ v \in L^2 \,\middle|\, \sum_{n=1}^{\infty} |\lambda_n|^2 |(v, e_n)|^2 < \infty \right\},$$

it is obvious that $\{e_n\} \subset D(A)$ and also that $H^\perp \subset D(A)$. Thus $D(A)$ is dense in L^2 and the range $R(A)$ is contained in H.

We may remark that in general the operator A is not bounded, because we may have $|\lambda_n| \to \infty$ as $n \to \infty$. However, it is true that *the linear integral operator $A: D(A) \to L^2$ defined above is closed.*

In fact, let $\{v_n\}$ be a sequence in $D(A)$ and let $u, v \in L^2$ be such that both $v_n \to v$ and $Av_n \to u$ in L^2. Since for every $n \in \mathbb{N}$ the image $Av_n \in H$ we have that $u \in H$. Moreover, since $\{e_n\}$ is a complete orthonormal basis for H, we can write

$$u = \sum_{i=1}^{\infty} (u, e_i) \quad \text{and} \quad \|u\|^2 = \sum_{i=1}^{\infty} |(u, e_i)|^2.$$

Now, for every $n \in \mathbb{N}$,

$$Av_n = \sum_{i=1}^{\infty} \lambda_i(v_n, e_i) e_i \quad \text{and} \quad (Av_n, e_j) = \lambda_j(v_n, e_j), \quad j \in \mathbb{N}.$$

Passing to limit as $n \to \infty$, we get $(u, e_j) = \lambda_j(v, e_j)$ for all $j \in \mathbb{N}$. Hence

$$\sum_{j=1}^{\infty} |\lambda_j|^2 |(v, e_j)|^2 = \sum_{j=1}^{\infty} |(u, e_j)|^2 = \|u\|^2.$$

This implies that $v \in D(A)$ and

$$Av = \sum_{j=1}^{\infty} \lambda_j(v, e_j)e_j = \sum_{j=1}^{\infty} (u, e_j)e_j = u,$$

that is, the operator A is closed.

Let us show further that *the linear integral operator* $A : D(A) \mapsto L^2$ *is self-adjoint*. In fact, by definition, $(Au, v) = (u, Av)$ for $u, v \in D(A)$, that is, A is symmetric. This means that $D(A) \subseteq D(A^*)$, where A^* restricted to $D(A)$ coincides with A. To prove that A is self-adjoint, it is sufficient to show that $D(A^*) \subseteq D(A)$. For this, let $v \in D(A^*)$. Then, there exists a constant $C > 0$ which depends on v such that

$$|(Au, v)| \leqslant C \|u\| \quad \text{for all } u \in D(A).$$

Take the element $u_n = \sum_{i=1}^{n} \lambda_i(v, e_i) e_i \in D(A)$ for every $n \in \mathbb{N}$. Then

$$\sum_{i=1}^{n} |\lambda_i|^2 |(v, e_i)|^2 = |(Au_n, v)| \leqslant C \|u_n\| = C \left\{ \sum_{i=1}^{n} |\lambda_i|^2 |(v, e_i)|^2 \right\}^{\frac{1}{2}}.$$

This implies that $\sum_{i=1}^{n} |\lambda_i|^2 |(v, e_i)|^2 \leqslant C^2$ for every $n \in \mathbb{N}$ and hence

$$\sum_{i=1}^{\infty} |\lambda_i|^2 |(v, e_i)|^2 \leqslant C^2. \text{ Thus } v \in D(A) \text{ and } D(A^*) \subseteq D (A).$$

If we assume that $\lambda_n \geqslant 0$ for all $n \in \mathbb{N}$, then

$$(Au, u) = \sum_{i=1}^{\infty} \lambda_i |(u, e_i)|^2 \geqslant 0, \quad u \in D(A).$$

Since A is also closed and self-adjoint. Theorem III.2.7 implies that it is maximal monotone.

Assume further that

$$\lambda_{2n-1} \geqslant 0 \text{ and } \lambda_{2n} \geqslant m \qquad \forall \, n \in \mathbb{N},$$

where m is a positive constant. The subspaces H_1 and H_2 spanned by the sequence $\{e_{2n-1}\}$ and $\{e_{2n}\}$ are orthogonal and complementary. Consider the operators $A_1: H_1 \mapsto H_1$ and $A_2: H_2 \mapsto H_2$, defined by

$$A_1 u = \sum_{i=1}^{\infty} \lambda_{2i-1}(u, e_{2i-1}) e_{2i-1} \qquad (\forall) \, u \in H_1,$$

and

$$A_2 u = \sum_{i=1}^{\infty} \lambda_{2i}(u, e_{2i})e_{2i} \qquad (\forall) \; u \in H_2,$$

which are maximal monotone, as we have proved above. Moreover, we have

$$(A_2 u, u) \geqslant m \, \|u\|^2 \qquad (\forall) \; u \in D(A_2).$$

Therefore the operators A_1 and A_2 satisfy the conditions of the theorem. Let $A = A_1 P_1 - A_2 P_2$, where P_i are the orthogonal projections of H onto H_i for $i = 1, 2$. Then A is a linear integral operator with indefinite kernel

$$k(x, y) = \sum_{n=1}^{\infty} \lambda_n e_n(x) \, e_n(y).$$

Now, let $f : [0,1] \times \mathbb{R} \mapsto \mathbb{R}$ be a function satisfying the conditions (j) $-$ (jj) in 2.6. Then the associated Nemitskyi operator $F : L^2 \mapsto L^2$ is strongly monotone.

By the theorem, we conclude that the Hammerstein equation

$$u(x) + \int_0^1 k(x, y) f(y, u(y)) \, \mathrm{d}y = w(x)$$

has a unique solution in $L^2(\Omega)$ for any given $w \in L^2(\Omega)$.

3. Angle-bounded mappings

We shall extend some of the previous results to Hammerstein equations in non-reflexive Banach spaces. Having this in view, we consider the angle-bounded operators as a subclass of monotone mappings. This is the main significant concept introduced recent by the study of Hammerstein equations into the theory of monotone operators.

Angle-bounded linear and nonlinear operators

3.1. Let X be a Banach space, let X^* be its dual space and let A be a monotone linear operator from X into X^*. We say that A is *angle-bounded* with the constant $a \geqslant 0$ if

$$|(Ax, y) - (Ay, x)| \leqslant 2a(Ax, x)^{\frac{1}{2}} (Ay, y)^{\frac{1}{2}}, \tag{1}$$

for all x and y in $D(A)$. The angle-boundedness of A with $a = 0$ corresponds to the symmetry of A, i.e.,

$$(Ax, y) = (Ay, x) \qquad (\forall) \ x, y \in D(A).$$

Consider a Hilbert space H with the inner product $<.,.>$. The triple of spaces (X, H, X^*) is said to be in *normal position* provided that
   (i) $X^* \subseteq H \subseteq X$ algebrically and topologically;
   (ii) $(f, x) = <f, x>$ for all $f \in X^*$ and $x \in H$;
   (iii) H is dense in X.
The most important examples of such spaces are the triples

$$(L^{p'}(\Omega), \ L^2(\Omega), \ L^p(\Omega)), \text{ where } \frac{1}{p} + \frac{1}{p'} = 1, \ p \geqslant 2,$$

where Ω is a domain on finite measure in $\mathbb{R}^N$.
   Suppose that (X, H, X^*) is in normal position. In this case an equivalent characterization of angle-boundedness can be given by means of the numerical range of A,

$$\{(Ax, x) \mid x \in D(A) \cap S(0,1)\},$$

where $S(0,1)$ denotes the unit sphere centred in the origin of X.
   Let $\mathbb{C}$ be the complex number field. Defining the usual multiplication by scalars on the product space $H \times H$, we obtain a linear space H_C of all elet ments of the form $z = x + iy$ with $x, y \in H$. If we take in H_C the inner produc-

$$\langle z_1, z_2 \rangle = \langle x_1, x_2 \rangle + \langle y_1, y_2 \rangle + i[\langle x_2, y_1 \rangle - \langle x_1, y_2 \rangle],$$

then H_C becomes a Hilbert space called the *complexification of H*. By definition, the complexification of a linear operator $A : H \mapsto H$ is the map $A_C z = Ax + iAy$ with $D(A_C) = \{z = x + iy \mid x, y \in D(A)\} \subseteq H_C$.
   If A is a monotone operator, then

$$\operatorname{Re} \langle A_C z, z \rangle = (Ax, x) + (Ay, y) \geqslant 0,$$

that is, *the numerical range of A_C lies in the right half-plane of* $\mathbb{C}$.
   If A is an angle-bounded operator and if we denote
$$\langle A_C z, z \rangle = (Ax, x) + (Ay, y) + i[(Ax, y) - (Ay, x)] = r (\cos \theta + i \sin \theta),$$
then in virtue of (1) we have

$$|\operatorname{tg} \theta| = \frac{|(Ax, y) - (Ay, x)|}{(Ax, x) + (Ay, y)} \leqslant 2a \frac{(Ay, y)^{\frac{1}{2}} (Ax, x)^{\frac{1}{2}}}{(Ax, x) + (Ay, y)} \leqslant a.$$

Setting $z = r e^{i\theta}$, we get that the numerical range of the operator A_C is contained in the sector

$$\left\{ z \in \mathbb{C} \mid |\arg z| \leqslant a \leqslant \frac{\pi}{2} \right\},$$

which justifies the name of angle-boundedness.
    Conversely, from $|\operatorname{tg} \theta| \leqslant a$ one obtains

$$\frac{|(Ax, y) - (Ay, x)|}{(Ax, x) + (Ay, y)} \leqslant a.$$

Replacing y by ρy with $\rho > 0$ we have

$$\rho \, \frac{|(Ax, y) - (Ay, x)|}{(Ax, x) + \rho^2(Ay, y)|} \leqslant a$$

or

$$|(Ax, y) - (Ay, x)| \leqslant a \left[\frac{(Ax, x)}{\rho} + \rho(Ay, y) \right],$$

for any real number $\rho > 0$. Taking, in particular, $\rho = \dfrac{(Ax, x)^{\frac{1}{2}}}{(Ay, y)^{\frac{1}{2}}}$, we get
the inequality (1).
    In order to extend the concept of angle-boundedness to nonlinear operators, we proceed from the trimonotonicity property of the Nemitskyi operator, i.e., from the fact that

$$(Fx - Fz, z - y) \leqslant (Fx - Fy, x - y),$$

for any three elements $x, y, z \in X$. A weaker property is the same inequality with a positive constant C multiplied in the right hand side, i.e.,

$$(Fx - Fz, z - y) \leqslant C(Fx - Fy, x - y).$$

    The axiomatization of this property leads to the following general definition: *A nonlinear mapping* $T: X \mapsto 2^{X^*}$ *is said to be* angle-bounded with the constant $C > 0$ *if*

$$(Tx - Tz, z - y) \leqslant C(Tx - Ty, x - y) \tag{2}$$

for any triple of elements $x, y, z \in D(T)$. For $y = z$ inequality (2) implies the monotonicity of T.
    Let us prove now the equivalence of the two definitions of angle-boundedness. We assume that the constants a and C in the relations (1) and (2) have minimal values.

THEOREM *If A is a linear angle-bounded operator, then inequalities* (1) *and* (2) *are equivalent and* $C = \dfrac{1}{4}(a^2 + 1)$.

Proof: Apply the translation $x - y = u$ and $z - y = v$. Since A is linear, inequality (2) is equivalent to

$$(Au, v) - (Av, v) \leqslant C(Au, u) \qquad (\forall) \, u, v \in D(A).$$

Replacing u by tu with $t \in \mathbb{R}$ one has

$$C(Au, u) \, t^2 - (Au, v)t + (Av, v) \geqslant 0 \qquad (\forall) \ t \in \mathbb{R}$$

and (2) becomes equivalent to

$$(Au, v) \leqslant 2C^{\frac{1}{2}} (Au, u)^{\frac{1}{2}} (Av, v)^{\frac{1}{2}} \qquad (\forall) \ u, v \in D(A). \tag{3}$$

As A is linear, we may define a positive semi-definite inner product on $D(A)$ by setting

$$\langle u, v \rangle = \frac{1}{2} [(Au, v) + (Av, u)].$$

Thus $\langle u, u \rangle \geqslant 0$ and $\langle u, v \rangle = \langle v, u \rangle$ for all $u, v \in D(A)$ and hence, by Schwartz generalized inequality

$$\langle u, v \rangle \leqslant \langle u, u \rangle^{\frac{1}{2}} \langle v, v \rangle^{\frac{1}{2}}.$$

Consider the set $N = \{u \in D(A) \, | \, \langle u, u \rangle = 0\}$. The last inequality implies that

$$N = \{u \in D(A) \, | \, \langle u, v \rangle = 0 \text{ for all } v \in D(A)\}$$

i.e., N is the null space of $\langle u, v \rangle$. The quotient space $H_0 = D(A)/N$ is a pre-Hilbert space with the norm

$$|u| = \langle u, u \rangle^{\frac{1}{2}}.$$

Define a linear operator T from H_0 into itself by

$$\langle Tu, v \rangle = (Au, v) \text{ for all } u, v \in H_0$$

and set $S = T - I$. Thus S is anti-symmetric since

$$\langle Su, v \rangle = (Au, v) - \langle u, v \rangle = \frac{1}{2} [(Au, v) - (Av, u)] = - \langle Sv, u \rangle.$$

Notice that $(Au, u) = \langle u, u \rangle$, so that (3) is equivalent to

$$|\langle Tu, v \rangle| \leqslant 2C^{\frac{1}{2}} |u| \, |v| \qquad (\forall) \ u, v \in H_0,$$

i.e., $|T| = 2C^{\frac{1}{2}}$, because C has minimal value.

On the other hand, (1) is equivalent to

$$\langle Su, v \rangle \leqslant a |u| \, |v| \qquad (\forall) \ u, v \in H_0,$$

and $|S| = a$ holds by the same argument. From the definition,

$$|Tu|^2 = \langle u + Su, u + Su \rangle = |u|^2 + |Su|^2$$

for all $u \in H_0$ and therefore $|T|^2 = 1 + |S|^2$, that is, $4C = 1 + a^2$ as required.

Example. Let Ω be the compact intreval $[a,b] \subset \mathbb{R}$ and let $k: \Omega \times \Omega \mapsto \mathbb{R}$ be a continuous kernel. Define the linear map $A: L^2(\Omega) \mapsto L^2(\Omega)$ by setting

$$(Ax)(t) = \int_\Omega k(t, s)x(s)\, ds \qquad (\forall)\, x \in L^2(\Omega). \tag{4}$$

Consider the functions $\varphi_j \in C(\Omega)$, $o \leqslant j \leqslant n$ and take

$$k(t, s) = \sum_{j=0}^n a_j\, \varphi_j(t)\, \varphi_j(s) + \sum_{j,k=0}^n b_{jk}[\varphi_j(t)\, \varphi_k(s) - \varphi_k(t)\, \varphi_j(s)],$$

with $a_j \geqslant 0$, $0 \leqslant j \leqslant n$. Suppose that $b_{jk} \neq 0$ yields $a_j a_k > 0$. Then the linear operator A is angle-bounded.

Indeed, by a double application of the Schwartz inequality, we obtain

$$|(Ax, y) - (Ay, x)|^2 = \left| 2 \sum_{j,k} b_{jk}[(\varphi_j, x)(\varphi_k, y) - (\varphi_j, y)(\varphi_k, x)] \right|^2 \leqslant$$

$$\leqslant 4 \left\{ \sum_{,k} |b_{jk}|\, [(\varphi_j, x)^2 + (\varphi_k, y)^2]^{\frac{1}{2}}\, [(\varphi_j, y)^2 + (\varphi_k, y)^2]^{\frac{1}{2}} \right\}^2 \leqslant$$

$$\tag{5}$$

$$\leqslant 4 \left\{ \sum_{j,k} |b_{jk}|\, [(\varphi_j, x)^2 + (\varphi_k, x)^2] \right\} \left\{ \sum_{j,k} |b_{jk}|\, [(\varphi_j, y)^2 + (\varphi_k, y)^2] \right\} \leqslant$$

$$\leqslant 4b^2 n^2 \sum_{j \in I} (\varphi_j, x^2)^2 \sum_{j \in I} (\varphi_j, y)^2,$$

where $b = \max\{|b_{jk}|\ |1 \leqslant j, k \leqslant n\}$ and

$$I = \{j \in (0,1, \dots, n)|\ \exists k \in (0,1, \dots, n) \text{ with } b_{jk} \neq 0 \text{ or } b_{kj} \neq 0\}.$$

On the other hand,

$$(Ax, x) = \sum_{j=0}^n a_j(\varphi_j, x)^2 \geqslant C \sum_{j \in J} (\varphi_j, x)^2 \geqslant 0, \tag{6}$$

where $J = \{j \in (0,1, \dots, n)|\ a_j > 0\}$ and $c = \min\{a_j | j \in J\}$. Hence, by (5) and (6) it follows that A is angle-bounded with the constant $a = \dfrac{2b^2 n^2}{c^2} > 0$

provided that $I \subseteq J$, which holds by hypothesis.

In particular, the linear map $A : L^2(\Omega) \mapsto L^2(\Omega)$ defined in (4) with

$$k(t, s) = 1 - t + s + 5ts \qquad (\forall)\ t, s \in [a, b],$$

is angle-bounded with constant $a = 1$.

The splitting of linear maps

3.2. Let us discuss some properties of angle-bounded mappings.

(P_1) Let $T : X \to X^*$ be an angle-bounded hemicontinuous operator. If $x, y \in X$ are such that $(Tx - Ty, x - y) = 0$, then $Tx = Ty$.

Indeed, definition (2) in 2.6 gives $(Tx - Tz, z - y) \leqslant 0$ for all $z \in X$. Setting $z = y + tu$ with $t > 0$, we get $(Tx - T(y + tu), u) \leqslant 0$. As $t \to 0$, if follows that $(Tx - Ty, u) \leqslant 0$ for all $u \in X$ and therefore $Tx = Ty$.

(P_2) Let $T : X \to X^*$ be a bounded operator. If T is angle-bounded with the constant C, then to every $r > 0$ there corresponds a constant $k(r) > 0$ such that

$$r\,\|Tx\| \leqslant C(Tx - T(0), x) + k(r) \qquad (\forall)\ x \in X. \tag{1}$$

Indeed, if we choose $y = 0$ in the definition of angle-boundedness, we obtain

$$r\,\|Tx\| = \sup_{\|z\| = r} |(Tx, z)| \leqslant C(Tx - T(0), x) + r \sup_{\|z\| = r} \|Tz\|.$$

To get (1) denote $k(r) = r \sup_{\|z\| = r} \|Tz\|$.

Let us now describe the structure of linear angle-bounded operators.

THEOREM *Let X be a Banach space, let X^* be its dual space and let $A : X \mapsto X^*$ be a linear operator which is angle-bounded with constant $a \geqslant 0$ and let $D(A) = X$. Then*

(a) *There exist a Hilbert space H, with norm $|\cdot|$, a bounded linear mapping S of X into H and a skew-adjoint bounded linear operator $B : H \mapsto H$ such that*

$$|S|^2 \leqslant \|A\|, \quad |B| \leqslant a \ \text{and}\ A = S^*(I + B)S,$$

where S^* is the adjoint map of S;

(b) *$(I + B)^{-1} : H \mapsto H$ is a bounded linear isomorphism which satisfies the inequality*

$$\langle (I + B)^{-1}f, f \rangle \geqslant (1 + a^2)^{-1}|f| \qquad (\forall) f \in H$$

(c) *The operator $S^* : H \mapsto X^*$ is one-to-one;*

(d) *If A is compact, then S is compact.*

Proof: (a) Since A is linear monotone and everywhere defined. Theorem III.2.2 implies that it is bounded. Let $A^* : X^{**} \mapsto X^*$ be the adjoint map of A. After identifying X with its canonical image in X^{**}, denote

$$A_1 x = \frac{1}{2}(Ax + A^*x) \text{ and } A_2 x = \frac{1}{2}(Ay - A^*x)$$

for each x in X. Let us apply a similar argument as in the proof of Theorem 2.6. We define a positive semi-definite inner product on $D(A) = X$ by setting

$$\langle x, y \rangle = (A_1 x, y) \qquad (\forall) x, y \in X,$$

and we introduce the pre-Hilbert space $H_0 = X/N$, where N is the null-space of the bilinear form $\langle x, y \rangle$. Denote by $\pi : X \mapsto H_0$ the canonical map and by H the completion of H_0 with respect to the norm $|\cdot|$ associated to the above inner product.

Define $S : X \mapsto H$ to be the composition of π with the injection of H_0 into H. By this definition, for each x, y in X,

$$\langle Sx, Sy \rangle = (A_1 x, y).$$

Thus S is linear and bounded since

$$|Sx|^2 = (A_1 x, x) \leqslant \|A_1\| \|x\|^2 \qquad (\forall) x \in X.$$

In particular, $|S|^2 \leqslant \|A\|$. As $D(A) = X$, we may consider its adjoint map $S^* : H \mapsto X^*$.

Since A is angle-bounded with constant a, we have

$$|(A_2 x, y)| = \frac{1}{2} |(Ay, y) - (Ay, x)| \leqslant a \langle x, x \rangle^{\frac{1}{2}} \langle y, y \rangle^{\frac{1}{2}} = a|Sx| \, |Sy|,$$

for all x, y in X. As $(A_2 x, y)$ is bounded on a dense subset of the Hilbert space H, there exists a bounded skew-adjoint linear operator $B : H \mapsto H$ such that

$$(A_2 x, y) = \langle BSx, Sy \rangle \qquad (\forall) x, y \in X.$$

Since $|\langle BSx, Sy \rangle| \leqslant a|Sx| \, |Sy|$ it follows that $|B| \leqslant a$.

Finally, for each x, y in X we have

$$(Ax, y) = (A_1 x, y) + (A_2 x, y) = \langle Sx, Sy \rangle + \langle BSx, Sy \rangle =$$

$$= \langle (I + B) Sx, Sy \rangle = \langle S^*(I + B) Sx, y \rangle$$

and therefore $A = S^*(I + B) S$, as required.

(b) Since $B^* = -B$ we obtain

$$\langle Bg, g \rangle = 0 \quad \text{and} \quad \langle (I + B) g, g \rangle = |g|^2 \qquad (\forall) g \in H.$$

Hence, $(I + B)^{-1}$ is linear and bounded. If $g = (I + B)^{-1}f$, then $f = (I + B) g$ and $\langle (I + B)^{-1}f, f \rangle = \langle g, (I + B)g \rangle = |g|^2$.

On the other hand,

$$|f|^2 = \langle (I + B) g, (I + B)g \rangle = |g|^2 + |Bg|^2 \leq (1 + |B|^2) |g|^2,$$

and since $|B| \leq a$, we deduce that

$$\langle (I + B)^{-1} f, f \rangle \geq (1 + |B|^2)^{-1} |f| \geq (1 + a^2)^{-1} |f|^2$$

for all $f \in H$. Therefore, $(I + B) : H \mapsto H$ is monotone and one-to-one.

(c) Since $R(S)$ is dense in H, it follows that S^* is also one-to-one.

(d) Let $\{u_n\}$ be a bounded sequence in X. If A is compact, then there exists a subsequence, denoted again by $\{u_n\}$, such that $\{Au_n\}$ is a Cauchy sequence in X^*. For this sequence we have

$$|Su_n - Su_m|^2 = |S(u_n - u_m)|^2 = (Au_n - Au_m, u_n - u_m) \to 0$$

as $m, n \to \infty$. Hence S is compact. Thus the theorem is proved.

In the case when $D(A)$ is not all of X we deduce the

COROLLARY *Let $A : X \mapsto X^*$ be a monotone symmetric linear operator densely defined in X. Then there exist a Hilbert space H and a linear map $S : X \mapsto H$ such that $A = S^*S$, where $S^* : H \mapsto X^*$ is the adjoint map of S. Moreover, S^* is one-to-one and $R(S) \subseteq D(S^*)$.*

Proof: In this case we construct the Hilbert space H starting with the positive semi-definite inner product

$$\langle x, y \rangle = (Ax, y) \qquad (\forall) \; x, y \in D(A).$$

The corresponding operator $S : X \mapsto H$, given by $\langle Sx, Sy \rangle = (Ax, y)$, for all $x, y \in D(A)$, is densely defined in X and it has a dense range $R(S)$ in H. Hence, the adjoint map $S^* : H \mapsto X^*$ is well-defined and one-to-one.

Let us show that $R(S) \subseteq D(S^*)$. This in particular will imply that S^* is densely defined in H. Indeed, Sx is defined provided $x \in D(A)$. Thus let $\varphi : D(A) \to \mathbb{R}$ be the linear function defined by $\varphi(y) = \langle Sx, Sy \rangle$. This function is bounded since

$$|\varphi(y)| = |\langle Sx, Sy \rangle| = |(Ax, y)| \leq \|Ax\| \, \|y\|.$$

It can be extended to all of X as a continuous function in X, which proves that

$$\langle Sx, Sy \rangle = (S^*Sx, y) \text{ for all } y \in X.$$

Hence $Sx \in R(S)$.

On the other hand, since $\langle Sx, Sy \rangle = (Ax, y)$ and $D(A)$ is dense in X, the equality $A = S^*S$ follows.

Equations with angle-bounded operators

3.3. Let X be a Banach space, not necessarily reflexive, let X^* be its dual space and let X^{**} be the bidual space of X. Unlike in the reflexive case, we consider now a homogeneous Hammerstein equation

$$(I + AF)\, u = 0 \tag{1}$$

in the dual space X^*, i.e., where the solution u lies in X^*, $F : X^* \mapsto X$ and $A : X \mapsto X^*$. In this situation we may use both the weak topologies $\sigma(X^*, X)$ and $\sigma(X^*, X^{**})$ and Alaoglu's theorem.

Suppose further that $A : X \mapsto X^*$ is a linear angle-bounded operator with $D(A) = X$. By virtue of the splitting given in Theorem 3.2, equation (1) becomes

$$u + S^*(I + B)\, SFu = 0.$$

Since S^* is one-to-one, any solution of this equation must be of the form $u = S^*v$ for some $v \in H$. Writing the equation in terms of v, we have

$$S^*[v + (I + B)\, SFS^*v] = 0.$$

Since the map S^* has a trivial null space, the latter equation is equivalent to

$$v + (I + B)\, SFS^*v = 0.$$

By Theorem 2.7, we know that $(I + B)$ is an isomorphism of H and we may write

$$(I + B)^{-1} v + SFS^*v = 0.$$

We strengthen further the conditions on A by assuming that it is strongly monotone on X, i.e., $(Ax, x) \geqslant m\|x\|^2$ for all $x \in X$ and some $m > 0$. Then A is also angle-bounded with $a = \dfrac{1}{m}\|A\|$, since

$$|(Ax, y) - (Ay, x)| \leqslant 2\|A\|\, \|x\|\, \|y\| \leqslant 2a(Ax, x)^{\frac{1}{2}} (Ay, y)^{\frac{1}{2}} \quad (\forall)\, x, y \in X.$$

In this case the linear injection $S : X \mapsto H$ is an invertible map.

LEMMA *Let $A : X \mapsto X^*$ be a strongly monotone linear operator. If $F : X^* \mapsto X$ is a quasi-bounded operator of type (M), then so is $Q = (I + B)^{-1} + SFS^* : H \mapsto H$.*

Proof: Let $\{v_n\}$ be any sequence in H such that $v_n \rightharpoonup v$ and $Qv_n \to f$ in H and $\lim \sup \langle Qv_n, v_n - v \rangle \leqslant 0$. It follows that $S^*v_n \to S^*v$ in X^*, $(I + B)^{-1} v_n \rightharpoonup \cdot (I + B)^{-1} v = h$ and

$$\langle (I + B)^{-1} v, v \rangle \leqslant \lim \inf \langle (I + B)^{-1} v_n, v_n \rangle.$$

We also have $FS^*v_n \rightharpoonup S^{-1}(f-h)$ in X and

$$\limsup (FS^*v_n, S^*v_n - S^*v) = \limsup \langle SFS^*v_n, v_n - v \rangle =$$

$$= \limsup \langle Qv_n, v_n - v \rangle - \liminf \langle (I+B)^{-1} v_n, v_n - v \rangle \leqslant 0.$$

As F is of type (M), we can conclude that $FS^*v = S^{-1}(f-h)$ and thus $Qv = f$.

Suppose that the sequences $\{v_n\}$ and $\{\langle Qv_n, v_n \rangle\}$ are bounded. From the equality

$$(FS^*v_n, S^*v_n) = \langle SFS^*v_n, v_n \rangle = \langle Qv_n, v_n \rangle + \langle (I+B)^{-1} v_n, v_n \rangle$$

and the boundedness of $(I+B)^{-1}$ it follows that the sequence $\{(FS^*v_n, S^*v_n)\}$ is bounded. Since F is quasi-bounded, $\{FS^*v_n\}$ is bounded in X and $\{SFS^*v_n\}$ is bounded in H. As $Qv_n = (I+B)^{-1}v_n + SFS^*v_n$, the sequence $\{Qv_n\}$ is bounded in H. This proves that Q is quasi-bounded.

THEOREM *Let* $A : X \mapsto X^*$ *be linear strongly monotone with constant* m *(therefore angle-bounded with constant* $a = \dfrac{1}{m} \|A\|$*) and let* $F : X \mapsto 2^{X^*}$ *be a quasi-bounded mapping of type* (M) *such that*

$$(F(g) - F(0), g) \geqslant - c\|g\|^2, \quad g \in X^*,$$

where $c(1 + a^2)\|A\| < 1$. *Then there exists a solution in* X *of equation* (1).

Proof: By the lemma, $Q = (I+B)^{-1} + SFS^* : H \mapsto H$ is quasi-bounded and of type (M). Moreover, we have

$$\langle Qv, v \rangle = \langle (I+B)^{-1} v, v \rangle + \langle FS^*v, S^*v \rangle \geqslant (1 + a^2)^{-1} |v| + c\|S^*v\|^2 +$$

$$+ (F(0), v) \geqslant (k|v| - \|F(0)\| \|A\|^{\frac{1}{2}}) |v|,$$

where $k = (1 + a^2)^{-1} - c\|A\| > 0$. Hence there exists an $R > 0$ such that $\langle Qv, v \rangle > 0$ for $v \in H$ with $|v| = R$. In virtue of theorem III.5.4, the equivalent equation $Qv = 0$ admits a solution.

3.4. Now let us extend the property of angle-boundedness upon the nonlinear operator F.

As we have noted in 1.4, under some suitable assumptions on the function $f(x, r)$, the corresponding Nemitskyi operator is weakly compact from $L^\infty(\Omega)$ into $L^1(\Omega)$. This justifies the following hypothesis:

(c) *The nonlinear operator* F *maps bounded sets of* X^* *into relatively weak compact sets of* X.

The existence and stability result is stated as follows:

THEOREM *Let* X *be a Banach space, let* $A : X \mapsto X^*$ *be a hemicontinuous bounded monotone operator and let* $F : X^* \mapsto X$ *be an angle-bounded hemi-*

continuous map satisfying (c). Then for every $w \in X^*$, there exists a unique solution $u \in X^*$ of the equation

$$u + AFu = w. \tag{1}$$

In addition, if A and F are continuous so, then is $T = (I + AF)^{-1}$.

Proof: For uniqueness, suppose that

$$u + AFu = w \quad \text{and} \quad v + AFv = w.$$

Then $u - v + AFu - AFv = 0$ and

$$(u - v, Fu - Fv) + (AFu - AFv, Fu - Fv) = 0.$$

Since A is monotone, $(AFu - AFv, Fu - Fv) \geqslant 0$ and the monotonicity of F implies $(u - v, Fu - Fv) = 0$. Now (P_1) implies that $Fu = Fv$. Hence,

$$u = w - AFu = w - AFv = v.$$

Further, we may always assume that $A(0) = 0$ and $F(0) = 0$. Indeed, by the shift

$$\tilde{F}u = Fu - F(0), \quad \tilde{A}u = A(u + F(0) - AF(0), \quad \text{and} \quad \tilde{w} = w - AF(0)$$

the Hammerstein equation becomes $u + \tilde{A}\tilde{F}u = \tilde{w}$.

If $w - u = AFu$, then $Fu \in A^{-1}(w - u)$ or

$$0 \in Fu - A^{-1}(w - u).$$

Denote by $A_w : X \mapsto 2^{X^*}$ the monotone mapping defined by

$$A_w v = - A^{-1}(w - v).$$

Apply Debrunner-Flor's lemma (see III.2.9) to the monotone set $A_w v$ and the mapping F defined on the weakly* compact subset

$$K = \{v \in X^* |\ \|v\| \leqslant r\},$$

where the positive number r will be determined later. Therefore there exists an element $u \in K$ such that

$$(Fu - x,\ f - u) \geqslant 0 \qquad (\forall)[f, - x] \in G(A_w|_K),$$

that is, for all $[f, x] \in X \times X^*$ with $\|f\| \leqslant r$ and $f + Ax = w$. Let us show that the element u is solution of equation (1). For that purpose, we rewrite the last inequality in the following form

$$(Fu - x,\ w - Ax - u) \geqslant 0 \text{ for all } x \in X \text{ with } \|Ax - w\| \leqslant r. \tag{2}$$

Take $x = 0$ in (2). Assuming that $r \geqslant \|w\|$, we obtain $(Fu, u) \leqslant (Fu, w)$. By hypothesis (c), the operator F is bounded. Moreover, we deduce from (P_2), that

$$R\|Fu\| \leqslant C(Fu, w) + K(R) \leqslant C\|Fu\| \|w\| + K(R).$$

For $R > C\|w\|$ fixed, we have

$$\|Fu\| \leqslant M = \frac{K(R)}{R - C\|w\|},$$

where M depends on $\|w\|$ and not on r. If we fix

$$r > \sup_{\|x\| \leqslant M+1} \|Ax - w\|,$$

then we can replace x by $x_t = Fu + tz$ with $0 \leqslant t \leqslant 1$ and $z \in X$, $\|z\| \leqslant 1$. Hence (2) becomes $(z, w - Ax_t - u) \geqslant 0$. Passing to the limit as $t \to 0$, we obtain $u + AFu = w$, which proves the existence of a solution for equation (1).

Now, let us assume the continuity of A and F. If

$$u + AFu = f \text{ and } v + AFv = g,$$

then using the fact that $(I + AF)^{-1}$ is one-to-one, we may write

$$u = (I + AF)^{-1}f \text{ and } v = (I = AF)^{-1} g.$$

On the other hand, set

$$\sup_{\|u-v\| \leqslant t} \|Fu - Fv\| \leqslant \omega(t) \text{ and } \sup_{\|x-y\| \leqslant r} \|Ax - Ay\| \leqslant \sigma(r),$$

where $\omega(t) \to 0$ as $t \to 0$ and $\sigma(r) \to 0$ as $r \to 0$. By the monotonicity of A,

$$(Fu - Fv, u - v) \leqslant (Fu - Fv, f - g)$$

and by the angle-boundedness of F,

$$(Fv - Fw, w - u) \leqslant C(Fu - Fv, u - v) \leqslant C\|Fu - Fv\| \|f - g\|$$

for all $w \in X^*$. Therefore,

$$(Fv - Fu, w - u) = (Fw - Fu, w - u) + (Fv - Fw, w - u) \leqslant$$

$$\leqslant \|Fw - Fu\| \|w - u\| + C\|Fu - Fv\| \|f - g\|.$$

If we choose $w = u + h$ with $\|h\| = t$, then

$$t\|Fu - Fv\| \leqslant t\omega(t) + C\|Fu - Fv\| \|f - g\| \qquad (\forall) t > 0.$$

When $\|f - g\| \leqslant \dfrac{t}{2C}$ we get that $\|Fu - Fv\| \leqslant 2\omega(t)$. This proves the continuity of the mapping $F(I + AF)^{-1}$. Finally, from

$$u - v = f - g - (AFu - AFv)$$

we deduce that

$$\|(I + AF)^{-1} f - (I + AF)^{-1} g\| \leqslant \frac{t}{2C} + \sigma(2\omega(t)).$$

Thereby we have established the continuous dependence of the solution of equation (1) on the inhomogeneous term.

An approximation method

3.5. The application of the continuous dependence theorem has permitted Brézis and Browder [3] to give a convergence result for suitably defined Galerkin approximations of Hammerstein equations.

For $w \in X^*$ we set $u = (I + AF)^{-1} w$.

PROPOSITION *Let* $\{A_n\}$ *and* $\{F_n\}$ *be sequences of mappings satisfying the assumptions of Theorem 3.4 with a uniform angle-boundedness constant C for F_n's. Further, we assume the boundedness of the operators A_n, F_n, uniformly for $n \in \mathbb{N}$, and that for every sequence $\{v_n\}$ such that $v_n \to u$ in X^* it follows that $F_n v_n \to Fu$ in X and $A_n Fu \to AFu$ in X^*. If $w_n \to w$ in X^* and $u_n = (I + AF)^{-1} w_n$, then $u_n \to u$ in X^*.*

Proof: By the monotonicity of A_n, we have

$$(A_n F_n u_n - A_n F_n(0), \; F_n u_n - F_n(0)) \geqslant 0,$$

and as $u_n + A_n F_n u_n = w_n$, we obtain

$$(w_n - u_n - A_n F_n(0), \; F_n u_n - F_n(0)) \geqslant 0.$$

Thus, in virtue of (P_1) we get

$$r\|F_n u_n\| \leqslant C(F_n u_n - F_n(0), u_n) + K(r) \leqslant C(w_n - A_n F_n(0), F_n u_n - F_n(0)) + K(r),$$

for all $r > 0$. Choosing $r > C\|w_n - A_n F_n(0)\|$, we see that $\{\|F_n u_n\|\}$ is bounded and, by our hypotheses, so is $\|u_n\|$.

Let us now write $(A_n F_n u_n - A_n Fu, F_n u_n - Fu) \geqslant 0$ in the form

$$(u_n - u - f_n, F_n u_n - F_n u + x_n) \leqslant 0,$$

where $x_n = F_n u - Fu$ and $f_n = w_n - A_n Fu - u$. Therefore

$$0 \leqslant (u_n - u, F_n u_n - F_n u) \leqslant \|f_n\| \, \|F_n u_n - F_n u\| + \|x_n\| \, \|u - u_n\| + \|f_n\| \, \|x_n\|.$$

Since $\|x_n\| \to 0$ and $\|f_n\| \to 0$, it follows that

$$\varepsilon_n = (u_n - u, F_n u_n - F_n u) \to 0.$$

If we denote $\omega_n(t) = \sup_{\|v - u\| \leqslant t} \|F_n v - F_n u\|$, then $\omega_n(t) \to 0$ as $t \to 0$ and $n \to \infty$, by the assumption on the convergence of $\{F_n u_n\}$. Similarly, as in Theorem 3.4, we get

$$t\|F_n u_n - F_n u\| \leqslant t\omega_n(t) + C(F_n u_n - F_n u, u_n - u) \qquad (\forall) \, t > 0.$$

If we take $t = \sqrt{\varepsilon_n}$ we obtain

$$\|F_n u_n - F_n u\| \leqslant \omega_n(\sqrt{\varepsilon_n}) + C\sqrt{\varepsilon_n}.$$

Hence $F_n u_n \to Fu$ in X and $w_n - A_n F_n u_n = u_n \to w - AFu = u$ in X^*. The proof is complete.

Next, let X be a separable Banach space and let Y be a closed subspace of X^* such that $A(X) \subseteq Y$. A scheme $\Gamma = \{X_n, P_n; Y_n, P_n\}$ is *projectionally complete* for (X, Y) provided that $\{X_n\} \subset X$ and $\{Y_n\} \subset Y$ are sequences of monotonically increasing finite-dimensional subspaces with dim $X_n =$ dim Y_n for each $n \in \mathbb{N}$ and $P_n : X \mapsto X_n$ and $P_n^* : Y \mapsto Y_n$ are adjoint linear projections such that $P_n x \to x$ in X and $P_n^* y \to y$ in Y for every $x \in X$ and $y \in Y$. A detailed study of classes of nonlinear mappings connected to these approximate schemes can be found in Petryšyn's works [1] – [2].

For a given $y \in Y$ consider the Hammerstein equation

$$(I + AF)u = y \tag{H}$$

and the n-th Galerkin approximation of this equation

$$(I + A_n F_n)u_n = P_n^* y, \tag{H_n}$$

where $A_n = P_n^* A P_n : X \mapsto Y_n$ and $F_n = P_n F P_n^* : Y \mapsto X_n$.

THEOREM *Let $A : X \mapsto Y$ be a bounded continuous monotone operator and let $F : Y \mapsto X$ be an angle-bounded and weakly compact mapping. Then for each $n \in \mathbb{N}$, the Galerkin approximation (H_n) admits a unique solution u_n in Y_n and u_n converges strongly in X^* to the unique solution $u \in Y$ of the equation (H).*

Proof: Since $\|P_n\| = \|P_n^*\| \leqslant 1$ for all $n \in \mathbb{N}$, the mappings A_n and F_n are uniformly bounded (independently of n). Also, each A_n is monotone

$$(A_n x - A_n y, x - y) = (A P_n x - A P_n y, P_n x - P_n y) \geqslant 0 \qquad (\forall) \, x, y \in X,$$

while each F_n is angle-bounded with the same constant as the mapping F itself

$$(F_n u - F_n w, w - v) = (FP_n^* u - FP_n^* w, P_n^* w - P_n^* v) \leqslant$$

$$\leqslant C(FP_n^* u - FP_n^* v, P_n^* u - P_n^* v) = C(F_n u - F_n v, u - v) \quad (\forall) u, v, w \in X^*.$$

The hypotheses of Theorem 3.4 being fulfilled, equation (H_n) admits the unique solution $u_n = P_n^* y - P_n^*(AF_n u_n)$ in Y_n.

For the convergence of the Galerkin approximations to the solution of the original equation, it is sufficient to verify only the convergence conditions of $\{A_n\}$ and $\{F_n\}$ in the proposition.

We note that

$$\| F_n u_n - Fu \| = \| P_n F P_n^* u_n - Fu \| \leqslant \| P_n [F P_n^* u_n - Fu] \| + \| P_n Fu - Fu \| \leqslant$$

$$\leqslant \| F P_n^* u_n - Fu \| + \| P_n Fu - Fu \| \to 0,$$

since $P_n^* u_n \to u$ in Y and F is continuous on Y.

Similarly, we have

$$\| A_n Fu - AFu \| = \| P_n^* AP_n Fu - AFu \| \leqslant$$

$$\leqslant \| P_n^* AP_n Fu - P_n^* AFu \| + \| P_n^* AFu - AFu \| \to 0,$$

since $AFu \in Y$ and $P_n^* AFu \to AFu$ in Y.

These observations complete the proof of the theorem.

A class of Urysohn equations

3.6. A nonlinear integral equation of Urysohn type has the form

$$u(x) + \int_\Omega K(x, y, u(y)) \, dy = w(x),$$

where Ω is a domain of σ-finite measure in $\mathbb{R}^N$. The inhomogeneous term w and the unknown function u are measurable on Ω. The function $K : \Omega \times \Omega \times \mathbb{R} \to \mathbb{R}$ is called the *Urysohn kernel*. When

$$K(x, y, r) = k(x, y) f(y, r),$$

the Urysohn equation becomes a Hammerstein one.

In this section we shall discuss some existence results for Urysohn equations, the kernels of which are of the form

$$K(x, y, r) = \sum_{i=1}^{m} k_i(x, y) f_i(y, r) \quad (\forall) x, y \in \Omega, r \in \mathbb{R}.$$

The integral equations with these kernels can be regarded as generalized equations of Hammerstein type and the previous results can be extended to the operational equation

$$u + \sum_{i=1}^{m} A_i F_i u = w, \tag{1}$$

where each A_i is a linear operator from a Banach space X to its dual space X^* while F_i is a nonlinear operator from X^* to X.

The study of these equations was initiated by Browder [12] and further developed by Joshi [1] and Gupta [5] — [6].

We shall show that by a suitable choice of some appropriate Banach spaces, we may assume without loss of generality that $m = 1$. In view of this, let $\mathscr{X} = X \times \ldots \times X$ and $\mathscr{X}^* = X^* \times \ldots \times X^*$ be the Carthesian products with m factors each. Then $U \in \mathscr{X}^*$ if and only if $U = (u_1, \ldots, u_m)$ with $u_i \in X^*$ and

$$\|U\|_{\mathscr{X}^*} = \left(\sum_{i=1}^{m} \|u_i\|_{X^*}^2 \right)^{\frac{1}{2}}.$$

A similar structure is used employing the elements $V \in \mathscr{X}$. Denote $\sum_{i=1}^{m} u_i$ by u and define the mappings $\mathscr{F} : \mathscr{X}^* \to \mathscr{X}$ and $\mathscr{A} : \mathscr{X} \to \mathscr{X}^*$ by

$$\mathscr{F} U = (F_1 u, \ldots, F_m u) \quad \text{and} \quad \mathscr{A} V = (A_1 v_1, \ldots, A_m v_m).$$

The maps $\mathscr{F}$ and $\mathscr{A}$ preserve all properties of the maps F_i and A_i.

Let us observe now that it is sufficient to prove that the equation

$$U + \mathscr{A}\mathscr{F} U = W$$

has a solution $U \in \mathscr{X}^*$ for any given $W = (w_1, \ldots, w_m) \in \mathscr{X}^*$. In fact, this implies that

$$u_i + A_i F_i u = w_i, \quad i = 1, \ldots, m,$$

and by addition we obtain

$$u + \sum_{i=1}^{m} A_i F_i u = \sum_{i=1}^{m} w_i.$$

Taking $W = (w, 0, \ldots, 0) \in \mathscr{X}^*$, we see that the element u is a solution of equation (1).

The following results related to Urysohn's equation (1) will be based on some extensions of the positivity and the angle-boundedness of linear operators.

3.7. Let X be a general Banach space and let X^* be its dual space. A linear bounded operator $A : X \mapsto X^*$ is said to be *quasi-positive* provided that

$$\mu_A = \inf \left\{ \frac{(Ax, x)}{\|Ax\|^2} \,\middle|\, x \in X, \ Ax \neq 0 \right\} > - \infty.$$

The relationship between this concept and the angle-boundedness is given by means of the following

PROPOSITION *Let $A: X \mapsto X^*$ be a linear angle-bounded operator with $D(A) = X$. Then there exists a constant $d > 0$ such that*

$$(Ax, y) \geqslant d\|Ax\|^2 \qquad (\forall)x \in X. \tag{*}$$

Proof: In fact, by Theorem 2.7, $A = S^*(I + B)S$ and

$$\langle (I + B)f, f \rangle \geqslant (1 + a^2)^{-1}|f|^2 \quad (\forall)f \in H,$$

where $a \geqslant 0$ is the constant of angle-boundedness of A. Then,

$$(Ax, x) = \langle (I + B)Sx, Sx \rangle \geqslant (1 + a^2)^{-1}|(I + B)Sx|^2 \geqslant$$

$$\geqslant (1 + a^2)^{-1}\|A\|^{-1}\|Ax\|^2,$$

because $\|Ax\|^2 = \|S^*(I + B)Sx\|^2 \leqslant \|S^*\|^2|(I + B)Sx|^2$ and $\|S^*\|^2 = \|S\|^2 \leqslant$ $\leqslant \|A\|$. Thus we get inequality (*) with $d = (1 + a^2)^{-1}\|A\|^{-1}$.

Therefore any linear angle-bounded operator A is quasi-positive with $\mu_A > 0$. The converse is not always true.

COROLLARY *Let $A : X \mapsto X^*$ be a linear monotone operator with $D(A) = X$. Then the following conditions are equivalent:*

(*) $(Ax, x) \geqslant d\|Ax\|^2 \quad (\forall) x \in X$;

(⁂) $(Ax, x - y) \geqslant -\dfrac{1}{4d}\|y\|^2 \quad (\forall) x, y \in X.$

Proof: In virtue of (*), we have

$$(Ax, y) \leqslant \|Ax\| \|y\| \leqslant \frac{1}{\sqrt{d}} (Ax, x)^{\frac{1}{2}}\|y\| \leqslant (Ax, x) + \frac{1}{4d}\|y\|^2,$$

and this implies (⁂).

Passing to the polar form $(x \mapsto tx, t \in \mathbb{R})$ in (⁂) one obtains (*).

The inequality (⁂) being invariant under passage to the adjoint operator, we deduce that

$$(A^*z, z) \geqslant d\|A^*z\|^2 \quad (\forall) z \in X,$$

that is, A^* verifies condition (*) with the same constant $d > 0$. Thus, condition (*) is an essential weakening of the angle-boundedness of monotone linear operators defined on all of X.

Now we can prove the

THEOREM *Let $\{A_1, \ldots, A_m\}$ be a family of compact quasi-positive linear operators from X into X^* such that $\bigcup\limits_{i=1}^{m} R(A_i) \subset Y$, where Y is a closed subspace* *of X^*, and let $\{F_1, \ldots, F_m\}$ be a family of bounded demicontinuous nonlinear mappings from Y into X. Suppose that there is a function $c: \mathbb{R}_+ \mapsto \mathbb{R}_+$ satisfying $\lim_{r \to \infty} r^{-2}c(r) = 0$ such that the inequality*

$$\sum_{i=1}^{m} (u_i, F_i u) + \lambda \sum_{i=1}^{m} \|u_i\|^2 y \geqslant -c \left(\left| \sum_{i=1}^{m} \|u_i\|^2 \right|^{\frac{1}{2}} \right)$$

holds for any m-tuple $\{u_1, \ldots, u_m\}$ in Y with $u = \sum_{i=1}^{m} u_i$ and some $\lambda < \mu =$

$= \inf \{\mu_{A_i} | 1 \leqslant i \leqslant m\}$. *Then the equation*

$$u + \sum_{i=1}^{m} A_i F_i u = 0$$

has at least one solution in Y.

Proof: By the remark in Section 3.5, it is sufficient to prove the theorem for $m = 1$. As $AF : Y \to Y$ is compact, we shall use the Leray-Schauder principle to show that there is an $r > 0$ such that $(I + tAF)u \neq 0$ for every $t \in [0,1]$ and $u \in Y$ with $\|u\| = r$. Now let $r > 0$ be such that $\mu - \lambda - r^{-2}c(r) > 0$; this is possible because $\alpha < \mu$ and $r^{-2}c(r) \to 0$ as $r \to \infty$. Clearly, $u + tAFu \neq 0$ for $t = 0$ and $\|u\| = r$. Assume that $u + tAFu = 0$ for some $t > 0$ and $\|u\| = r$. By the above hypotheses we have

$$0 = (u, Fu) + t(AFu, Fu) \geqslant (u, Fu) + t\mu\|AFu\|^2$$

$$= (u, Fu) + \frac{1}{t} \mu\|u\|^2 \geqslant (\mu - \lambda - r^{-2}c(r))r^2 > 0.$$

This contradiction proves the theorem.

3.8. Let us now investigate Urysohn integral equations in Banach spaces which are in normal position in the sense defined in Section 3.1.

Let (X, H, X^*) be a triple in normal position. Denote by $\langle .,. \rangle$ and $|.|$ the inner product and the norm in H. A linear bounded operator $A : X \mapsto X^*$ is said to be *quasi-accretive* provided that

$$v_A = \inf \left\{ \frac{(Ax, x)}{|Ax|^2} \,\middle|\, x \in X, Ax \neq 0 \right\} > -\infty.$$

In general, the concepts of quasi-positivity and quasi-accretivity are different. The quasi-accretive operators are easier to investigate than quasi-positive ones because in their study one can make use of the spectral theory of linear maps in Hilbert spaces.

Denote by $\hat{A}$ the restriction of A to H.

LEMMA *Let $A : X \mapsto X^*$ be a quasi-accretive operator and let $\lambda < v_A$. Then*

(a) *$I - \lambda\hat{A}$ is one-to-one;*

(b) *$R(I - \lambda\hat{A})$ is closed in H.*

Proof: (a) If $u \in H$ and $u - \lambda \hat{A}u = 0$, then

$$0 = \langle \hat{A}u, u - \hat{A}u \rangle = (\hat{A}u, u) - \lambda |Au|^2 \geqslant (v_A - \lambda)|\hat{A}u|^2.$$

We deduce that $\hat{A}u = 0$ and $u = 0$.

(b) Let $v_n = (I - \lambda \hat{A})u_n \in R(I - \lambda \hat{A})$ and $v_n \to v$ in H. From

$$\langle \hat{A}u_n, v_n \rangle = \langle \hat{A}u_n, u_n - \lambda \hat{A}u_n \rangle = (Au_n, u_n) - \lambda |Au_n|^2 \geqslant (v_A - \lambda)|Au_n|^2$$

we can see that $\{Au_n\}$ is bounded in H. We may assume that $Au_n \rightharpoonup w$ in H. Then $u_n \to v + \lambda w = z$ and $\hat{A}u_n \rightharpoonup \hat{A}z$ in H. This implies that $z \in H$ and $(I - \lambda \hat{A})z = v$, that is, $R(I - \lambda \hat{A})$ is closed in H.

Moreover, we have the

PROPOSITION *Let* $A : X \mapsto X^*$ *be a quasi-accretive operator and assume that it is either compact or that it satisfies condition* (*). *Then*

(c) $R(I - \lambda \hat{A}) = H$ *for every* $\lambda < v_A$ *and thus* $\dfrac{1}{\lambda} \in \rho(\hat{A})$ *(the resolvent of* $\hat{A}$*)*;

(d) $A_\lambda = (I - \lambda \hat{A})^{-1}A : X \mapsto X^*$ *is a bounded monotone linear operator for every* $\lambda < v_A$.

Proof: (c) If A satisfies condition (*), then $v_A > 0$ and, by Corollary 3.7, we have

$$v_A = \inf \left\{ \frac{(A^*x, x)}{|A^*x|^2} \mid x \in X, A^*x \neq 0 \right\}.$$

In virtue of (a) we see that $\ker (I - \lambda A^*) = \{0\}$ for $\lambda < v_A$. Now it follows from (b) that $R(I - \lambda \hat{A})^{\perp} = \ker (I - \lambda \hat{A}^*)$. If A is compact, then (c) is an obvious conclusion from the Fredholm alternative for compact linear operators.

(d) By (c), the maps $(I - \lambda \hat{A})^{-1} \colon H \mapsto H$ and $\hat{A}_\lambda \colon X \mapsto H$ are continuous. Setting $v = (I - \lambda \hat{A})u$ for $u \in H$, we have

$$\langle A_\lambda u, u \rangle = ((I - \lambda \hat{A})^{-1}\hat{A}u, u) = (\hat{A}(I - \lambda \hat{A})^{-1}u, u) = (\hat{A}v, (I - \lambda \hat{A})v) =$$

$$= \langle Av, v \rangle - \lambda |Av|^2 \geqslant 0.$$

This inequality remains valid for $u \in X$ because A_λ is continuous and H is dense in X.

The following version of Theorem 2.1 is interesting from the methodological standpoint.

THEOREM *Let (X, H, X^*) be in normal position, let $\{A_1, \ldots, A_m\}$ be a family of compact quasi-accretive operators from X into X^*, such that $\bigcup\limits_{i=1}^{m} R(A_i) \subset Y$, where Y is a closed subspace of X^*, and let $\{F_1, \ldots, F_m\}$ be a family of bounded demicontinuous mappings from Y into X. Suppose that there is an $r > 0$ such that*

$$\sum_{i=1}^{m} (u_i, F_i u) + \lambda \sum_{i=1}^{m} |u_i|^2 > 0,$$

for any m-tuple $\{u_1, \ldots, u_m\}$ in Y, $u = \sum\limits_{i=1}^{m} u_i$ with $\sum\limits_{i=1}^{m} |u_i|^2 \geqslant r^2$, and for some $\lambda < \min \{v_{A_i} | 1 \leqslant i \leqslant m\}$. Then the equation

$$u + \sum_{i=1}^{m} A_i F_i u = 0$$

has at least one solution in Y.

Proof: As above, we take $m = 1$. Setting $F_\lambda u = (F + \lambda I)u$ for $u \in H$, by hypothesis, we have $\langle F_\lambda u, u \rangle \geqslant 0$ for $|u| \geqslant r$. Consider the equation

$$u + A_\lambda F_\lambda u = 0, \tag{1}$$

which is obtained by adding $\lambda A u$ to and subtracting it from the equation $u + A F u = 0$ and then multiplying it by $(I - \lambda \hat{A})^{-1}$. Since A_λ is a linear monotone compact operator, in view of Theorem, 2.1, equation (1) has a solution in X^*, which lies in Y because $R(A) \subset Y$.

3.9. A similar result for linear operators satisfying condition (∗) can be given by means of a variant of the angle-boundedness.

Let (X, H, X^*) be a triple in normal position. A quasi-accretive operator $A : X \mapsto X^*$ is said to be *quasi-angle-bounded* if there exist constants $\alpha < v_A$ and $\gamma \geqslant 0$ such that

$$|(Ax, y) - (Ay, x)| \leqslant \gamma[(Ax, x) - \alpha |Ax|^2]^{\frac{1}{2}}[(Ay, y) - \alpha |Ay|^2]^{\frac{1}{2}},$$

for all $x, y \in X$. Every symmetric quasi-accretive linear operator $A : X \mapsto X^*$ is quasi-angle-bounded.

PROPOSITION *Let $A : X \mapsto X^*$ be a quasi-angle-bounded operator and assume that it is either compact or it satisfies condition (∗). Then the operator $A_\lambda = (I - \lambda \hat{A})^{-1} A : X \mapsto X^*$ is angle-bounded for every $\lambda < v_A$.*

Proof: By Proposition 3.8, the maps $(I - \lambda \hat{A})^{-1} : H \mapsto H$ and $A_\lambda : X \mapsto H$ are continuous. Set $x = (I - \lambda \hat{A})^{-1} u$ and $y = (I - \lambda \hat{A})^{-1} v$ for $u, v \in H$.

Then

$$|\langle A_\lambda u, v\rangle - \langle A_\lambda v, u\rangle| = |(\hat{A}_\lambda u, v) - (\hat{A}_\lambda v, u)| =$$

$$= |\hat{A}(I - \lambda\hat{A})^{-1}u, v) - (\hat{A}(I - \lambda\hat{A})^{-1}v, u)| = |(\hat{A}x, y) - (\hat{A}y, x)| \leqslant$$

$$\leqslant \gamma(\hat{A}x, (I - \lambda\hat{A})x)^{\frac{1}{2}}(\hat{A}y, (I - \lambda\hat{A})y)^{\frac{1}{2}} = \gamma(A_\lambda u, u)^{\frac{1}{2}}(A_\lambda v, v)^{\frac{1}{2}}.$$

This inequality remains valid for $u, v \in X$ because A_λ is continuous and H is dense in X. Hence, A_λ is angle-bounded with the constant $\gamma > 0$. In particular, by Proposition 3.7, A satisfies condition (∗).

Clearly, a map A which satisfies condition (∗) is bounded and maximal monotone. As a consequence of condition (∗), the maximal monotone inverse mapping $A^{-1} : X^* \mapsto 2^X$ is coercive.

Using these facts, we can prove the following:

THEOREM *Let* (X, H, X^*) *be in normal position, where* X *is a reflexive space. Let* $\{A_1, \ldots, A_m\}$ *be a family of quasi-angle-bounded operators from* X *into* X^*, *such that condition* (∗) *holds for each of them, and let* $\{F_1, \ldots, F_m\}$ *be a family of demicontinuous mappings from* X^* *into* X. *Assume that*

$$\sum_{i=1}^m (u_i - v_i, F_i u - F_i v) + \lambda \sum_{i=1}^m |u_i - v_i|^2 \geqslant 0$$

for some $\lambda < \min\{v_{A_i}| 1 \leqslant i \leqslant m\}$ *and for each pair of* m-tuples $\{u_1, \ldots, u_m\}$, $\{v_1, \ldots, v_m\}$ *in* X^*, *with* $u = \sum_{i=1}^m u_i$ *and* $v = \sum_{i=1}^m v_i$. *Then the equation*

$$u + \sum_{i=1}^m A_i F_i u = w$$

has a unique solution in X^* *for any given* w *in* X^*.

Proof: As in the proof of Theorem 3.8, it is sufficient to show that the equation

$$u + A_\lambda F_\lambda u = f \tag{1}$$

has a unique solution in X^* for each given $f \in X^*$. In view of our hypotheses $A_\lambda = (I - \lambda\hat{A})^{-1}A : X \mapsto H$ verifies condition (∗) and $F_\lambda : X^* \mapsto X$ is demicontinuous and satisfies $(F_\lambda u - F_\lambda v, u - v) \geqslant 0$ for all $u, v \in X^*$. Therefore the mappings A_λ and F_λ are both maximal monotone.

Equation (1) is equivalent to the equation

$$0 \in A_\lambda^{-1}(u - f) + F_\lambda u.$$

Let $v = u - f$ and let $F_{\lambda f}$ denote the maximal monotone mapping defined for any $v \in X^*$ by $F_{\lambda f} = F_\lambda(v + f)$. Then equation (1) holds if and only if

$$0 \in (A_\lambda^{-1} + F_{\lambda f})v. \tag{2}$$

By Theorem III.3.7, thes um $A_\lambda^{-1} + F_{\lambda f}$ is maximal monotone. Moreover, $A_\lambda^{-1} + F_{\lambda f}$ is coercive. The solvability of equation (2) follows now from the surjectivity of coercive maximal monotone mappings (Theorem III.2.10).

We may also disregard the assumption on the reflexivity of X and use the splitting of the angle-bounded linear map A_λ (see Browder and Gupta [1]).

3.10. For the purpose of applications, we need the following variant of Theorem 3.7:

THEOREM *Let* (X, H, X^*) *be in normal position. Let* $\{A_1, \ldots, A_m\}$ *be a family of compact quasi-angle-bounded operators from X into X^* such that* $\bigcup\limits_{i=1}^{m} R(A_i) \subset Y$, *where Y is a closed subspace of X^*, and let* $\{F_1, \ldots, F_m\}$ *be a family of bounded demicontinuous nonlinear mappings from Y into X. Assume that there is a function* $c : \mathbb{R}_+ \mapsto \mathbb{R}_+$ *satisfying* $\lim\limits_{r \to \infty} r^{-2}c(r) = 0$ *such that the inequality*

$$\sum_{i=1}^{m} (u_i, F_i u) + \lambda \sum_{i=1}^{m} \|u_i\|^2 \geqslant - c\left(\left|\sum_{i=1}^{m} \|u_i\|^2\right|^{\frac{1}{2}}\right)$$

holds for any m-tuple $\{u_1, \ldots, u_m\}$ *in Y with* $u = \sum\limits_{i=1}^{m} u_i$ *and some* $\lambda < \mu =$
$= \inf \{\mu_{A_i} \mid 1 \leqslant i \leqslant m\}$. *Then the equation*

$$u + \sum_{i=1}^{m} A_i F_i u = 0$$

has at least one solution in Y.

Proof: As in the proof of Theorem 3.8, it is sufficient to show the solvability in X^* of the equation $u + A_\lambda F_\lambda u = 0$ where $A_\lambda = (I - \lambda \hat{A})A$ and $F_\lambda =$
$= F + \lambda I$. By Proposition 3.9, we have $(A_\lambda u, u) \geqslant d\|Au\|^2$. Our assumptions impliy that $(u, F_\lambda u) \geqslant - c(\|u\|)$, for all $u \in Y$. Next we apply the argument in the proof of Theorem 3.7.

Hammerstein integral equations

3.11. We shall apply Theorem 3.9 to the nonlinear integral equations. For the sake of simplicity, we take $m = 1$.

Let Ω be a bounded measurable domain in $\mathbb{R}^N$, let $k : \bar{\Omega} \times \bar{\Omega} \to \mathbb{R}$ be a continuous kernel and let $f : \bar{\Omega} \times \mathbb{R} \to \mathbb{R}$ be a continuous function. Then, by Theorem II.1.2, the linear integral operator

$$Av(x) = \int_\Omega k(x, y) v(y) \, dy$$

maps $L^1(\Omega)$ compactly into $L^\infty(\Omega)$ and the Nemitskyi operator

$$Fu(x) = f(x, u(x)), \quad x \in \Omega,$$

is bounded continuous from $L^\infty(\Omega)$ into $L^1(\Omega)$.

PROPOSITION *If A satisfies condition* (∗), *i.e.,*

$$(Au, u) \geqslant d \|Au\|_\infty^2 \qquad (\forall)\, u \in L^1(\Omega),$$

and $f(x, t)$ has the sign property

$$f(x, t)\, t \geqslant 0 \qquad (\forall)\, x \in \Omega, |t| > r,$$

where r is a positive number, then the integral equation

$$u(x) + \int_\Omega k(x, y) f(y, u(y))\, dy = 0, \quad x \in \Omega,$$

has at least one continuous solution.

Proof: Obviously, $R(A) \subset C(\bar\Omega)$. For any $u \in C(\bar\Omega)$ we have

$$(u, Fu) = \int_\Omega f(x, u(x)) u(x)\, dx =$$

$$= \int_{\{x\in\Omega \mid |u(x)| > r\}} f(x, u(x))\, u(x)\, dx + \int_{\{x\in\Omega \mid |u(x)| \leqslant r\}} f(x, u(x))\, u(x)\, dx \geqslant -c_0,$$

where $c_0 = r\mu(\Omega)\sup\{f(x, t) \mid x \in \Omega, |t| \leqslant r\}$ and $\mu(\Omega) = $ meas Ω. Since A is compact and F is bounded and continuous, the conditions of Theorem 3.7 are fulfilled with $\mu = d > 0$ and $\lambda = 0$. Hence there exists at least one solution $u(x)$ of equation (1) which is necessarily continuous on Ω.

Let us call a non-zero real number μ a *characteristic value* of the kernel k provided that there exists a function $u \in L^2(\Omega)$ which does not vanish almost everywhere in Ω and which satisfies

$$u(x) = \mu \int_\Omega k(x, y)\, u(y)\, dy \qquad \text{a.e. } x \in \Omega.$$

In other words, μ is a characteristic value of the corresponding linear integral operator A in $L^2(\Omega)$.

THEOREM *Let the kernel k be continuous and symmetric and assume that it has a finite number of negative characteristic values, the smallest of which is denoted by μ_0. If the function $f: \Omega \times \mathbb{R} \mapsto \mathbb{R}$ is continuous and there exist constants $\lambda < \mu_0$ and $r > 0$ such that*

$$f(x, t)t + \lambda |t|^2 \geqslant 0 \quad \text{for all} \quad |t| \geqslant r,$$

then the integral equation (1) has at least one continuous solution.

Proof: Since Ω is bounded, the triple $(L^1(\Omega), L^2(\Omega), L^\infty(\Omega))$ is in normal position and $R(A) \subset C(\bar{\Omega})$. Denote by $\hat{A}$ the restriction of A to $L^2(\Omega)$ and by $\langle . , . \rangle$ the inner product in $L^2(\Omega)$. Clearly $\hat{A}$ is compact and symmetric. Denote by $\mu_0 = v_0 \leqslant v_1 \leqslant \cdots$ the characteristic values of $\hat{A}$ and by $\{e_i\}$ the orthonormal system of corresponding eigenfunctions. Thus, for each $v \in L^2(\Omega)$, we have the Fourier representation

$$\hat{A}v = \sum_{i \geqslant 0} \frac{\langle v, e_i \rangle}{v_i} e_i.$$

If $\hat{A}v \neq 0$, then this representation implies that

$$\frac{\langle \hat{A}v, v \rangle}{\|\hat{A}v\|_2^2} = \frac{\displaystyle\sum_{i \geqslant 0} \frac{\langle v, e_i \rangle^2}{v_i}}{\displaystyle\sum_{i \geqslant 0} \frac{\langle v, e_i \rangle^2}{v_i^2}} = \frac{\displaystyle\sum_{i \geqslant 0} v_i \alpha_i^2}{\displaystyle\sum_{i \geqslant 0} \alpha_i^2} \geqslant \mu_0,$$

where $\alpha_i = \dfrac{\langle v, e_i \rangle}{v_i}$, while $\dfrac{\langle \hat{A}e_0, e_0 \rangle}{\|\hat{A}e_0\|_2^2} = \mu_0$. By the continuity of $A: L^1(\Omega) \mapsto L^\infty(\Omega)$

and the density of $L^2(\Omega)$ in $L^1(\Omega)$, it follows that

$$\inf \left\{ \frac{(Au, u)}{\|Au\|_2^2} \,\middle|\, u \in L^1(\Omega),\ Au \neq 0 \right\} = \inf \left\{ \frac{\langle \hat{A}u, u \rangle}{\|\hat{A}u\|_2^2} \,\middle|\, u \in L^2(\Omega),\ Au \neq 0 \right\}.$$

Since $A: L^1(\Omega) \mapsto L^\infty(\Omega)$ is symmetric and quasi-accretive, it is quasi-angle-bounded and $\mu_A = \mu_0$.

The Nemitskyi operator $F: L^\infty(\Omega) \mapsto L^1(\Omega)$, which corresponds to $f(x, t)$, is bounded and continuous. For all $u \in C(\bar{\Omega})$, we have

$$\|u\|_2^2 = \int_\Omega |u(x)|^2 \, dx \leqslant \int_{\{x \in \Omega \,|\, |u(x)| > r\}} |u(x)|^2 \, dx + r^2 \mu(\Omega)$$

and just as in the previous proposition

$$(u, Fu) \geqslant \lambda \int_{\{x \in \Omega \,|\, |u(x)| > r\}} |u(x)|^2 \, dx - r\mu(\Omega) \sup \{|f|(x, t)| \,\big|\, x \in \Omega, |t| \leqslant r\}.$$

Hence

$$(u, Fu) + \lambda \|u\|_2^2 \geqslant -r\mu(\Omega) [\lambda r + \sup \{|f(x, t)| \,\big|\, x \in \Omega, |t| \leqslant r\}$$

for all $u \in C(\bar{\Omega})$. Since moreover $\lambda < \mu_A$, all assumptions of Theorem 3.10 are satisfied and our proof is complete.

4. Variational methods

Russian mathematicians have used variational methods in the treatment of Hammerstein equations involving unbounded linear operators in Hilbert spaces (see Vainberg [2]). The extension of this approach to general Banach spaces, due to De Figueiredo and Gupta [3], is based on the splitting property of linear symmetric mappings given in Corollary 3.2.

Banach space case

4.1. Let X be a Banach space with dual X^* and let $\varphi : X^* \to \overline{\mathbb{R}}$ be a proper function. We set

$$\partial\varphi(u) = \{x \in X \mid \varphi(v) - \varphi(u) \geqslant (x, v - u) \text{ for all } v \in X^*\}.$$

The elements x which verify this inequality are called *subgradients** of φ at the point u. The set of all subgradients* of φ at u defines a monotone mapping (generally multivalued) $\partial\varphi : X^* \to 2^X$, called the *subdifferential** of φ. When X is reflexive, the subgradients* and subgradients are the same. In non-reflexive Banach spaces the subgradients are elements of X^{**}, hence, in general, a subgradient is not a subgradient*.

Assume that X has an equivalent norm such that the unit ball in X^* is weak* sequentially compact. This condition is fulfilled when X is either a separable or a reflexive Banach space. Denote by "$\to$" the convergence in the $\sigma(X^*, X)$-topology.

We now proceed to prove the following general

THEOREM *Let $A : X \to X^*$ be a symmetric monotone linear operator, densely defined in X, and let $\partial\varphi : X^* \to 2^X$ be the subdifferential* of a weakly* lower semicontinuous function $\varphi : X^* \to \overline{\mathbb{R}}$. Suppose that φ is coercive, i.e., there exists a function $c : \mathbb{R}_+ \to \mathbb{R}$ bounded from below, such that $c(r) \to \infty$ as $r \to \infty$ and*

$$\varphi(v) \geqslant c(\|v\|) \text{ for all } v \in X^*. \tag{1}$$

Then there exists at least one solution $u \in X^*$ of the equation

$$(I + A \partial\varphi) u \ni w,$$

for every $w \in X^*$.

Proof: As usually, replacing $\varphi(u)$ by $\varphi(u + w)$, we may assume that $w = 0$.

Let H be the Hilbert space with the inner product $\langle . , . \rangle$ and the norm $|.|$ introduced in Corollary 3.2. Then $A = S^*S$, where $S: X \mapsto H$ and $S^*: H \mapsto X^*$ are adjoint maps. Define the function

$$\Phi(u) = \frac{1}{2} |u|^2 + \varphi(S^*u) \quad \text{with} \quad D(\Phi) = D(S^*).$$

If we denote $M = \inf \{c(r) | r > 0\}$, we get

$$\Phi(u) \geqslant \frac{1}{2} |u|^2 + c(\| S^*u \|) \geqslant \frac{1}{2} |u|^2 + M, \text{ for all } u \in D(\Phi).$$

Let $d = \inf \{\Phi(u) \mid u \in D(\Phi)\} > - \infty$ and let $\{u_n\}$ be a minimizing sequence of Φ. Then for sufficiently large n, we have

$$d + 1 \geqslant \Phi(u_n) \geqslant \frac{1}{2} |u|^2 + c(\| S^*u_n \|) \geqslant M.$$

Consequently the sequences $\{u_n\}$ and $\{S^*u_n\}$ are both bounded. Since we have assumed the weak* sequential compactness of the closed balls on X^*, passing if need be to subsequences, we have $u_n \rightharpoonup u_0$ in H and $S^*u_n \rightharpoonup g$ in X^*.

We claim that $u_0 \in D(S^*)$. The linear functional $f: D(A) \mapsto \mathbb{R}$ defined by $f(x) = \langle u_0, Sx \rangle$ is bounded because

$$\langle u_0, Sx \rangle = \lim \langle S^*u_n, x \rangle = (g, x) \leqslant \|g\| \, \|x\|$$

and therefore $S^*u_0 = g$.

Let us show now that $\Phi(u)$ *realizes its infimum at* u_0. Indeed, the norm of H and φ being both weakly * lower semicontinuous, it follows that

$$\frac{1}{2} |u_0|^2 + \varphi(S^*u_0) \leqslant \liminf \left\{ \frac{1}{2} |u_n|^2 + \varphi(S^*u_n) \right\} = \lim \Phi(u_n) = d,$$

and hence $\Phi(u_0) = d$.

Finally, Φ attains a minimal value at u_0 and, necessarily, $0 \in \partial \Phi(u_0)$. As $\partial \Phi(u_0) = u_0 + S \partial \varphi(S^*u_0)$, we obtain

$$0 \in S^*u_0 + A\varphi(S^*u_0).$$

Therefore S^*u_0 is a solution of the equation $0 \in (I + A \, \partial\varphi)u$, and this proves the theorem.

In the case of Hilbert spaces we use the particular estimate of type (1) with

$$c(r) = a_1 r^2 + a_2 r^\theta + a_3, \qquad r \in \mathbb{R},$$

where $a_1 > 0$, $0 < \theta < 2$ and a_2, a_3 are arbitrary constants.

Hilbert space case

4.2. Consider now the Hilbert space case where, unlike in Banach spaces, we can reduce the variational problem to the minimization of a function on a subspace.

Let H be a Hilbert space with inner product $(.,.)$ and the induced norm $\|\cdot\|$, and let $A: H \mapsto H$ be a self-adjoint linear monotone mapping. Assume that the latter is densely defined in H but not necessarily bounded. By Theorem III.2.7, A is maximal monotone and thus there exists the square root $|A|^{\frac{1}{2}}$ which is self-adjoint and monotone and satisfies $D(A) \subseteq D(|A|^{\frac{1}{2}})$ (for details, see Kato [2, p. 281]).

Let H_1, H_2 be closed orthogonal subspaces of H such that $H = H_1 \oplus H_2$ and let $P_i : H \mapsto H_i$, $i = 1, 2$, denote the corresponding orthogonal projections. Consider a densely defined self-adjoint linear mapping $A : H \mapsto H$ such that $P_i(D(A)) \subset D(A)$ for $i = 1, 2$ and $A = AP_1 + AP_2$ with $P_i A = AP_i$ for $i = 1, 2$. Next, suppose that

$$(AP_1 x, x) \geqslant 0 \quad \text{and} \quad (AP_2 x, x) \leqslant 0 \quad (\forall)\, x \in D(A).$$

Since the projections P_i, $i = 1, 2$ are bounded, the linear operators

$$AP_1, \quad -AP_2 \quad \text{and} \quad |A| = AP_1 - AP_2,$$

both defined on $D(A)$, are self-adjoint, monotone and densely defined. It follows immediately from the definition that

$$|A|^{\frac{1}{2}} = (AP_1)^{\frac{1}{2}} + (-AP_2)^{\frac{1}{2}} \tag{1}$$

$$[(AP_1)^{\frac{1}{2}} - (-AP_2)^{\frac{1}{2}}](P_1 - P_2) = |A|^{\frac{1}{2}} \tag{2}$$

$$[(AP_1)^{\frac{1}{2}} - (-AP_2)^{\frac{1}{2}}]|A|^{\frac{1}{2}} = |A|^{\frac{1}{2}}[(AP_1)^{\frac{1}{2}} - (-AP_2)^{\frac{1}{2}}] = A. \tag{3}$$

Within this framework, we have the following special case of the preceding result

THEOREM *Let* H, H_i, P_i *and* A *be as above. Assume further that* dim $H_2 < \infty$ *and that there exists an* $m > 0$ *such that*

$$(AP_2 x, x) \leqslant -m\|x\|^2 \quad (\forall)\, x \in D(A).$$

Suppose that the function $\varphi : H \mapsto \mathbb{R}$ *is Gâteaux differentiable (weakly lower semicontinuous) and satisfying*

$$\varphi(x) \geqslant \frac{1}{2} a_1\|x\|^2 + a_2\|x\|^\theta + a_3,$$

where $a_1 m > 1$, $a_2 \leqslant 0$, $a_3 \leqslant 0$ and $0 \leqslant \theta < 2$. *Then the equation* $u + AFu = w$, *where* $F = \text{grad } \varphi$, *has a solution in H for each given* $w \in H$.

Proof: Without loss of generality we may assume that $w = 0$. Define the inner product in $D(|A|^{\frac{1}{2}})$ by

$$[x, y] = (x, y) + (|A|^{\frac{1}{2}} x, |A|^{\frac{1}{2}} y) \qquad (\forall) \, x, y \in D(|A|^{\frac{1}{2}}).$$

This generates the norm

$$|x| = (\|x\|^2 + |A|^{\frac{1}{2}}x\|^2)^{\frac{1}{2}}, \qquad x \in D(|A|^{\frac{1}{2}}).$$

$D(|A|^{\frac{1}{2}})$ *is complete* with respect to this norm and consequently a Hilbert space which will be denoted by H'. Indeed, if $\{u_n\}$ is a Cauchy sequence in H', then $\{u_n\}$ and $\{|A|^{\frac{1}{2}} u_n\}$ are both Cauchy sequences in H. Therefore there exist $u, v \in H$ such that $u_n \to u$ and $|A|^{\frac{1}{2}} u_n \to v$ in H. Since $|A|^{\frac{1}{2}}$ is closed, it follows that $u \in D(|A|^{\frac{1}{2}})$ and $|A|^{\frac{1}{2}} u = v$. Hence $u \in D(|A|^{\frac{1}{2}})$ and thus $u_n \to u$ in H'.

Consider the function $\Phi : H' \to \mathbb{R}$ defined by

$$\Phi(u) = \frac{1}{2} \|P_1 u\|^2 - \frac{1}{2} \|P_2 u\|^2 + \varphi(|A|^{\frac{1}{2}} u), \quad u \in H'.$$

We assert that Φ *is weakly lower semicontinuous and coercive on* H'. In fact, let $u_n \rightharpoonup u$ in H'. Then $[u_n, v] = (u_n, v) + (|A|^{\frac{1}{2}} u_n, |A|^{\frac{1}{2}} v) = (u_n, v+|A|v)$ for all $v \in D(|A|^{\frac{1}{2}})$. Since $D(A) \subseteq D(|A|^{\frac{1}{2}})$ and $R(I+|A|) = H$, we obtain $u_n \rightharpoonup u$ in H. Similarly, using the relation

$$[u_n, |A|^{\frac{1}{2}} v] = (u_n, |A|^{\frac{1}{2}} v) + (|A|^{\frac{1}{2}} u_n, |A|v) = (|A|^{\frac{1}{2}} u_n, v + |A|v)$$

we also deduce that $|A|^{\frac{1}{2}} u_n \rightharpoonup |A|^{\frac{1}{2}} u$ in H. Therefore it follows that

$$\|u\| \leqslant \lim \inf \|u_n\| \quad \text{and} \quad \varphi(|A|^{\frac{1}{2}} u) \leqslant \lim \inf \varphi(|A|^{\frac{1}{2}} u_n)$$

whenever $u_n \rightharpoonup u$ in H'. In virtue of the fact that $\dim H_2 < \infty$, we have $\|P_2 u_n\| \to \|P_2 u\|$, and consequently

$$\Phi(u) \leqslant \lim \inf \Phi(u_n),$$

that is, we obtain the weak lower semicontinuity of Φ.

Under the above assumptions on φ we have

$$\Phi(u) \geqslant \frac{1}{2} \|P_1 u\|^2 - \frac{1}{2} \|P_2 u\|^2 + \frac{1}{2} a_1 \||A|^{\frac{1}{2}} u\|^2 + a_2 \||A|^{\frac{1}{2}} u\|^\theta + a_3.$$

Denote $a_1 - \dfrac{1}{m} = 2\varepsilon$. Then

$$\Phi(u) \geqslant \frac{1}{2}\|P_1u\|^2 - \frac{1}{2}\|P_2u\|^2 + \frac{1}{2m}\||A|^{\frac{1}{2}}u\|^2 + \frac{\varepsilon}{2}\||A|^{\frac{1}{2}}u\|^2$$

$$+ \frac{\varepsilon}{2}\||A|^{\frac{1}{2}}u\|^2 + a_2\|u\|^{\theta} + a_3 \geqslant \frac{1}{2}\|P_1u\|^2 - \frac{1}{2}\|P_2u\|^2$$

$$+ \frac{1}{2m}\|(-AP_2)^{\frac{1}{2}}u\|^2 + \frac{\varepsilon}{2}\|(-AP_2)^{\frac{1}{2}}u\|^2 + \frac{\varepsilon}{2}\||A|^{\frac{1}{2}}u\|^2 + a_2|u|^{\theta} + a_3 .$$

Since

$$\|(-AP_2)^{\frac{1}{2}}u\|^2 = (-AP_2u, u) \geqslant m\|u\|^2 \geqslant m\|P_2u\|^2, \quad (\forall)\ u \in D(|A|^{\frac{1}{2}}),$$

we conclude that

$$\Phi(u) \geqslant k|u|^2 + a_2|u|^{\theta} + a_3 \quad (\forall)\ u \in H',$$

where $k = \min(1, m\varepsilon, \varepsilon)$. This proves the coerciveness of Φ.

Therefore there exists a $u_0 \in H'$ at which Φ attains a minimum and consequently at this point the gradient of Φ vanishes:

$$(P_1u_0 - P_2u_0, v) + (F|A|^{\frac{1}{2}}u_0, |A|^{\frac{1}{2}}v) = 0 \quad (\forall)\ v \in D(|A|^{\frac{1}{2}}).$$

Since $|A|^{\frac{1}{2}}$ is self-adjoint, the above equation implies that $-F|A|^{\frac{1}{2}}u_0 \in D(|A|^{\frac{1}{2}})$ and

$$P_1u_0 - P_2u_0 + |A|^{\frac{1}{2}}F|A|^{\frac{1}{2}}u_0 = 0.$$

Apply $(AP_1)^{\frac{1}{2}} - (-AP_2)^{\frac{1}{2}}$ to both sides of this equation. It follows by a simple calculation that

$$|A|^{\frac{1}{2}}u_0 + AF|A|^{\frac{1}{2}}u_0 = 0,$$

i.e., $u = |A|^{\frac{1}{2}}u_0$ is a solution of the the equation $(I + AF)u = 0$.

COROLLARY *Let H, H_i, P_i and A be as in the theorem. Suppose that* $\dim H_2 < \infty$ *and*

$$(AP_2x, x) \leqslant -m\|x\|^2 \quad (\forall)\ x \in D(A)$$

for some $m > 0$. If $F = \operatorname{grad} \varphi$, where $\varphi : H \mapsto \overline{\mathbb{R}}$ is a weakly lower semi-continuous function such that F is strongly monotone, and

$$(Fx - Fy, x - y) \geqslant c\|x - y\|^2 \quad (\forall)\ x, y \in H$$

with the constant c satisfying cm > 1, then the equation u + AFu = w is solvable for each w ∈ H.

Proof: Just as in the theorem we see that the function $\Phi(u)$ is weakly lower semicontinuous on H'. Next, for all $u \in H'$,

$$[\text{grad } \Phi(u), u] = (P_1 u, u) - (P_2 u, u) + (F|A|^{\frac{1}{2}} u, |A|^{\frac{1}{2}} u) =$$

$$= \|P_1 u\|^2 - \|P_2 u\|^2 + (F|A|^{\frac{1}{2}} u, |A|^{\frac{1}{2}} u) \geqslant$$

$$\geqslant \|P_1 u\|^2 - \|P_2 u\|^2 + c\| |A|^{\frac{1}{2}} u \|^2 - \| F(0) \| \| |A|^{\frac{1}{2}} u \| \geqslant$$

$$\geqslant k |u|^2 - \|F(0)\| \, |u|,$$

where $k = \min(1, m\varepsilon, \varepsilon)$ and $a - \dfrac{1}{m} = 2\varepsilon$. Hence

$$\Phi(u) - \Phi(0) = \int_0^1 [\text{grad } \Phi(tu), u] \, dt = \int_0^1 [\text{grad } \Phi(tu), tu] \, \frac{dt}{t}$$

$$\geqslant \frac{k}{2} |u|^2 - \|F(0)\| \, |u|.$$

This implies the coerciveness of $\Phi(u)$. Thus there is an $r > 0$ such that

$$\Phi(u) - \Phi(0) > 0 \quad \text{for } |u| \geqslant r$$

and therefore there exists a $u_0 \in H'$ with $|u_0| < r$ at which Φ attains its minimum. As in the proof of the theorem, $\text{grad } \Phi(u_0) = 0$ implies that $u = |A|^{\frac{1}{2}} u_0$ is a solution of the equation $(I + AF) u = 0$.

5. Related topics and exercises

5.1. Let Ω be a domain of σ-finite measure in $\mathbb{R}^N$ and let $H = L^2(\Omega)$. Suppose that $f(x, r)$ verifies the Caratheodory conditions, $f(.,0) \in H$ and $f(x,r)$ is nondecreasing in r for each $x \in \Omega$. Let F be the Nemitskyi operator with domain $D(F) = \{u \in H \mid Fu \in H\}$. Then F is a single-valued maximal monotone mapping (Browder [14]).

5.2. Let Ω be a domain of σ-finite measure in $\mathbb{R}^N$. Assume that $A: L^1(\Omega) \to L^\infty(\Omega)$ is a bounded hemicontinuous monotone mapping and $Fu = f(x, u(x))$, where $f(x, r): \Omega \times \mathbb{R} \to \mathbb{R}$ is continuous and nondecreasing in r for a.e. $x \in \Omega$, and integrable in x for all $r \in \mathbb{R}$. Then the equation $u + AFu = w$ has a unique solution in $L^{\infty}(\Omega)$ for every $w \in L^{\infty}(\Omega)$ (Brézis and Browder [1]).

5.3. Suppose that $f(x,r)$ satisfies the conditions (i), (ii) and (iv)' in Section 2.2 and the sign condition

$$f(x, r) \cdot r \geqslant c \, | f(x, r)| \, |r| \quad \text{for} \quad |r| \geqslant R.$$

Prove that for each $k > 0$ there exists a constant $q(k)$ such that

$$(Fu, u) \geqslant k \, \| Fu \|_1 - q(k),$$

where F is the corresponding Nemitskyi operator.

HINT The proof follows the method used in Lemma 2.2 (see Brézis and Browder [2]).

5.4. Let F be a bounded continuous mapping of $L^{\infty}(\Omega)$ into $L^1(\Omega)$ and let A be a compact operator of $L^1(\Omega)$ into $L^{\infty}(\Omega)$, where Ω is a domain of σ-finite measure in $\mathbb{R}^N$. Suppose that

(a) $(v, Av) \geqslant 0 \quad$ for all $\quad v \in L^1(\Omega)$;

(b) for each $k > 0$ there exists a $q(k) > 0$ such that

$$(Fu, u) \geqslant k \, \|Fu\|_1 - q(k) \quad (\forall) \, u \in L^{\infty}(\Omega).$$

Prove that for each $w \in L^{\infty}(\Omega)$ the equation $(I + AF) \, u = w$ has at least one solution in $L^{\infty}(\Omega)$.

HINT Apply the Leray-Schauder principle (see Brézis and Browder [2]).

5.5. Let X be a reflexive Banach space and let $A : X^* \mapsto X$ be a linear monotone operator such that A^{-1} is single-valued and defined on all of X. Suppose that $F : X \mapsto X^*$ is a quasi--bounded coercive mapping of type (M). Then $R(I + AF) = X$ (de Figueiredo and Gupta [1]).

5.6. Let X be a reflexive Banach space, let $A : X^* \mapsto X$ be a bounded pseudo-monotone operator and let $F : X \mapsto X^*$ be a hemicontinuous monotone mapping. Suppose that $k(r)$ and $c(r)$ are real-valued functions on $\mathbb{R}_+$ such that $k(r) + c(r) \to \infty$, as $r \to \infty$, and the following two inequalities are satisfied:

(a) $(v, Av) \geqslant k(\|v\|) \, \|v\| \quad (\forall) \, v \in X^*$;

(b) $(v, u) \geqslant c(\|v\|) \, \|v\| \quad (\forall) \, v \in X^*$, for all $u \in F^{-1}v$.

Prove that $R(I + AF) = X$.

HINT Since $(I + AF)F^{-1} = A + F^{-1}$, it is sufficient to prove that $R(A + F^{-1}) = X$ (see Browder, de Figueiredo and Gupta [1]).

5.7. Let X be a reflexive Banach space, let $A : X \mapsto 2^{X^*}$ be a maximal monotone linear mapping and let $F : X \mapsto X^*$ be a bounded coercive monotone mapping of type (S_+). Suppose that a continuous family $\{P_t\}$ of compact mappings of X into X^* is given such that

$P_0 = 0$. Let Ω be a bounded open subset of X and let w be an element of X^*, such that the equation $w \in (A + F)u$ has a solution in Ω. Assume further that for each $t \in [0, 1]$ the equation $w \in (A + F + P_t) u$ has no solution on $\partial\Omega$. Then for each $t \in [0,1]$ the equation $w \in (A + F + P_t) u$ has a solution in Ω (Gupta [2]).

5.8. Let (X, H, X^*) be a triple of spaces in normal position and let L be the restriction to H of a quasi-accretive operator from X into X^*. Denote by $S = \dfrac{1}{2}(L + L^*)$ and $A = \dfrac{1}{2}(L - L^*)$ the symmetric and anti-symmetric part of L. If $R(A) \subset R(S)$, then L is angle-bounded (Amann [1]).

5.9. Consider the linear uniformly elliptic differential operator $\sum\limits_{|\alpha|, |\beta| \leqslant m} D^\alpha(a_{\alpha\beta}(x)D^\beta)$, $x \in \Omega$, where Ω is a domain in $\mathbb{R}^N$. Suppose that the associated Dirichlet form $a(u, v)$ verifies Gårding's inequality

$$a(u, u) \geqslant \gamma_0 |u|^2_{m,2} - \lambda_0 |u|^2_2 \qquad (\forall) u \in W_0^{m,2}(\Omega)$$

with $\gamma_0 > 0$, $\lambda_0 \geqslant 0$ and that it is bounded in $W_2^{m,2}(\Omega)$, i.e., there exists a $\delta > 0$ such that

$$|a(u, v)| \leqslant \delta |u|_{m,2} |v|_{m,2} \qquad (\forall) u, v \in W_0^{m,2}(\Omega).$$

Then the linear operator $L: L^2(\Omega) \mapsto L^2(\Omega)$ given by

$$(Lu, v) = a(u, v) \qquad (\forall) v \in W_0^{m,2}(\Omega),$$

is bounded, continuous and monotone on

$$D(L) = \{u \in W_0^{m,2}(\Omega) \mid |a(u, v)| \leqslant \gamma(u) \|v\|_2, \ \gamma(u) < \infty \quad (\forall) v \in W_0^{m,2}(\Omega)\}.$$

Moreover, for every $\lambda \geqslant \lambda_0$, the operator $(L + \lambda I)^{-1}: L^2(\Omega) \mapsto L^2(\Omega)$ is angle-bounded (Amann [1]–[2]).

5.10. Consider in a Banach space X the *generalized Hammerstein equation*

$$u + A(u) F(u) = 0, \tag{1}$$

where $A: X \mapsto L(X^*, X)$ is compact and $F: X \mapsto X^*$ is a nonlinear bounded continuous mapping satisfying the inequality

$$(F(u) - F(v), u - v) \geqslant -c \|u - v\|^2, \qquad (\forall) u, v \in X.$$

Suppose that for each $x \in X$ the operator $A(x): X^* \mapsto X$ is angle-bounded, with constant of angle-boundedness a independent of x, and suppose that there exists a $\gamma > 0$ such that $\gamma < (1 + a^2)^{-1}c^{-1}$ and $|A(x)|_{L(X^*, X)} \leqslant \gamma$, for each $x \in X$. Then equation (1) has at least one solution in X. The result remains valid if X is replaced by X^*.

Apply this result to the boundary-value problem for the singular quasi-linear differential equation:

$$\frac{d}{dx}\left\{\left[\frac{1}{x} + a(x, u(x))\right]^{-1}\frac{du}{dx}\right\} + \left[\frac{1}{x} + a(x, u(x))\right]u(x) = 2f(x, u(x)), \text{ a.a. } x \in (0, 1]$$

with boundary conditions

$$[u(0)] < \infty \quad \text{and} \quad u(1) + \frac{du}{dx}(1) = 0.$$

Assume that

(a) $f: [0, 1] \times \mathbb{R} \mapsto \mathbb{R}$ satisfies the Caratheodory conditions;

(b) f is bounded on bounded subsets of $[0, 1] \times \mathbb{R}$;

(c) There is a $c < 1$ such that

$$[f(x, u) - f(x, v)] (u - v) \leqslant c (u - v)^2 \qquad (\forall) \ u, v \in \mathbb{R} \text{ for all } x \in [0, 1];$$

(d) $a: [0,1] \times \mathbb{R} \mapsto \mathbb{R}_+$ is continuous and $a(1, \xi) = 0$ for each $\xi \in \mathbb{R}$.

Under these conditions there exists a solution $u \subset C^1(0, 1]$ of the above boundary-value problem (Backwinkel—Schillings [1]).

5.11. Let $\gamma: \mathbb{R}_+ \mapsto \mathbb{R}_+$ be a monotone increasing continuous function such that $\gamma(r) \to \infty$ as $r \to \infty$. Assume that $T: X \mapsto X^*$ is angle-bounded with constant C and

$$\| Tu - Tv \| \leqslant \gamma(\|u - v\|) \qquad (\forall) \ u, v \in X.$$

Then the inequality

$$C(Tu - Tv, u - v) \geqslant \frac{1}{2} \|Tu - Tv\| \gamma^{-1}\left(\frac{1}{2} \|Tu - Tv\|\right) \qquad (\forall) \ u, v \in X,$$

holds. In particular, if T is Lipschitz continuous with constant k, then

$$C (Tu - Tv, u - v) \geqslant \frac{1}{4k} \|Tu - Tv\|^2 \qquad (\forall) \ u, v \in X.$$

These inequalities are extensions of condition $(*)$ in Section 3.6 to nonlinear mappings (Brézis and Browder [3]).

5.12. Let X be a Banach space, let $A: X \mapsto X^*$ be an angle-bounded linear operator and let $F: X^* \to X$ be a locally bounded from below pseudo-monotone mapping. Prove (using the notations in Theorem 3.2) that the mapping $Q = (I + B)^{-1} + SFS^*$ from H into itself is locally bounded from below and pseudo-monotone.

HINT Use similar arguments as in Lemma 3.3 (see Petryshyn and Fitzpatrick [1]).

5.13. Let Ω be a domain of finite measure in $\mathbb{R}^N$ and let X be a reflexive Banach space such that $L^\infty(\Omega) \subseteq X, X^* \subseteq L^1(\Omega)$. Let $F: X \to X^*$ be a hemicontinuous monotone angle-bounded mapping with $0 \in \text{Int } R(F)$ and let $A: L^1(\Omega) \to L^1(\Omega)$ be a bounded linear operator with $(Av, v) \geqslant 0$ for all $v \in L^\infty(\Omega)$. Then for each $w \in X$, the equation $(I + AF)u = w$ admits solutions in X such that

$$(AFu - Av, Fu - v) \geqslant 0 \quad \text{for all} \quad v \in L^\infty(\Omega) \text{ with } Av \in X,$$

(see Brézis and Browder [4]).

Bibliographical comments

The first systematic study of the Nemitskyi operator was made in Vainberg's book [1]. The proofs of its properties given here are due to Browder [12] and Prodi — Ambrosetti [1]. We note that condition (1) in 1.1 is not only sufficient but also necessary for the continuity of the operator F from $L^p(\Omega)$ into $L^q(\Omega)$ (see Krasnoselskii [1, p. 27]). For a detailed study of the Nemitskyi operator in $L^p(\Omega)$ and in Sobolev spaces we refer to Krasnoselskii—Zabreyko—Pustylnik—Sobolevski [1] and Marcus — Mizel [1]—[2]. The perturbations of second order elliptic operators in $L^p(\Omega)$ by Nemitskyi operators have been studied by Calvert [2]. Brézis — Browder [3] showed the trimonotonicity of F.

In the theory of abstract Hammerstein equations in a Banach space X we distinguish between the regular case and the singular one by means of the usual criterion applied to linear equations: we have regularity when the product AF is defined on all of X and the singular case otherwise. We have dwelt upon the recent significant results in the regular case. For the singular equations see Browder [14].

Several mathematicians have established sufficient conditions for the existence and uniqueness of solutions of nonlinear integral equations of Hammerstein type based on the compactness of the linear integral operator A. In this respect, we refer to Dolph — Minty [1] and Martin Jr. [1]. The general results contained in 2.1 are due to Browder [12], while the case $L^1(\Omega)$ was given by Brézis — Browder [1]—[4].

The first attempt to use monotone operators in order to solve integral equations has been implicitly used 40 years ago by Golomb [1]. Further studies in Hilbert spaces have been carried out by Dolph — Minty [1], Kolodner [1] and Kolomy [1]—[2]. The extensions of such results to Banach spaces have been performed by Amann [1]—[3], Brézis [1], Browder — De Figueiredo — Gupta [1], Petryshyn — Fitzpatrick [1], Vainberg [2] and Zini [2]. A unified treatment of monotonicity methods for the study of abstract Hammerstein equations is given in the survey papers by Browder [12] and Gupta [1]. The results have been sharpened in the Hilbert space case by De Figueiredo — Gupta [1]—[2]. These works were made use of in writing Sections 2.3—2.7.

The splitting of angle-bounded linear operator is due to Browder — Gupta [1] and Amann [2]. The extension of the concept of angle-boundedness to nonlinear mappings and the general existence Theorem 3.4 were given by Brézis—Browder [3]. A recent study of properties of angle-bounded mappings and an extension of Theorem 3.1 are due to Baillon — Haddad [1].

The approximate method follows Brézis — Browder [3] and Pascali [2].

The approach to Uryshon equations follows the paper of Gupta [6], which extends the results of Amann [4] for Hammerstein equations. Here the theory of Hammerstein equations is connected with the spectral theory of linear operators. Condition (*) pointed out by Hess [3], [6] will be studied throughout the next section.

Finally, in presenting the variational method, we have used the papers by DeFigueiredo—Gupta [3], Gupta [1] and Pascali [5]. Under more restrictive conditions this method has been investigated by Rall [1], Vainberg—Lavrentiev [1], Kosiskyi [1] and Vainberg [2].

Petrovanu [1] studied discrete Hammerstein equations. Existence results for the generalized Hammerstein equation $u + A(u)F(u) = w$ have been obtained by Avramescu [1], Petry [1]—[2] and Backwinkel-Schillings [1].

Chapter V

Homotopy arguments

Let us go deeper into the investigation of the existence of solutions of the nonlinear functional equation $Tu=f$. In Chapter III, this topic has been studied by the extension of Brouwer's theorem about fixed points to coercive pseudo-monotone mappings in a reflexive Banach space X. It is the purpose of this chapter to investigate the equation $Tu=f$ on a bounded domain of X which is symmetric about the origin. We begin with some extensions of Borsuk's antipodal theorem to nonlinear mappings of monotone type which are homotopic to odd operators.

In order to avoid any auxiliary considerations, we shall deal only with the case when the bounded domain is a ball about the origin. This is sufficient for most applications.

Next, we shall make only mild continuity assumptions on the mappings involved and we shall not assume that they are bounded. By using the asymptotically homogeneous operators we can obtain some nonlinear Fredholm alternatives.

In the latter part of this chapter we establish some necessary and sufficient conditions for certain significant nonlinear boundary-value problems.

1. Odd mappings

Let X be a reflexive Banach space with its dual space X^*, let $B = B(0, r)$ be the open ball in X with center at the origin and radius r, and let ∂B and $\bar B$ be its boundary and closure respectively. A mapping $T : X \to 2^{X^*}$ is said to be *odd* if $[-x, -f] \in G(T)$ provided that $[x, f] \in G(T)$. In particular an operator $T : X \to X^*$ is odd on ∂B if $T(-x) = -Tx$ for all $\|x\| = r$.

Infinite-dimensional variant of Borsuk's theorem

1.1. A basic extension of Borsuk's theorem is given by the following

PROPOSITION *Let* $T : \bar B \mapsto X^*$ *be a demicontinuous operator of type* (S) *which is odd on* ∂B. *Then there exists a solution in* $\bar B$ *of the equation* $Tx = 0$.

Proof: Let Λ be the set of all finite-dimensional subspaces F of X with the norms $\|u\| = \|u\|_F = \|u\|_X$ for any $u \in F$. For every $F \in \Lambda$ we denote by $j_F : F \mapsto X$ the canonical injection and by $j_F^* : X^* \mapsto F^*$ its adjoint map. Then $T_F = j_F^* T j_F$ is a continuous mapping from $\bar{B} \cap F$ into F^* which is odd on $\partial B \cap F$. Hence, by Borsuk's theorem, the equation $T_F x = 0$ has at least one solution $x_F \in \bar{B} \cap F$.

For any $F_0 \subset \Lambda$ consider the set of these solutions $W_{F_0} = \bigcup \{x_F \mid F \supseteq F_0\}$. The family of all such sets has the finite intersection property and the same holds for their weak closures. Hence there exists an element $x_0 \in \bar{B}$ which belongs to the intersection of the weak closure of all the W_{F_0}, i.e. $x_0 \in \bigcap \{\tilde{W}_{F_0} \mid F_0 \in \Lambda\}$. We shall prove that $T x_0 = 0$.

Given $y \in X$, we may choose a finite-dimensional subspace $F_1 \subset X$ such that $\{x_0, y\} \subset F_1$. As $\tilde{W}_{F_1}$ is weakly relatively compact Proposition I.1.4 implies that there exists a sequence $\{x_n\} = \{x_{F_n}\}$ in W_{F_1} such that each $x_n \in F_n$, $F_n \supseteq F_1$ and $x_n \rightharpoonup x_0$. Since $x_n - x_0 \in F_n$, we have

$$(T x_n - T x_0, x_n - x_0) = (j_{F_n}^* T j_{F_n} x_n - T x_0, x_n - x_0)$$

$$= (T_{F_n} x_n - T x_0, x_n - x_0) = -(T x_0, x_n - x_0) \to 0.$$

and $x_n \to x_0$ because T is of type (S). By the demicontinuity of T it follows that $T x_n \rightharpoonup T x_0$. Since $(T x_n, y) = (T_{F_n} x_n, y) = 0$ holds for all $y \in X$, it follows that $T x_n \rightharpoonup 0$, and therefore $T x_0 = 0$.

This result remains invariant under homotopy:

THEOREM *Let $A_t u = A(u, t) : \bar{B} \times [0, 1] \mapsto X^*$ be a mapping with the following properties:*

(i) *For fixed $t \in [0, 1]$, the operator $A_t : \bar{B} \mapsto X^*$ is demicontinuous and of type (S);*

(ii) *$A(u, t)$ is continuous in t uniformly for $x \in \partial B$;*

(iii) *$A(u, t) \neq 0$ for all $u \in \partial B$ and all $t \in [0, 1]$;*

(iv) *A_1 is odd on ∂B.*

Then there exists a $u_0 \in B$ such that $A_0 u_0 = 0$.

Proof: Without loss of generality we may assume that $r = 1$. Define the operator $T : \bar{B} \mapsto X^*$ by setting

$$Tx = \begin{cases} A(2x, 0) & \text{if} \quad \|x\| \leq \dfrac{1}{2}, \\[2ex] A\left(\dfrac{x}{\|x\|}, 2\|x\| - 1\right) & \text{if} \quad \dfrac{1}{2} \leq \|x\| \leq 1. \end{cases}$$

By (i), T is demicontinuous and by (iv), T is odd on ∂B. Let us prove that T is also of type (S). By the first part in the proof of the above proposition, there exists a sequence $\{x_n\} \subset \bar{B}$ such that $Tx_n = T_{F_n} x_n = 0$, $x_n \to x_0$ and $\lim (Tx_n, x_n - x_0) = 0$. We shall show that $x_n \to x_0$.

Let $\{x_k\}$ and $\{x_j\}$ be subsequences of $\{x_n\}$ such that $\|x_k\| \leq \dfrac{1}{2}$ and

$\dfrac{1}{2} \leq \|x_j\| \leq 1$. Since A_0 is of type (S), it follows that $x_k \to x_0$. On the other hand, $\{x_j\}$ has a subsequence, denoted below again by $\{x_j\}$, such that $\|x_j\| \to \alpha \in \left[\dfrac{1}{2}, 1\right]$. Since $Tx_j = 0$, we conclude that

$$\left(A\left(\frac{x_j}{\|x_j\|}, 2\|x_j\| - 1 \right), x_j - x_0 \right) = 0.$$

Therefore assumption (ii) implies that

$$\left\| A\left(\frac{x_j}{\|x_j\|}, 2\alpha - 1 \right) - A\left(\frac{x_j}{\|x_j\|}, 2\|x_j\| - 1 \right) \right\| \to 0 \quad \text{as } j \to \infty$$

and hence

$$\lim \left(A\left(\frac{x_j}{\|x_j\|}, 2\alpha - 1 \right), x_j - x_0 \right) = 0.$$

Consequently,

$$\lim_{j \to \infty} \left(A\left(\frac{x_j}{\|x_j\|}, 2\alpha - 1 \right), \frac{x_j}{\|x_j\|} - \frac{x_0}{\alpha} \right)$$

$$= \lim_{j \to \infty} \left(A\left(\frac{x_j}{\|x_j\|}, 2\alpha - 1 \right), \frac{x_j}{\|x_j\|} - \frac{x_0}{\|x_j\|} \right)$$

$$+ \lim \left(A\left(\frac{x_j}{\|x_j\|}, 2\alpha - 1 \right), x_0\left(\frac{1}{\|x_j\|} - \frac{1}{\alpha} \right) \right) = 0.$$

Since $A_{2\alpha-1}$ is of type (S), we obtain that $\dfrac{x_j}{\|x_j\|} \to \dfrac{x_0}{\alpha}$ and $x_n \to x_0$. As in the latter part of the proof of the proposition we infer that $Tx_0 = 0$. Thus assumption (iii) implies that $\|x_0\| < \dfrac{1}{2}$ and hence $u_0 = 2x_0$ satisfies $A_0 u_0 = 0$.

COROLLARY *If assumption* (iv) *is replaced by*

(iv') $(A_1 u, u) \geq 0$ *for all* $u \in \partial B$,

then the conclusion of the theorem remains valid.

Proof: Assuming (iv'), we apply the topological degree argument to the linear homotopy

$$H_{F_n}(\,.\,,t) = (1-t)\, T_{F_n} + t I_{F_n},$$

where I_{F_n} is the identity map on F_n. Then there exists an $x_n \in B \cap F_n$ such that $T x_n = T_{F_n} x_n = 0$.

We can regard assumption (iv') as the coercivity condition with respect to the origin in X. Consider now the trivial homotopy

$$A_t x = T x - f$$

for any $f \in X^*$; we can conclude that *any demicontinuous coercive operator of type (S) is surjective.*

1.2. A mapping $T: X \mapsto X^*$ is said to be *quasi-monotone* or satisfying *condition (P)*, if for every sequence $\{x_n\} \subset D(T)$ such that $x_n \rightharpoonup x$ in X it follows that

$$\lim \sup\,(T x_n, x_n - x) \geqslant 0.$$

This condition is an extremely mild monotonicity assumption because the class of quasi-monotone maps contains all operators of monotone type considered so far, for example, *any pseudo-monotone operator is quasi-monotone.* Indeed, let T be pseudo-monotone. Consider $x_n \rightharpoonup x$ and assume that $\lim \sup\,(T x_n, x_n - x) < 0$. Then for any $y \in X$ we have

$$\lim \inf\,(T x_n, x_n - y) \geqslant (Tx, x - y).$$

In particular, for $x = y$ we obtain $\lim \inf\,(T x_n, x_n - x) \geqslant 0$, which contradicts our assumption.

In virtue of Trojanski's theorem let us assume that X and X^* are endowed with locally uniform convex norms. In this case the duality map J is single-valued and of type (S).

The relation between quasi-monotone mappings and operators of type (S) is given by the following

LEMMA *A demicontinuous operator* $T: X \mapsto X^*$ *is quasi-monotone if and only if for each* $\varepsilon > 0$ *the mapping* $T + \varepsilon J$ *is of type* (S_+).

Proof: The "if" part. Let T be a quasi-monotone operator and assume that

$$\lim \sup\,((T + \varepsilon J)\, x_n - (T + \varepsilon J)\, x, x_n - x) \leqslant 0$$

whenever $x_n \rightharpoonup x$ in X. According to the quasi-monotonicity of T, $\lim \sup\,(J x_n - J x, x_n - x) \leqslant 0$ and, by Theorem III.2.6, we deduce that $x_n \to x$ in X, i.e., $T + \varepsilon J$ is of type (S_+).

The "only if" part. Assume that $T + \varepsilon J$ is of type (S_+) for each $\varepsilon > 0$. If T is not quasi-monotone, then there exists a sequence $\{x_n\}$ such that $x_n \rightharpoonup x$ and $\lim \sup (Tx_n - Tx, x_n - x) = -\delta$ with $\delta > 0$. Since T is demicontinuous, $x_n \nrightarrow x$. By the boundedness of $\{x_n\}$, there exists an $M > 0$ such that $\|x_n\| \leq M$ and

$$|\varepsilon(Jx_n - Jx, x_n - x)| < 4\varepsilon M^2.$$

Take $\varepsilon < \dfrac{\delta}{8M^2}$. It follows that

$$\lim \sup ((T + \varepsilon J) x_n - (T + \varepsilon J) x, x_n - x) \leq -\frac{1}{2} \delta < 0$$

and $x_n \nrightarrow x$, which contradicts our assumption.

It is easy to check that the set of all quasi-monotone mappings forms a convex cone.

THEOREM *Let X be a reflexive Banach space and let $T : X \mapsto X^*$ be a demi-continuous quasi-monotone mapping. Assume that $T(\overline{B})$ is closed for each $B = \overline{B}(0, \rho) \subset X$ and that T satisfies one of the following two conditions:*
  (a) *there is an $r > 0$ such that T is odd on $X - B(0,r) = \{x \in X \mid \|x\| \geq r\}$;*
  (b) *there is an $r > 0$ such that $(Tx, x) \geq 0$ when $\|x\| \geq r$.*
If T^{-1} is locally bounded on X^*, then $R(T) = X^*$.

Proof: For each $f \in X^*$, by the local boundedness of T^{-1}, we can choose $R \geq r$ and $c > 0$ such that

$$\|Tx - tf\| \geq c, \quad \text{for all } t \in [0, 1] \text{ and } \|x\| = R.$$

Indeed, suppose that on the contrary there exists a sequence $\{x_n\} \subset X$ with $\|x_n\| \geq n$ and $\{t_n\} \subset [0, 1]$ such that $\|Tx_n - t_n f\| < \dfrac{1}{n}$. Assume that $t_n \to t \in [0, 1]$. Then

$$\|Tx_n - tf\| \leq \|Tx_n - t_n f\| + |t_n - t| \|f\|,$$

i.e., $Tx_n \to tf$. This together with $\|x_n\| \to \infty$ contradicts the local boundedness of T^{-1}.

Now, choose $\varepsilon_0 > 0$ such that $\varepsilon_0 R < \dfrac{1}{2} c$ and

$$\|(T + \varepsilon J) x - tf\| \geq \frac{1}{2} c \qquad (\forall) t \in [0, 1], \ \|x\| = R, \tag{1}$$

for all $\varepsilon \in (0, \varepsilon_0)$.

Case (*a*): The mapping $A_t x = (T + \varepsilon J) x - (1 - t) f$ verifies the condition of Theorem 1.1. Hence there exists an $x_\varepsilon \in B(0, R)$ such that $(T + \varepsilon J) x_\varepsilon = f$, for all $\varepsilon \in (0, \varepsilon_0)$.

Case (*b*): Let F be a finite-dimensional subspace of X. As $(Tx, x) \geqslant 0$ when $\|x\| \geqslant r$, the homotopy $H_F: B(0, R) \cap F \times [0, 1] \mapsto F^*$ defined by

$$H_F(x, t) = t(T + \varepsilon J)_F x + (1 - t) x,$$

is continuous in x and $(H_F(x, t), x) > 0$ for all $t \in [0, 1]$ and $\|x\| \geqslant R$. Hence, $\deg((T + \varepsilon J), 0, F \cap B(0, R)) = \deg(I, 0, F \cap B(0, R)) = 1$. Let $f_F = j_F^* f$. On the other hand, inequality (1) assures that $0 \neq (T + \varepsilon J) x - t f_F$ for all $x \in \partial B(0, R)$ and all $t \in [0, 1]$. By property (d2) of topological degree, (see II.3.4), there exists at least one solution x_F in $F \cap B(0, R)$ of the equation $(T + \varepsilon J) x_F - f_F = 0$. By the lemma, $(T + \varepsilon J)$ is of type (S_+). Thus, as in the last part in the proof of Proposition 1.1, we see that there exists an $x_\varepsilon \in B(0, R)$ such that $x_F \to x_\varepsilon$ and $(T + \varepsilon J) x_\varepsilon = f$.

We conclude that $Tx_\varepsilon \to f$ as $\varepsilon \to 0$ in both cases. Since $T(\overline{B}(0, R))$ is closed, there exists an $x_0 \in \overline{B}(0, R)$ such that $Tx_0 = f$. The proof is complete.

The assumption that $T(\overline{B})$ is closed in X^* for any closed ball $\overline{B}$ in X is satisfied when T has the generalized pseudo-monotone property. In fact, let $\{f_n\}$ be a sequence in $T(\overline{B})$ such that $f_n \to f$. As $f_n \in Tu_n$ and $\{u_n\} \subset \overline{B}$, we may suppose that $u_n \rightharpoonup u \in \overline{B}$. Then $\lim (f_n, u_n - u) = 0$ and $f \in Tu$.

The above result generalizes Theorem III.4.2.

A mapping $T: X \mapsto 2^{X^*}$ is said to be *k-homogeneous* if $[x, f] \in G(T)$ implies that $[\lambda x, \lambda^k f] \in G(T)$ for all $\lambda \geqslant 0$.

COROLLARY *Let* $T: X \mapsto X^*$ *be a quasi-monotone operator such that* $T(\overline{B})$ *is closed in* X^* *for each* $\overline{B} = \overline{B}(0, \rho) \subset X$. *Suppose that for some* $r > 0$, T *is odd and k-homogeneous outside of* $B(0, r)$. *If* $0 \notin T(\partial B(0, r))$, *then* T *is surjective.*

Proof: It is sufficient to check that T^{-1} is locally bounded on X^*. By our hypotheses, there exists an open ball $B^*(0, d) \subset X^*$ such that

$$T(\partial B(0, r)) \cap B^*(0, d) = \emptyset.$$

By the *k*-homogeneity of T, we have that $T(\partial B(0, tr)) \cap B^*(0, t^k d) = \emptyset$ with $t \geqslant 1$. Thus, for $R \geqslant d$ and $t = \left(\dfrac{R}{d}\right)^{\frac{1}{k}}$ we conclude that $T^{-1}(B^*(0, R)) \subseteq$ $\subseteq B(0, tr)$, i.e., the local boundedness of T^{-1} on X^*.

Perturbations homotopic to odd operators

1.3. We shall discuss the solvability of nonlinear equations defined by the sum of a maximal monotone mapping and an odd pseudo-monotone operator.

PROPOSITION *Let $B = B(0, r)$ be an open ball in a reflexive Banach space X and let $T: B \mapsto 2^{X^*}$ be a maximal monotone mapping which is odd on ∂B. Suppose that $P_t u = P(u, t): B \times [0, 1] \mapsto X^*$ is an operator with the following properties:*

    (j) *For each fixed t, the map P_t is demicontinuous and quasi-monotone;*
    (jj) *$P_t u$ is uniformly continuous in $t \in [0, 1]$ for $u \in \partial B$;*
    (jjj) *$0 \notin (T + P_t)(\partial B)$ for all $t \in [0, 1]$;*
    (jv) *P_0 is pseudo-monotone and P_1 is odd on ∂B.*
Then there exists a $u_0 \in B$ such that $0 \in (T + P_0) u_0$.

Proof: By Lemma 1.2, the map $A_t^\varepsilon = T + P_t + \varepsilon J$ satisfies all conditions of Theorem 1.1 for each $\varepsilon > 0$. Then to $\varepsilon = \dfrac{1}{n}$ there corresponds a $u_n \in B$ such

that $\left(T + P_0 + \dfrac{1}{n} J\right) u_n = 0$. We assume that $u_n \to u_0 \in \bar{B}$ and

$$(T + P_0) u_n = -\frac{1}{n} J u_n \to 0.$$

Since Proposition III.2.5 ensures the pseudo-monotonicity of $S = T + P_0$ for $f_n \in S u_n$ and $f_0 \in S u_0$, we have that

$$0 = \lim (f_n, u_n - v) \geq (f_0, u_0 - v) \quad \text{for all } v \in X.$$

This implies that $0 \in S u_0 = (T + P_0) u_0$. Because of assumption (jjj), it follows now that $u_0 \in B$. This completes the proof.

As in Corollary 1.1, we can replace the oddness of T and P_1 on ∂B by the coercivity condition

$$((T + P_1) u, u) \geq 0 \quad \text{for all } u \in \partial B.$$

COROLLARY *Let $T: B \mapsto 2^{X^*}$ be maximal monotone and odd on ∂B, let $P: B \mapsto X^*$ be demicontinuous of type (S_+), also odd on ∂B, and let $Q: B \mapsto X^*$ be demicontinuous and pseudo-monotone. If*

$$0 \notin (T + P + (1 - t) Q)(\partial B) \quad \text{for all } t \in [0, 1],$$

then the equation $0 \in (T + P + Q) u$ has at least one solution in B.

Proof: Let $P_t = P + (1 - t) Q$. Since any demicontinuous operator of type (S_+) is pseudo-monotone, the corollary follows as a special case of the above proposition.

We apply these results to obtain a nonlinear Fredholm alternative. The Fredholm alternative guarantees the solvability of equation $Tu = f$ for any

$f \in X^*$ provided that the corresponding homogeneous equation admits only the null solution.

An operator $Q: X \mapsto X^*$ is called *k-asymptotically zero* with $k > 0$ provided that

$$\lim_{\|u\| \to \infty} \frac{\|Qu\|}{\|u\|^k} = 0.$$

THEOREM *Let* $T: X \mapsto 2^{X^*}$ *be an odd, k-homogeneous, maximal monotone operator with* $D(T) = X$, *let* $P: X \mapsto X^*$ *be a demicontinuous, odd, k-homogeneous operator of type* (S_+) *and let* $Q: X \mapsto X^*$ *be a bounded demicontinuous pseudo-monotone operator which is k-asymptotically zero with the same* $k > 0$. *If either T or P are bounded and* $0 \in (T + P)u$ *implies that* $u = 0$, *then the range of* $T + P + Q$ *is all of* X^*.

Proof: Without loss of generality, we may suppose that T is single-valued. First, we show the existence of a constant $r > 0$ such that any solution of the equation

$$(T + P + (1 - t) Q) u = 0 \quad (\forall) \; t \in [0, 1]$$

lies in $B(0, r)$.

Assume that, on the contrary, there exist $t_n \in [0, 1]$ and $u_n \in D(T)$ with $\|u_n\| \geqslant n$ and such that $(T + P + (1 - t_n) Q) u_n = 0$. Set $v_n = \dfrac{u_n}{\|u_n\|}$. Passing if necessary to a subsequence, we may suppose that $v_n \to v_0$. By the k-homogeneity of $T + P$ we have

$$Tv_n + Pv_n = - (1 - t_n) \|u_n\|^{-k} Qu_n \to 0.$$

The boundedness of T or of P implies the existence of a subsequence of $\{v_n\}$, denoted below again by $\{v_n\}$, such that $Pv_n \rightharpoonup g$, $Tv_n \rightharpoonup -g$ and

$$(Tv_n, v_n - v_0) + (Pv_n, v_n - v_0) \to 0.$$

Since T is monotone, we have $(Tv_n, v_n - v_0) \geqslant (Tv_0, v_n - v_0) \to 0$. Then $\lim (Pv_n, v_n - v_0) \leqslant 0$ and thus $v_n \to v$ because P is of type (S_+). Therefore $v_0 \neq 0$. By the demiclosedness of T and the demicontinuity of P we have that $Tv_0 = -g$ and $Pv_0 = g$. Thus, we arrive to the contradiction $(T + P) v_0 = 0$ and $\|v_0\| = 1$.

Now, for any $f \in X^*$ the mapping $Q_f u = Qu - f$ is still k-asymptotically zero and the sum $P + Q_f$ is pseudo-monotone. Thus, by the first part of the proof, there exists an $r > 0$ such that

$$(T + P + (1 - t) Q_f) u \neq 0 \text{ for all } t \in [0, 1] \text{ and } u \in \partial B(0, r).$$

The corollary ensures that $(T + P + Q_f) u_0 = 0$ for some $u_0 \in B(0, r)$, whence $f \in (T + P + Q) u_0$. This completes the proof.

Clearly, the *l*-homogeneity (being *l*-asymptotically zero) is the usual homogeneity (being asymptotically zero).

We apply the above Fredholm alternative to the abstract Hammerstein equation

$$u + T(P + Q) u = w \quad \text{with} \quad u, w \in X. \tag{1}$$

Here $T: X^* \mapsto X$ is a linear maximal monotone operator. (It is sufficient that T be symmetrical and positive semidefinite), $P: X \mapsto X^*$ is a bounded odd homogeneous demicontinuous map of type (S_+) and $Q: X \mapsto X^*$ is bounded, demicontinuous pseudo-monotone and asymptotically zero.

Equation (1) is equivalent to

$$0 \in T^{-1} v + (P + Q)(v + w),$$

where $v = u - w$ and $T^{-1}: X \mapsto 2^{X^*}$ is a linear maximal monotone mapping. Therefore, if $(I + TP) u = 0$ only for $u = 0$, then $0 \in (T^{-1} + P) v$ implies that $v = 0$ and by the above theorem, equation (1) is solvable for each $w \in X$.

The conditions of the theorem are ensured by a Hammerstein integral equation in $L^2(\Omega)$ (Ω is a domain in $\mathbb{R}^N$ of σ-finite measure)

$$u(x) + \int_\Omega k(x, y) f(y, u(y)) = w(x), \quad x \in \Omega \subset \mathbb{R}^N, \tag{2}$$

if the following assumptions hold:
(k) The kernel $k \in L^2(\Omega \times \Omega)$ is symmetric and

$$\int_\Omega \int_\Omega k(x, y) v(y) v(x) \, \mathrm{d}y \, \mathrm{d}x \geq 0 \quad (\forall) \; v \in L^2(\Omega);$$

(kk) The function $f: \Omega \times \mathbb{R} \mapsto \mathbb{R}$ satisfies the Caratheodory conditions and the growth condition

$$\left| \frac{1}{t} f(x, tu) - a(x) u \right| \leq c(t) (1 + |u|) \quad (\forall) \; u \in L^2(\Omega), \; t \geq 1,$$

where $a \in L^\infty(\Omega)$ is a non-negative function and $c(t) \to 0$ as $t \to \infty$;
(kkk) $f(x, u)$ is strictly increasing in u for fixed $x \in \Omega$.

Set $Tv = \int_\Omega k(., y) v(y) \, \mathrm{d}y$, $Pu = a(.)u$ and $Qu = f(., u) - a(.)u$.

We can easily see that T and P are linear monotone operators which are symmetrical and positive semidefinite, P is of type (S_+) and Q is a bounded monotone map (see IV.1.1). Further, Q is asymptotically zero since

$$\lim_{|u|\to\infty} \frac{\|Qu\|}{\|u\|} = \lim_{t\to\infty} \frac{\|Q(tu)\|}{t\|u\|} = \frac{1}{\|u\|}\lim_{t\to\infty}\frac{1}{t}\left(\int_\Omega |f(.,tu)-ta(.)u|^2\,dx\right)^{\frac{1}{2}}$$

$$= \frac{1}{\|u\|}\lim_{t\to\infty}\left(\int_\Omega \left|\frac{1}{t}f(.,tu)-a(.)u\right|^2\,dx\right)^{\frac{1}{2}} = 0.$$

Hence, equation (2) has a solution for each $w\in L^2(\Omega)$.

1.4. In this section we study some consequences of the foregoing results in the case of coercive mappings.

Let X be a reflexive Banach space and let X^* be its dual space.

PROPOSITION *Any demicontinuous coercive pseudo-monotone mapping P of X into X^* is surjective.*

Proof: Let $f\in X^*$ be arbitrarily chosen. The coercivity condition of $P: X\mapsto X^*$ ensures the existence of an $r>0$ such that

$$(Pu-f,u)>0 \quad \text{for all } u\in\partial B(0,r).$$

Consider the trivial homotopy

$$P_t u = Pu - f \quad (\forall)\, t\in[0,1].$$

Then $0\notin P_t(\partial B)$ and we can apply Proposition 1.3 with the maximal monotone map $T: X\mapsto 2^{X^*}$ defined by

$$Tu = \{0\}, \quad \text{for all } u\in X.$$

Hence there exists a $u_0\in B(0,r)$ such that $Pu_0 = f$, i.e., $R(P)=X^*$.

THEOREM *Let $P: X\mapsto X^*$ be a pseudo-monotone operator such that*

$$\|Pu\| + \frac{(Pu,u)}{\|u\|} \to \infty \quad as \quad \|u\| \to \infty. \tag{1}$$

Then $R(P) = X^*$.

Proof: For any $f\in X^*$ the map $Su = Pu - f$ still satisfies (1) since

$$\|Su\| + \frac{(Su,u)}{\|u\|} \geqslant \|Pu\| + \frac{(Pu,u)}{\|u\|} - 2\|f\| \to \infty \quad as \quad \|u\| \to \infty.$$

Hence, we can take $f=0$ and prove the solvability of $Pu=0$.

Choose $r>0$ such that

$$\|Pu\| + \frac{(Pu,u)}{\|u\|} > 0 \quad \text{for all } u\in\partial B(0,r) \tag{2}$$

and consider the homotopy

$$P_t u = (1 - t)\, Pu + tJu, \quad u \in X, \quad t \in [0, 1].$$

Then $0 \notin P_t(\partial P)$ for all $t \in [0, 1]$.

Indeed, suppose that for some t and

$$(1 - t)\, Pu + tJu = 0. \quad u \in \partial B(0, r).$$

Then $t < 1$ and hence $Pu = -\dfrac{t}{1 - t}\, Ju$. We infer that

$$\| Pu \| = \frac{1}{1 - t}\, \| u \| \quad \text{and} \quad \frac{(Pu, u)}{\| u \|} = -\frac{1}{1 - t}\, \| u \|$$

which contradicts (2). Thus, $P_t u \neq 0$ for all $u \in \partial B(0, r)$ and all $t \in [0, 1]$.

Now, as in the proof of the proposition we can apply Proposition 1.3 to show that there exists a $u_0 \in B(0, r)$ such that $P_0 u_0 \equiv Pu_0 = 0$. The proof is now complete.

We remark that we have removed the condition on the boundedness of the mappings involved and these results generalize the similar results obtained in Section III.2.9.

2. Eigenvalue problems for maximal monotone operators

Let H be a real Hilbert space, with inner product $(.\,,\,.)$ and norm $\|\cdot\|$, and let $T : H \to 2^H$ be a maximal monotone mapping. As we have seen in Section III.3.1, the mapping $\mathscr{J}_\lambda = (I + \lambda T)^{-1}$ is monotone and non-expansive on H for all $\lambda > 0$.

Given $f \in H$ and $\lambda \in \mathbb{R}$, consider the equation

$$f + \lambda u \in Tu. \tag{1}$$

When $\lambda < 0$ we can easily show that

$$u = \left(I - \frac{1}{\lambda}\, T \right)^{-1} \left(-\frac{1}{\lambda}\, f \right) = \mathscr{J}_{-\frac{1}{\lambda}} \left(-\frac{1}{\lambda}\, f \right)$$

is the unique solution of equation (1). For $\lambda > 0$, we denote $\mathscr{J}_1$ by P and equation (1) becomes equivalent to

$$v - \mu Pv = f,$$

where $\mu = 1 + \lambda$ and $v = \mu u + f$.

Mappings with a compact resolvent

2.1. A real number $\mu \geqslant 1$ is a *characteristic value* of P provided that the equation $(I - \mu P)v = 0$ has nonzero solutions. We denote by $E(P)$ the set of all characteristic values of P. Obviously, $\mu \in E(P)$ if and only if $\lambda = \mu - 1$ is an eigenvalue of T, i.e., $\ker(\lambda I - T) \neq 0$. Consequently $E(P)$ is a closed set in $\mathbb{R}$.

Assume further that

(i) $[0, 0] \in G(T)$, which is equivalent to $P(0) = 0$;

(ii) $T(tu) = tTu$ for all $u \in D(T)$ and $t > 0$, which implies that $P(tu) = tPu$ for all $u \in H$ and $t > 0$;

(iii) P is a compact mapping.

Since P is also maximal monotone, the compactness of P ensures its complete continuity, i.e., $u_n \rightharpoonup u$ implies that $Pu_n \to Pu$.

LEMMA *Under these conditions, if $\mu \notin E(P)$, then*

$$\lim_{\|u\| \to \infty} \|(I - \mu P)u\| = \infty,$$

that is, the (multivalued) mapping $(I - \mu P)^{-1}$ is bounded.

Proof: Let us assume the contrary. Then there exist a sequence $\{u_n\} \subset X$ and a number $M > 0$ such that $\|u_n\| \to \infty$ and $\|(I - \mu P)u_n\| \leqslant M$. Let $v_n = \dfrac{u_n}{\|u_n\|}$ and suppose that $v_n \rightharpoonup v$. Since $\|(I - \mu P)v_n\| \leqslant \dfrac{M}{\|u_n\|}$, we also have $(I - \mu P)v_n \to 0$ as $n \to \infty$. As $Pv_n \to Pv$, we have $v = \mu Pv$ and $v \neq 0$. This contradicts the fact that $\mu \notin E(P)$.

We can establish now the following spectral representation:

THEOREM *Suppose that the conditions (i)—(iii) are fulfilled. Then for any $\mu \in \mathbb{R}$ there exists a finite-dimensional subspace E_μ of H such that*

$$H = E_\mu \oplus R(I - \mu P) = E_\mu \oplus R(\lambda I - T),$$

where $\lambda = \mu - 1$.

Proof: Choose $\varepsilon > 0$ arbitrarily. By Theorem II.1.1, there exists an ε-approximation P_ε of P on $B(0, 1)$ of finite rank. The operator

$$R_\varepsilon x = \|x\| P_\varepsilon\left(\frac{x}{\|x\|}\right)$$

is homogeneous and of finite rank in all of X and

$$\|Px - R_\varepsilon x\| < \varepsilon \|x\| \qquad (\forall) \ x \in X.$$

Write $I - \mu P = I - \mu(P - R_\varepsilon) - \mu R_\varepsilon$ and observe that, by Corollary III.5.3, $Q = I - \mu(P - R_\varepsilon)$ is of type (M). For $\varepsilon < \dfrac{1}{\mu}$, Q is also coercive. In virtue of Corollary III.5.4, the range of Q is all of H, and this proves the theorem.

We shall show at the end of this section by means of a counterexample that the condition $\mu \notin E(P)$ is not sufficient for the surjectivity of $I-\mu P$. This fact justifies giving here the

COROLLARY *Let $\mu \notin E(P)$ and $r > 0$ be such that*

$$0 \notin (I - \mu P) S_r \quad \text{and} \quad \deg(I - \mu P, 0, B_r) \neq 0,$$

where $B_r = B(0, r)$ and $S_r = \partial B_r$. Then $R(I - \mu P) = H$.

Proof: Since $(I - \mu P) S_r$ is closed, there exists a $\rho < r$ such that $B_\rho \cap (I - \mu P) S_r = \emptyset$. Hence, for $f \in B_\rho$ we have $\deg(I - \mu P, f, B_r) = \deg(I - \mu P, 0, B_r)$ and thus $B_\rho \subset (I - \mu P)(B_r)$. Therefore $R(I - \mu P) = H$ because of the homogeneity of P.

In particular, the above assumptions are fulfilled whenever P is odd in H. In this case $\deg(I - \mu P, 0, B_r) \neq 0$ by Borsuk's theorem.

The distribution of non-characteristic values of P is illustrated by the following

PROPOSITION *Suppose that* (i)—(iii) *are satisfied and*

$$a = \sup\{\|Pu\| \mid u \in S_1\} < 1.$$

Then $\left[1, \dfrac{1}{a}\right] \cap E(P) = \emptyset$ and $R(I - \mu P) = H$ for all $\mu \in \left[1, \dfrac{1}{a}\right)$.

Proof: Since $P(0) = 0$ and $a < 1$, we have $\|Pu\| \leq \|u\|$ for all $u \in H$. Set $H(t, u) = (I - t_\mu P) u$, $t \in [0, 1]$ and let $f \in B(0, 1 - \mu a)$. If there exists a $t_0 \in [0, 1]$ such that $f = (I - t_0 \mu P)u$ with $\|u\| = 1$, then

$$\mu a = 1 - (1 - \mu a) < 1 - \|f\| = \|u\| - \|f\| \leq t_0 \mu \|Pu\| \leq \mu a,$$

which is a contradiction. By the homotopy property,

$$\deg(I - \mu P, f, B_1) = \deg(I, f, B_1) = 1$$

and $B(0, 1 - \mu a) \subset (I - \mu P)(B_1)$. Using now the same argument as in the proof of the above corollary we obtain the proposition.

Let $H = L^2(\Omega)$, where Ω is a bounded domain in $\mathbb{R}^N$, such that the Sobolev imbedding theorem holds. Consider a proper convex l.s.c. function $j: \mathbb{R} \mapsto \overline{\mathbb{R}}$ with $j(0) = 0$. Its subdifferential $\beta = \partial j: \mathbb{R} \mapsto 2^{\mathbb{R}}$ is a maximal monotone mapping. Assume that $[0, 0] \in G(\beta)$. Let $\varphi: H \mapsto \overline{\mathbb{R}}$ be the function defined by

$$(*_*^*) \quad \varphi(u) = \begin{cases} \dfrac{1}{2} \displaystyle\int_\Omega |\nabla u|^2 \, dx + \int_{\partial\Omega} j(u) \, d\sigma, & \text{if } u \in H^1(\Omega) \text{ and } j(u) \in L^1(\Omega), \\ +\infty & \text{otherwise.} \end{cases}$$

By virtue of III.6.13, the function φ is convex l.s.c., $\varphi(0) = 0$ and $\varphi(u) \geqslant 0$ for all $u \in H$. Further,

$$Tu = \partial\varphi(u) = -\Delta u \text{ for } u \in D(T) = \left\{ u \in H^2(\Omega) \, \middle| \, -\frac{\partial u}{\partial n} \in \beta(u) \text{ a.e. on } \partial\Omega \right\},$$

where $\dfrac{\partial}{\partial n}$ is the outward normal derivative.

Given $f \in L^2(\Omega)$, the associated eigenvalue problem requires a solution $u \in H^2(\Omega)$ of the equation

$$-\Delta u - \lambda u = f \text{ in } \Omega, \text{ with } -\frac{\partial u}{\partial n} \in \beta(u) \quad \text{a.e. on } \partial\Omega. \tag{1}$$

Clearly, T verifies (i). In view of the compactness of $P = (I + T)^{-1}$, we have that $Pu \in D(T)$ and

$$\begin{cases} -\Delta Pu + Pu = u & \text{in } \Omega, \\ -\dfrac{\partial Pu}{\partial n} \in \beta(Pu) & \text{a.e. on } \partial\Omega. \end{cases} \tag{2}$$

Then integrating by parts we have

$$\int_\Omega u \, Pu \, dx = -\int_\Omega \Delta Pu \, Pu \, dx + \int_\Omega |Pu|^2 \, dx$$

$$= \int_\Omega |\nabla Pu|^2 \, dx - \int_{\partial\Omega} \frac{\partial Pu}{\partial n} Pu \, d\sigma + \int_\Omega |Pu|^2 \, dx.$$

Since $0 \in \beta(0)$ and $-\dfrac{\partial Pu}{\partial n} \in \beta(Pu)$ a.e. on $\partial\Omega$ yield

$$-\int_{\partial\Omega} \frac{\partial Pu}{\partial n} Pu \, d\sigma \geqslant 0,$$

we deduce that

$$\|Pu\|_{1,2}^2 = \|\nabla Pu\|^2 + \|Pu\|^2 \leqslant (Pu, u). \tag{3}$$

Thus, $\|Pu\|_{1,2} \leqslant \|u\|$ implies that P is compact in H.

Moreover, if $\|Pu\| = \|u\|$, by (3) we get that

$$\|\nabla Pu\|^2 + \|Pu\| \, \|u\| \leqslant (Pu, u) \leqslant \|Pu\| \, \|u\|.$$

Then $\|\nabla Pu\| = 0$ and $(Pu, u) = \|Pu\| \, \|u\|$. Therefore $Pu = u$ and $u = c \in \beta^{-1}(0)$. In particular, $T^{-1}(0) = \beta^{-1}(0)$.

Assume further that

$$\beta(tr) = t\beta(r), \qquad (\forall)\ r \in D(\beta),\ t > 0,$$

and that $T = \partial\varphi$ verifies condition (ii). If $\beta^{-1}(0) = \{0\}$, then $P(0) = 0$ and, by the proposition, problem (1) is solvable for each $\lambda \in \left[0,\ \dfrac{1}{a} - 1\right)$.

Evidently, one has the null solution if $f = 0$.

Example. Let $k \geqslant 0$, and let $\beta: \mathbb{R} \mapsto 2^{\mathbb{R}}$ be defined by

$$\beta(r) = \begin{cases} kr & \text{if } r > 0, \\ (-\infty, 0] & \text{if } r = 0, \\ \varnothing & \text{if } r < 0. \end{cases}$$

The corresponding functions is

$$j(r) = \begin{cases} \dfrac{1}{2}\, kr^2 & \text{if } r \geqslant 0, \\ \infty & \text{if } r < 0. \end{cases}$$

If $k > 0$ then the above conclusion is valid.

When $k = 0$ (Signorini's boundary conditions) we obtain the desired counterexample. In fact, put $\Omega = (0,1)$ and $\lambda_n = \left[\left(2n + \dfrac{1}{2}\right)\pi\right]^2$, $n \in \mathbb{N}$.

It is readily seen that problem (1) with $f = 0$ admits uniquely the null solution; hence $\mu_n = \lambda_n + 1 \notin E(P)$. However, the associated homogeneous problem (1) with $f(x) = x$ has no solution.

The subdifferential case

2.2. We shall give now some existence results for a class of nonlinear eigenvalue problems.

Let $T = \partial\varphi$ be the subdifferential of a proper convex l.s.c. function $\varphi: H \mapsto \mathbb{R}$. Then the equation $f + \lambda u \in Tu$ with $f \in H$ and $\lambda \in \mathbb{R}$ can be written in the equivalent form

$$u \in D(\varphi),\ (f - \lambda u, v - u) \leqslant \varphi(v) - \varphi(u) \qquad (\forall)\ v \in D(\varphi),$$

and related to the prox mapping.

As in Section I.2.7, for any $u \in H$, let $\psi_u: H \mapsto \overline{\mathbb{R}}$ be the proper convex l.s.c. function defined by

$$\psi_u(v) = \frac{1}{2}\|u - v\|^2 + \varphi(v), \quad v \in H.$$

Its minimum value is attained at a unique point $P_\varphi u = (I + \partial\varphi)^{-1}u$.

Moreover, as we have seen in Corollary III.3.2, the convex function

$$\Phi(u) = \frac{1}{2} \|u\|^2 - \frac{1}{2} \|u - P_\varphi u\|^2 - \varphi(P_\varphi u)$$

$$= (u, P_\varphi u) - \frac{1}{2} \|P_\varphi u\|^2 - \varphi(P_\varphi u), \quad u \in H,$$

is Fréchet differentiable and $d\Phi(u) = P_\varphi u$ for all $u \in H$.

In this case, the equation

$$\lambda u \in Tu \quad \text{with } \lambda > 0 \text{ and } u \in D(T)$$

is equivalent to

$$u = (I + \partial\varphi)^{-1} (u + \lambda u) = P_\varphi (\lambda + 1) u. \tag{1}$$

THEOREM *Let* $T = \partial\varphi$, *where* $\varphi : H \mapsto \overline{\mathbb{R}}$ *is a proper convex l.s.c. function with* $\varphi(0) = 0$. *Suppose that*

(j) $[0. 0] \in G(T)$;

(jj) $T(0) \neq H$ *and* $T^{-1}(0)$ *is bounded*;

(jjj) Φ *is strongly continuous, i.e.,* $\Phi(v_n) \to \Phi(v)$ *whenever* $v_n \rightharpoonup v$ *in* H.

Then for each $r \in \Phi(H) - \Phi(T^{-1}(0))$, *there exists a* $\lambda > 0$ *and* $0 \neq u \in H$ *such that* $\Phi ((1 + \lambda)u) = r$ *and* $\lambda u \in Tu$.

Proof: Set $\mu = \lambda + 1$ and $v = \mu u$. Then equation (1) is equivalent to

$$0 \neq v \in H, \ \mu \geqslant 1, \ v - \mu P_\varphi v = 0. \tag{2}$$

Observe that $T(0) = \{u \in H \mid P_\varphi u = 0\}$ and since $\varphi(0) = 0$, we have

$$\Phi(u) = 0 \quad \text{for all} \ u \in T(0).$$

By (jj) there exists an $r > 0$ such that

$$M_r = \{u \in H \mid \Phi(u) = r\} \neq \varnothing \ \text{and} \ M_r \cap T^{-1}(0) = \varnothing.$$

Define $\Psi(u) = \frac{1}{2} \|u\|^2$ and denote $d = \inf \{\Psi(u) \mid u \in M_r\}$. Let $\{v_n\}$ be a sequence in M_r such that $\Psi(v_n) \to d$. Since $\{v_n\}$ is a bounded sequence, it contains a subsequence $\{v_k\}$ with $v_k \rightharpoonup v_0$. Thus $d = \Psi(v_0)$, because Ψ is l.s.c. On the other hand, by (jjj) we have $\Phi(v_n) \to \Phi(v_0)$ and hence $v_0 \in M_r$.

Since $d\Phi(v_0) = P_\varphi v_0 \neq 0$, we can apply Theorem I.1.9 to the functions Φ and Ψ on M_r. Thus, there exists a $\theta \in \mathbb{R}$ such that

$$d\Psi(v_0) = \theta \, d\Phi(v_0), \text{ i.e., } v_0 = \theta \, P_\varphi v_0.$$

As $v_0 \notin T(0)$, $\|v_0\|^2 = \theta(P_\varphi v_0, v_0)$ and $(P_\varphi v_0, v_0) \geqslant 0$, we deduce that $v_0 \neq 0$ and $\theta > 0$. By virtue of $\|P_\varphi v_0\| = \|v_0\|$ we get $\theta \geqslant 1$. Hence $\mu = \theta$ and $v = v_0$ verify (2). The conclusions of the theorem follow easily.

Let us consider again the convex function $(_*{}^*{}_*)$ in the foregoing section and let us give an interesting application of this function. Using the notation of the quoted example we have the

COROLLARY Suppose that $\beta^{-1}(0) = \{0\}$; Then for each $r \in \Phi(H) - \{0\}$ there exist $\lambda > 0$ and $0 \neq u \in H$ such that $\Phi((1 + \lambda)u) = r$ and $\lambda u \in \partial\varphi(u)$, i.e.,

$$u \in H^2(\Omega), \quad -\Delta u = \lambda u \text{ in } \Omega, \quad -\frac{\partial u}{\partial n} \in \beta(u) \quad \text{a.e on } \Gamma. \tag{3}$$

Proof: By III.6.13,

$$\|u\|_{1,2} \leqslant c_1 \|-\Delta u + u\| + c_2$$

and since the equation holds for $P_\varphi u$ in Ω, we obtain

$$\|P_\varphi u\|_{1,2} \leqslant c_1 \|u\| + c_2 \qquad (\forall) \ u \in H = L^2(\Omega).$$

Now let $u_n \to u$ in H. By the above inequality and the compact imbedding of $H^2(\Omega)$ in $H^1(\Omega)$, we conclude, passing if necessary to a subsequence, that $P_\varphi u_n \to g$ in $H^1(\Omega)$. As P_φ is also maximal monotone, we have $P_\varphi u_n \to Pu$ in $H^1(\Omega)$. In virtue of the properties of $\Phi(u)$, we get

$$\Phi(u_n) = \int_0^t (P_\varphi t u_n, u_n) \, \mathrm{d}t \to \int_0^t (P_\varphi t u, u) \, \mathrm{d}t = \Phi(u).$$

Thus, condition (jjj) is fulfilled.

The above application suggests the replacement of (jjj) by the equivalent condition

$$\{u \in H \mid \|u\| \leqslant 0 \ \text{and} \ \varphi(u) \leqslant c\} \ \text{is compact for all } c > 0. \tag{4}$$

This is obvious for the function $(_*{}^*{}_*)$.

The relation between the subdifferentiable case and the results of the previous section is further elucidated by the following

PROPOSITION Let $\varphi: H \mapsto \mathbb{R}$ be a proper convex l.s.c. function with $\varphi(0) = 0$, $\varphi(u) \geqslant 0$ for all $u \in H$ and assume condition (4). Then the operator $P_\varphi = (I + \partial\varphi)^{-1}$ is compact.

Proof: The function $\Phi: H \mapsto \mathbb{R}$ defined above is Fréchet differentiable with $\mathrm{d}\Phi(u) = P_\varphi u$ and $(P_\varphi u, u) \geqslant 0$ for all $u \in H$.

Since $\Phi(0) = 0$, we have

$$\Phi(u) = \Phi(u) - \Phi(0) = \int_0^1 (P_\varphi tu, \, tu) \, \frac{dt}{t} \geqslant 0 \quad (\forall) \; u \in H.$$

Hence $\varphi(P_\varphi u) \leqslant (P_\varphi u, u) \leqslant \|u\|^2$ because $\|P_\varphi u\| \leqslant \|u\|$. The result follows now by condition (4).

A Fredholm alternative

2.3. Suppose there are given two mappings T and P from a reflexive Banach space X into the dual space X^* and an element $f \in X^*$; the Fredholm alternative deals with the role of the real parameter λ in solving the equation $(T - \lambda P)u = f$. Theorem 1.1 provides a result on surjectivity for unbounded mappings in X, which lies at the basis on a nonlinear Fredholm alternative.

We recall that T is weakly coercive in X provided that

$$\lim_{\|x\| \to \infty} \|Tx\| = \infty.$$

We denote $T - \lambda P$ by S_λ. Let us prove the following

PROPOSITION *Let* $T: X \mapsto X^*$ *be a demicontinuous operator and let* $P: X \mapsto X^*$ *be a completely continuous operator such that* S_λ *is weakly coercive. Assume that* $Q: X \mapsto X^*$ *is a demicontinuous operator such that* $Q + S_\lambda$ *is odd and*

$$\lim_{\|u\| \to \infty} \frac{\|Qu\|}{\|S_\lambda u\|} = 0. \tag{1}$$

If $T + tQ$ *is of type* (S), *then for each* $t \in [0,1]$ *the range of* S_λ *is all of* X^* *and* $(T - \lambda P)^{-1}$ *is a bounded mapping.*

Proof: For any $f \in X^*$, we consider the operator $A_t: X \mapsto X^*$ defined by

$$A_t u = S_\lambda u + tQu - (1 - t)f.$$

From the weak coerciveness of S_λ and by (1), we have
$$\lim_{\|u\| \to \infty} \|S_\lambda u + tQu - (1-t)f\| = \infty \text{ uniformly for } t \in [0,1].$$
Hence we can choose an $r > 0$ such that

$$\|S_\lambda u + tQu - (1-t)f\| \geqslant 0, \text{ for } \|u\| = r \text{ and } t \in [0,1].$$

In virtue of Theorem 1.1, there exists a solution $u \in B(0, r)$ of the equation $A_0 u = 0$ so that $S_\lambda u = f$. Since r depends on f, the inverse mapping $(T - \lambda P)^{-1}$ is bounded.

We shall find a sufficient condition for weak coerciveness by means of the *k*-asymptote at infinity of an operator. A *k*-homogeneous operator $T_0 \colon X \mapsto X^*$ is called the *k-asymptote* of $T \colon X \mapsto X^*$ provided that

$$\lim_{\|u\| \to \infty} \frac{\|Tu - T_0 u\|}{\|u\|^k} = 0. \qquad (k > 0).$$

The real number $\lambda \neq 0$ is not an *eigenvalue* for the couple of operators (T, P) provided that $(T - \lambda P)u = 0$ implies that $u = 0$. The element $u \neq 0$ of X which is a solution of $(T - \lambda P)u = 0$ is called the *eigenvector* corresponding to the *eigenvalue* λ. The mapping inverse to $(T - \lambda P)$ is said to be *k-bounded* if there exists a constant $C > 0$ such that

$$\|u\| \leqslant C \, (1 + \|(T - \lambda P)u\|)^{\frac{1}{k}} \text{ for all } u \in X.$$

THEOREM *Let T and P be two mappings with the odd k-asymptotes T_0 and P_0. Suppose that T and T_0 are demicontinuous, P and P_0 are completely continuous and that $tT + (1 - t) T_0$ is of type (S) for each $t \in [0,1]$. If $\lambda \neq 0$ is not an eigenvalue for the couple $\{T_0, P_0\}$, then $R(T - \lambda P) = X^*$.*

Proof: First, we remark that if λ is not an eigenvalue for $\{T_0, P_0\}$, then there exists a constant $C > 0$ such that

$$\|u\| \leqslant C \, \|(T_0 - \lambda P_0) \, u\|^{\frac{1}{k}}. \qquad (2)$$

In virtue of the *k*-homogeneity of T_0 and P_0 it is sufficient to show that

$$\inf_{\|u\|=1} \|(T_0 - \lambda P_0) \, u\| > 0.$$

If this is not so, then there exists a sequence $\{u_n\}$ with $\|u_n\| = 1$ such that $u_n \rightharpoonup u$, $\|(T_0 - \lambda P_0)u_n\| \to 0$ and $P_0 u_n \to P_0 u$. Since T_0 is of type (S), we have that $u_n \to u$, and this contradicts the fact that λ is an eigenvalue for $\{T_0, P_0\}$.

Next, in order to apply the proposition, we remark that $T - \lambda P = (T - T_0) - \lambda(P - P_0) + (T_0 - \lambda P_0)$. From

$$\lim_{\|u\| \to \infty} \frac{\|(T - T_0)u\|}{\|u\|^k} = 0, \quad \lim_{\|u\| \to \infty} \frac{\|(P - P_0)u\|}{\|u\|^k} = 0 \qquad (3)$$

and (2), we obtain that $\|(T - \lambda P)u\| \geqslant \dfrac{1}{C} \, \|u\|^k$ for large $\|u\|$, i.e., the operator S_λ is weakly coercive. Set $Q = (T - \lambda P) - (T_0 - \lambda P_0)$; then the operator $T - \lambda P - Q = (T_0 - \lambda P_0)$ is odd and

$$\lim_{\|u\| \to \infty} \frac{\|Qu\|}{\|(T - \lambda P)u\|} \leqslant \lim_{u \to \infty} \frac{\|(T - T_0)u\|}{\frac{1}{C}\|u\|^k} + \lim_{u \to \infty} \frac{\|(P - P_0)u\|}{\frac{1}{C}\|u\|^k} = 0.$$

By the proposition, the range of S_λ is all of X^* and S_λ^{-1} is a k-bounded mapping.

We complete this nonlinear Fredholm alternative by the following a priori estimate:

COROLLARY *Under the assumptions of the theorem, the number $\lambda \neq 0$ is not an eigenvalue of the couple (T_0, P_0) if and only if there exists a constant $C > 0$ such that*

$$\|u\| \leqslant C(1 + \|S_\lambda u\|)^{\frac{1}{k}}, \text{ for every } u \in X. \tag{4}$$

Proof: We put $S_\lambda^0 = T_0 - \lambda P_0$. If (4) does not hold and $\lambda \neq 0$ is an eigenvalue, then there exists a $u_0 \neq 0$ with $S_\lambda^0 u_0 = 0$. For $u = tu_0$ we also have $S_\lambda^0 u = 0$. By (3) we deduce that $\lim\limits_{|u| \to \infty} \dfrac{\|S_\lambda^0(tu_0)\|}{(t\|u_0\|)^k} = 0$. Hence we get from (4) that

$$\|u_0\| \leqslant C\left[\frac{1}{(t\|u_0\|)^k} + \frac{\|S_\lambda^0(tu_0)\|}{(t\|u_0\|)^k}\right]^{\frac{1}{k}} \|u_0\| \to 0 \text{ as } t \to \infty.$$

This contradicts the assumption that $u_0 \neq 0$ and, therefore λ is not an eigenvalue.

Conversely, if $\lambda \neq 0$ is not an eigenvalue for the couple $\{T_0, P_0\}$ and (4) does not hold, then there exists a sequence $\{u_n\} \subset X$ with $\|u_n\| \to \infty$ such that

$$\|u_n\|^k > n(1 + \|S_\lambda^0 u_n\|) \text{ for all } n \in \mathbb{N}.$$

Hence we have $\lim\limits_{n \to \infty} \dfrac{\|S_\lambda^0 u_n\|}{\|u_n\|^k} = 0$. From (3) we can easily see that $\lim\limits_{n \to \infty} \dfrac{\|S_\lambda^0 u_n\|}{\|u_n\|^k} = 0$. Set $v_n = \dfrac{u_n}{\|u_n\|}$ and suppose that $v_n \rightharpoonup v$. By the k-homogeneity of S_λ^0, it follows that $\lim\limits_{n \to \infty} \|S_\lambda^0 v_n\| = 0$, i.e., $S_\lambda^0 v_n \to 0$. Since the operator P_0 is completely continuous and T_0 is of type (S), it follows that $v_n \to v$ and $S_\lambda^0 v = 0$. This contradicts the assumption that $\lambda \neq 0$ is not an eigenvalue for the couple $\{T_0, P_0\}$. Consequently inequality (4) is true.

2.4. We shall apply the above Fredholm alternative to the study of the Dirichlet problem for divergence equations. For the sake of simplicity we shall consider only the second order equations.

Let Ω be a bounded domain in $\mathbb{R}^N$ for which the Sobolev imbedding theorem holds.

Find a solution $u \in W_0^{1,p}(\Omega)$ for the equation

$$\sum_{i=1}^{N} D^i A_i(x, u, D^j u) - \lambda A_0(x, u, D^j u) = f, \tag{1}$$

where $f \in W^{-1,p'}(\Omega)$, $1 < p < \infty$ and $\frac{1}{p} + \frac{1}{p'} = 1$. Denote $\zeta = \{D^i u \mid i = 1, \ldots, N\}$.

Concerning the coefficients $A_i \in C(\bar{\Omega}) \times \mathbb{R}^{N+1}$, $i = 0, 1, \ldots, N$, we make the following assumptions:

$$\sum_{i=0}^{N} A_i(x, u, \zeta) \leqslant c_0 (1 + |u| + |\zeta|)^{p-1}, \quad c_0 > 0; \tag{2}$$

$$\sum_{i=1}^{N} A_i(x, u, \zeta) \zeta_i \geqslant c_1 |\zeta|^p - c_2 |u|^p, \quad c_1 > 0, c_2 \in \mathbb{R}; \tag{3}$$

$$\sum_{i=1}^{N} (A_i(x, u, \zeta') - A_i(x, u, \zeta)) (\zeta_i' - \zeta_i) > 0, \tag{4}$$

for each $x \in \Omega$, $u \in \mathbb{R}$, $\zeta', \zeta \in \mathbb{R}^N$, $\zeta' \neq \zeta$.

Further, we assume the existence of functions $a_i \in C(\Omega \times \mathbb{R}^{N+1})$, $i = 0, 1, \ldots, N$, satisfying the conditions:

$$a_i(x, -u, -\zeta) = -a_i(x, u, \zeta); \quad a_i(x, tu, t\zeta) = t^{p-1} a_i(x, u, \zeta); \tag{5}$$

$$\left| \frac{1}{t^{p-1}} A_i(x, tu, t\zeta) - a_i(x, u, \zeta) \right| \leqslant c(t) (1 + |u| + |\zeta|)^{p-1}, \tag{6}$$

for each $x \in \Omega$, $u \in \mathbb{R}$, $\zeta \in \mathbb{R}^N$, where $c(t) \to 0$ as $t \to \infty$. Moreover, we suppose the validity of inequality (3) for a_i.

THEOREM *Under the above assumptions, equation* (1) *has a solution for each f provided that the equation*

$$-\sum_{i=1}^{N} D^i a_i(x, u, D^j u) - \lambda a_0(x, u, D^j u) = 0$$

has only the null solution in $W_0^{1,p}(\Omega)$.

Proof: Define the operators

$$(Tu, v) = \sum_{i=1}^{N} \int_{\Omega} A_i(x, u, D^j u) D^i v \, dx; \quad (T_0 u, v) = \sum_{i=1}^{N} \int_{\Omega} a_i(x, u, D^j u) D^i v \, dx;$$

$$(Pu, v) = \int_{\Omega} A_0(x, u, D^j u) v(x) \, dx; \quad (P_0 u, v) = \int_{\Omega} a_0(x, u, D^j u) v(x) \, dx,$$

for each $u, v \in W_0^{1,p}(\Omega)$. They map $W_0^{1,p}(\Omega)$ into $W^{-1,p'}(\Omega)$.

Condition (6) shows that T_0 and P_0 are $(p-1)$-asymptotes of T and P. Inequality (2) assures the demicontinuity of T and T_0. Also, P and P_0 are completely continuous maps from $W_0^{1,p}(\Omega)$ into $W^{-1,p'}(\Omega)$ because of their continuity in $L^p(\Omega)$ and the compact imbedding of $W_0^{1,p}(\Omega)$ in $L^p(\Omega)$.

We infer that T and T_0 are also of type (S). In fact, if $u_n \rightharpoonup u$ in $W_0^{1,p}(\Omega)$, then $u_n \to u$ in $L^p(\Omega)$ and, choosing if need be a subsequence, we have $u_n(x) \to u(x)$ a.e. $x \in \Omega$. Denote

$$g_n(x) = \sum_{i=1}^{N} (A_i(x, u_n, D^j u_n) - A_i(x, u, D^j u))(D^i u_n - D^i u).$$

Then $g_n(x) > 0$ by (4) and $\int_{\Omega} g_n(x)\,dx \to 0$ as $n \to \infty$. This implies that $D^i u_n(x) \to D^i u(x)$ a.e. in Ω. From (3) one deduces that

$$\int_{\Omega'} \sum_{i=1}^{N} |D^i u_n(x)|^p\,dx \quad are \ bounded \ uniformly \ in \ n \in \mathbb{N}$$

for any measurable subset Ω' of Ω. By Vitali's theorem, $D^i u_n \to D^i u$ in $L^p(\Omega)$. Hence, $u_n \to u$ in $W_0^{1,p}(\Omega)$, i.e. T and T_0 are of type (S).

Thus all assumptions of Theorem 2.3 are fulfilled. Moreover, if $u \in W_0^{1,p}(\Omega)$ is a solution of equation (1), then by Corollary 2.3, $\lambda \neq 0$ is not an eigenvalue of the corresponding homogeneous equation (1) if and only if there exists a constant $C > 0$ such that

$$\|u\|_{1,2} \leqslant C(1 + |f|)^{\frac{1}{p-1}}.$$

In particular, let us take

$$- \Delta u - \lambda u \, \frac{|u|^s}{1 + |u|^s} = f \text{ in } \Omega \text{ and } u|_{\partial\Omega} = 0, \tag{7}$$

where $s > 0$. If $\lambda \neq 0$ is not an eigenvalue of the null Dirichlet problem for the Laplace equation, then the equation (7) has a solution in $W_0^{1,p}(\Omega)$ for each $f \in L^2(\Omega)$.

3. Range of sums of monotone mappings

Let T and S be two multivalued mappings of a Hilbert space H into itself. Let us compare the sets

$$R(T + S) = \bigcup_{u \in H} (T + S)u \text{ and } R(T) + R(S) = \bigcup_{v,w \in H} (Tv + Sw).$$

Clearly, $R(T) + R(S)$ is much larger than $R(T + S)$. For instance, if one takes on $H = \mathbb{R}^2$ the maps

$$T = \text{rotation by } \frac{\pi}{2} \quad \text{and} \quad S = \text{rotation by } \left(-\frac{\pi}{2} \right),$$

then $R(T + S) = \{0\}$ and $R(T) + R(S) = H$.

We say that two sets M_1 and M_2 of H are *almost equal* and write $M_1 \simeq M_2$ provided that

$$\overline{M_1} = \overline{M_2} \quad \text{and} \quad \text{Int } M_1 = \text{Int } M_2.$$

In particular, we have almost equality $R(T + S) \simeq R(T) + R(S)$, i.e.,

$$\overline{R(T + S)} = \overline{R(T) + R(S)} \quad \text{and} \quad \text{Int } R(T + S) = \text{Int } [R(T) + R(S)],$$

when both T and S are maximal monotone and satisfy some additional conditions. We apply this concept to the problem of solvability of the non-coercive equation of the form $(T + S)u \ni f$, where $f \in [R(T) + R(S)]$.

3.1. The simplest case of almost equality is given by the following

PROPOSITION *If* $\partial \varphi : H \to 2^H$ *is the subdifferential of a proper convex l.s.c. function* $\varphi : H \to \mathbb{R}$, *then*

$$D(\partial \varphi) \simeq D(\varphi).$$

Proof: It is sufficient to prove that

$$D(\varphi) \subseteq \overline{D(\partial \varphi)} \quad \text{and} \quad \text{Int } D(\varphi) \subseteq D(\partial \varphi).$$

To show the first inclusion, let $v \in D(\varphi)$ and let $u_\lambda \in D(\partial \varphi)$ be the solution of the equation

$$(I + \lambda \partial \varphi) u_\lambda \ni v, \quad \text{with } \lambda > 0. \tag{1}$$

Then $\left(\dfrac{v - u_\lambda}{\lambda}, v - u_\lambda \right) \leqslant \varphi(v) - \varphi(u_\lambda)$ and thus

$$\| v - u_\lambda \|^2 \leqslant \lambda \varphi(v) - \lambda(a, u_\lambda) - \lambda b,$$

because $\varphi(u) \geqslant (a, u) + b$, (see Proposition I.2.5). Hence, $u_\lambda \to v$ as $\lambda \to 0$ and $v \in \overline{D(\partial \varphi)}$.

To show the second inclusion, let $v \in \text{Int } D(\varphi)$ and let u_λ be the solution of equation (1) such that $u_\lambda \to v$. We assert that $\dfrac{1}{\lambda} \| u_\lambda - v \|$ remains bounded as $\lambda \to 0$. In fact, there exists an $r > 0$ such that $B(v, r) \subset D(\varphi)$. Thus for any $w \in H$ with $\| w \| < r$, we have

$$\varphi(v + w) - \varphi(u_\lambda) \geqslant \left(\frac{v - u_\lambda}{\lambda}, v + w - u_\lambda \right) \geqslant \left(\frac{v - u_\lambda}{\lambda}, w \right).$$

245

Hence, $\left(\dfrac{v - u_\lambda}{\lambda}, w\right) \leqslant C(w)$. Thus $\dfrac{1}{\lambda}(v - u_\lambda)$ is weakly bounded and also

strongly bounded. It follows that $\dfrac{1}{\lambda}(v - u_\lambda) \to f$ in H and thus $f \in D(\partial\varphi)$,

by the maximal monotonicity of $\partial\varphi$. Hence $v \in D(\partial\varphi)$.

We recall that the conjugate function of φ is

$$\varphi^*(f) = \sup\{(f, u) - \varphi(u) \mid u \in H\}$$

and $\partial\varphi^* = (\partial\varphi)^{-1}$ (see 1.3.9).

THEOREM *Let* $T = \partial\varphi$ *and* $S = \partial\psi$ *where* $\varphi, \psi : H \mapsto \overline{\mathbb{R}}$ *are proper convex l.s.c. functions. Suppose that* $T + S$ *is maximal monotone. Then*

$$R(T + S) \simeq R(T) + R(S).$$

Proof: First, we remark that

$$D(\varphi^*) + D(\psi^*) \subseteq D((\varphi + \Psi)^*).$$

Indeed, if $f \in D(\varphi^*)$ and $g \in D(\psi^*)$, then

$$(f, u) - \varphi(u) \leqslant \varphi^*(f) \quad \text{and} \quad (g, u) - \psi(u) \leqslant \psi^*(g)$$

for all $u \in H$. Then $(\varphi + \psi)^*(f + g) \leqslant \varphi^*(f) + \psi^*(g)$.

We deduce that

$$R(T) + R(S) = D(\partial\varphi^*) + D(\partial\psi^*) \subseteq D((\varphi + \psi)^*) \qquad (2)$$

and, in virtue of the proposition,

$$\overline{R(T) + R(S)} \subseteq \overline{D((\varphi + \psi)^*)} = \overline{D(\partial(\varphi + \psi)^*)}.$$

Consequently, by Proposition I.2.6, we have that

$$\overline{R(T) + R(S)} \subseteq \overline{R(\partial(\varphi + \psi))} = \overline{R(T + S)} \subseteq \overline{R(T) + R(S)}.$$

Similarly, from (2),

$$\text{Int}\,[R(T) + R(S)] \subseteq \text{Int}\,[R(T + S)] \subseteq \text{Int}\,[R(T) + R(S)].$$

Condition (∗) for nonlinear mappings

3.2. In IV.3.6 we have introduced condition (∗) on linear (monotone) operators. This condition is an extension of the angle-boundedness of linear operators. Here we shall extend condition (∗) to nonlinear mappings.

Let X be a Banach space and let X^* be its dual space. A monotone mapping $T : X \mapsto 2^{X^*}$ verifies *condition* (∗) provided that to each $[z, h] \in D(T) \times R(T)$ there corresponds a constant $k(h, z)$ such that

$$(f - h, x - z) \geq k(h, z) \qquad (\forall) [x, f] \in G(T).$$

A monotone operator $T : X \mapsto X^*$ verifies condition (∗) provided that for each pair $[y, z] \in D(T)$ there exists a constant $k(y, z)$ such that

$$(Tx - Ty, x - z) \geq k(y, z) \qquad (\forall) x \in D(T).$$

Obviously, the map T verifies condition (∗) if and only if T^{-1} has the same property.

By Corollary IV.3.5, condition (∗) for a linear operator $A : X \mapsto X^*$ is equivalent to the existence of a constant $d > 0$ such that

$$(Au, u - v) \geq - \frac{1}{4d} \|v\|^2 \qquad (\forall) u, v \in X.$$

Then, for all $x \in D(A)$, we have

$$(Ax - Ay, x - z) = (A(x - y), (x - y) - (z - y)) \geq - \frac{1}{4d} \|z - y\|^2$$

and A verifies condition (∗) in the above sense.

We indicate some classes of monotone mappings which verify condition (∗):

Example 1. We can readily see that *any nonlinear mapping $T : X \mapsto 2^{X^*}$ which is angle-bounded with constant $C > 0$, i.e., such that*

$$(h - f, x - z) \leq C(h - g, y - z) \qquad (\forall) [x, f] \in G(T),$$

for each $[y, g], [z, h] \in G(T)$, verifies condition (∗). In particular, the subdifferential of a proper convex l.s.c. function verifies condition (∗).

Example 2. Any monotone mapping $T : X \mapsto 2^{X^*}$ which is coercive in the sense that

$$\lim_{\substack{[x, f] \in G(T) \\ \|x\| \to \infty}} \frac{(f, x - z)}{\|x\|} = \infty$$

for each $z \in D(T)$, verifies condition (∗). Indeed, for any pair $[z, h] \in D(T) \times R(T)$ by the above coerciveness, one deduces that there exists an $R > 0$ such that $(f - h, x - z) \geq 0$ for all $[x, f] \in G(T)$ with $\|x\| > R$. Moreover, if $\|x\| < R$, we have

$$(f - h, x - z) \geq (g - h, x - z) \geq - (\|g\| + \|h\|)(R + \|z\|),$$

where $g \in Tz$.

In particular, let T be a monotone operator, let $\lambda > 0$ and let J be the duality map. Then $T + \lambda J$ satisfies condition (∗).

In the remainder of this section, we restrict the discussion to Hilbert spaces. In this case, *if T is maximal monotone in H, then $\mathscr{J}_\lambda = (I + \lambda T)^{-1}$ and $T_\lambda = (T^{-1} + \lambda I)^{-1}$ verify condition (∗) for all $\lambda > 0$*

By Theorem 3.1 and the fact that the subdifferentials verify condition (∗), we can foresee a connection between condition (∗) and the almost equality of the ranges of sums of maximal monotone mappings. Indeed, we have the following

LEMMA *Let $Q : H \mapsto 2^H$ be a maximal monotone mapping and let M be a subset of H. If to each $h \in M$ there correspond $z \in H$ and $k \in \mathbb{R}$ such that*

$$(Qx - h, x - z) \geqslant k \qquad (\forall)\, x \in D(Q),$$

then conv $M \subset \overline{R(Q)}$ *and* Int [conv M] $\subset R(Q)$.

Proof: First, we remark that one can replace here M by $M' = $ conv M. In fact, if $h \in M'$, then $h = \Sigma\, t_i h_i$ with $h_i \in M$, $t_i > 0$ and $\Sigma\, t_i = 1$. For all $x \in D(Q)$ let us write $(Qx - h_i,\, x - z_i) \geqslant k_i$ in the form

$$(Qx, x) - (Qx, z_i) - (h_i, x) \geqslant k_i - (h_i, z_i) = k_i'.$$

Multiplying by t_i and adding, we get

$$(Qx, x) - (Qx, z) - (h, x) \geqslant t_i k_i',$$

where $z = \Sigma\, t_i z_i \in D(Q)$, and thereby $(Qx - h, x - z) \geqslant k$.

Therefore it remains to establish that $M \subset \overline{R(Q)}$ and Int $M \subset R(Q)$. Let $h \in M$ and let $u_\lambda \in D(Q)$ be the solution of equation

$$h \in (Q + \lambda I)\, u_\lambda, \text{ with } \lambda > 0, \tag{1}$$

which exists because $R(Q + \lambda I) = H$. It is sufficient to prove that $\lambda u_\lambda \to 0$ as $\lambda \to 0$. Since $(Qu_\lambda - h, u_\lambda - z) \geqslant k$, or $(h - \lambda u_\lambda - h, u_\lambda - z) \geqslant k$, we deduce that $\lambda\|u_\lambda\|^2 \leqslant k + \lambda\|u\|\,\|z\|$ and thus $\sqrt{\lambda}\|u_\lambda\| < k_1$. Hence $h \in \overline{R(Q)}$.

Let now $h \in$ Int M and $r > 0$ be such that $B(h, r) \subset M$. To any $g \in H$ with $\|g\| < r$ there correspond $z_g \in H$ and $k_g \in \mathbb{R}$ such that

$$(Qx - h - g, x - z_g) \geqslant k_g \qquad (\forall)\, x \in D(Q).$$

Using the solution of (1) we can write

$$\lambda\|u_\lambda\|^2 + (g, u_\lambda) \leqslant \lambda\|u_\lambda\|\,\|z_g\| + \|g\|\,\|z_g\| - k_g.$$

This estimate shows the boundedness of the set $\{u_\lambda\}$ as $\lambda \to 0$. Let $u_{\lambda_n} \rightharpoonup u$ in H as $\lambda_n \to 0$. Since $h - \lambda_n u_{\lambda_n} \in Qu_{\lambda_n}$ and $h - \lambda_n u_{\lambda_n} \to h$, by the demiclosedness of Q we may conclude that $h \in Qu$, i.e., $h \in R(Q)$.

We can now prove the following

THEOREM *Let T and S be monotone mappings such that* $T + S$ *is maximal monotone. Suppose that S satisfies condition* (∗) *and* $D(T) \subseteq D(S)$. *Then*

$$R(T + S) \simeq R(T) + R(S).$$

Proof: It is sufficient to establish that

$$R(T) + R(S) \subseteq \overline{R(T + S)} \text{ and Int } [R(T) + R(S)] \subset R(T + S).$$

For this purpose we apply the lemma with $Q = T + S$ and $M = R(T) + R(S)$. If $h \in M$, then $h \in Tv + Sw$. The monotonicity of T yields

$$(Tx - Tv, x - v) \geqslant 0.$$

Since S verifies condition (∗) and $v \in D(T) \subseteq D(S)$, we have

$$(Sx - Sw, x - v) \geqslant k(v, w).$$

Adding these inequalities, we obtain

$$((T + S)x - h, x - v) \geqslant k(v,w).$$

The conclusion of the theorem follows now by the lemma.

REMARK The assumption that $D(T) \subset R(S)$ is indispensable. Take for instance on $H = \mathbb{R}^2$ the maps $T = $ rotation by $\dfrac{\pi}{2}$ and $S = \partial I_W$, where I_W is the indicator function of $W = \mathbb{R} \times \{0\}$. It is easy to see that $R(T) = \mathbb{R}^2$, $R(S) = \{0\} \times \mathbb{R}$ and hence $R(T) + R(S) = \mathbb{R}^2$ while $R(T + S) = \{0\} \times \mathbb{R}$.

COROLLARY *Let T and S be two maximal mappings such that* $D(S) = H$. *Then* $R(T + S_\lambda) \simeq R(T) + R(S_\lambda) = R(T) + R(S)$ *for all* $\lambda > 0$.

Proof: Indeed $R(S) = R(S_\lambda)$ since $f \in Su$ if and only if $S_\lambda(u + \lambda f) = f$.

PROPOSITION *Let T and S be maximal monotone mappings such that* $R(S)$ *is bounded. Then* $R(T + S) \simeq R(T) + R(S)$.

Proof: In virtue of Theorem III.4.2, the boundedness of $R(S)$ implies that $D(S) = H$.

Now, for any pair $[z, h] \in D(S) \times R(S)$ there exist $[y, h], [z, g] \in G(S)$ such that

$$(f - h, x - z) = (f - h, x - y) + (f - h, y - z) \geqslant (f - h, y - z),$$

for all $[x, f] \in G(S)$. Thus S satisfies condition (∗).

In our first application, let A and F be two maximal monotone mappings with $D(A) = D(F) = H$. Suppose that either A or F satisfies condition (∗). Then the Hammerstein equation

$$u + AFu \ni w \qquad (2)$$

has a solution $u \in H$ for each $w \in H$.

In fact, if A satisfies condition (∗) we take $v = Fu$ and write (2) in the form $F^{-1}v + Av \ni w$. Then we apply the theorem with $T = F^{-1}$ and $S = A$, whence,

$$R(F^{-1} + A) \simeq R(F^{-1}) + R(A) = D(F) + R(A) = H.$$

When F satisfies condition (∗), write (2) in the form $-A^{-1}(w - u) + Fu \ni 0$. Apply again the theorem with $Tu = A^{-1}(w - u)$ and $Su = Fu$. Then

$$R(T + S) \simeq R(T) + R(S) = -D(A) + R(F) = H.$$

Thus equation (2) is solvable in both cases.

The case $(Tu, S_\lambda u) \geqslant 0$.

3.3. For the purpose of applications we shall need some criteria of maximality for sums of maximal monotone mappings.

THEOREM *Let T and S be two maximal monotone maps such that $(Tu, S_\lambda u) \geqslant 0$ for all $u \in D(T)$. Then $R(T + S) \simeq R(T) + R(S)$.*

Proof: By Theorem III.3.8, the condition $(Tu, S_\lambda u) \geqslant 0$ implies the maximal monotonicity of $T + S$ and thus for each $h \in H$, there exists a u_ε such that

$$(\varepsilon I + T + S)u_\varepsilon \ni h \qquad (1)$$

for any fixed $\varepsilon > 0$. We prove first that $\{Tu_\varepsilon\}$ and $\{Su_\varepsilon\}$ remain bounded as $\varepsilon \to 0$.

Let $u_{\varepsilon\lambda} \in D(T)$ be such that

$$(\varepsilon I + T + S_\lambda)u_{\varepsilon\lambda} \ni h \text{ with } \lambda > 0. \qquad (2)$$

Corollary III.3.3 asserts that $u_{\varepsilon\lambda} \to u_\varepsilon$ as $\lambda \to 0$. Taking the second power of (2) we get

$$\varepsilon^2\|u_{\varepsilon\lambda}\|^2 + \|Tu_{\varepsilon\lambda}\|^2 + \|S_\lambda u_{\varepsilon\lambda}\|^2 + 2\varepsilon(u_{\varepsilon\lambda}, Tu_{\varepsilon\lambda} + S_\lambda u_{\varepsilon\lambda}) \leqslant \|h\|^2. \qquad (3)$$

For a fixed element $v \in D(T)$, we can write

$$(Tu_{\varepsilon\lambda} + S_\lambda u_{\varepsilon\lambda} - Tv - S_\lambda v, \ u_{\varepsilon\lambda} - v) \geqslant 0.$$

This implies that

$$-(Tu_{\varepsilon\lambda} + S_\lambda u_{\varepsilon\lambda}, \ u_{\varepsilon\lambda}) \leqslant C(\|u_{\varepsilon\lambda}\| + \|Tu_{\varepsilon\lambda}\| + \|S_\lambda u_{\varepsilon\lambda}\| + 1). \tag{4}$$

Combining (3) and (4) we obtain

$$\varepsilon^2\|u_{\varepsilon\lambda}\|^2 + \|Tu_{\varepsilon\lambda}\|^2 + \|S_\lambda u_{\varepsilon\lambda}\|^2 \leqslant \|h\|^2 + 2\varepsilon C(\|u_{\varepsilon\lambda}\| + \|Tu_{\varepsilon\lambda}\| + \|S_\lambda u_{\varepsilon\lambda}\| + 1),$$

and thus,

$$\varepsilon\|u_{\varepsilon\lambda}\| + \|Tu_{\varepsilon\lambda}\| + \|S_\lambda u_{\varepsilon\lambda}\| \leqslant c(h).$$

Hence, passing to the limit as $\lambda \to 0$, we deduce the boundedness of $\{Tu_\varepsilon\}$. By (1), the same reasoning applies to $\{Su_\varepsilon\}$, because $\{\varepsilon u_\varepsilon\}$ is bounded.

Now, if $h \in R(T) + R(S)$, then $h \in Tv + Sw$. We have

$$(Tu_\varepsilon - Tv, u_\varepsilon - v) \geqslant 0 \quad \text{and} \quad (Su_\varepsilon - Sw, u_\varepsilon - w) \geqslant 0,$$

where u_ε is the solution of (1). Adding these inequalities, we obtain

$$(h - \varepsilon u_\varepsilon - h, u_\varepsilon) \geqslant (Tu_\varepsilon - Tv, v) + (Su_\varepsilon - Sw, w) \geqslant k$$

and thus $\varepsilon\|u_\varepsilon\|^2$ remains bounded as $\varepsilon \to 0$. Since $\varepsilon u_\varepsilon \to 0$, it follows from (1) that $h \in \overline{R(T + S)}$.

If $h \in \text{Int } [R(T) + R(S)]$, then there exists an $r > 0$ such that $B(h, r) \subset R(T) + R(S)$. For all $g \in H$ with $\|g\| < r$ we have $h + g \in Tv_g + Sw_g$. Thus

$$(Tu_\varepsilon - Tv_g, u_\varepsilon - v_g) \geqslant 0 \quad \text{and} \quad (Su_\varepsilon - Sv_g, u_\varepsilon - v_g) \geqslant 0.$$

Adding these inequalities we get

$$(h - \varepsilon u_\varepsilon - h - g, u_\varepsilon) \geqslant (Tu_\varepsilon - Tv_g, v_g) + (Su_\varepsilon - Sw_g, w_g) \geqslant k(g),$$

where $k(g)$ is independent of ε. Consequently $(g, u_\varepsilon) \leqslant -k(g)$ and $\{u_\varepsilon\}$ remains bounded. Thus for a subsequence we have $u_\varepsilon \rightharpoonup u$ and $h - \varepsilon u_\varepsilon \to h$. By the demiclosedness of $T + S$, we conclude now that $h \in (T + S)u$, i.e., $h \in R(T + S)$. The proof is complete.

The assumptions of the theorem can be easily verified to hold in certain significant problems.

1) Let Ω be a bounded domain in $\mathbb{R}^N$ such that Green's formula holds and let $\beta : \mathbb{R} \mapsto 2^{\mathbb{R}}$ be a maximal monotone mapping. Given $f \in H^2(\Omega)$, we shall establish the existence conditions for the solution in $H^2(\Omega)$ of the

nonlinear problem of Neumann type

$$
\begin{cases}
-\Delta u + \beta(u) \ni f & \text{in } \Omega, \\[2mm]
\dfrac{\partial u}{\partial n} = 0 & \text{on } \partial\Omega,
\end{cases}
\tag{5}
$$

where $\dfrac{\partial}{\partial n}$ is the outward normal derivative. We can readily see that for any solution u of (5) we must have

$$
\int_\Omega \beta(u)\, dx = \int_\Omega f\, dx \quad \text{and thus} \quad \frac{1}{\mu(\Omega)} \int_\Omega f\, dx \in R(\beta).
$$

Now les us prove that the condition

$$
\frac{1}{\mu(\Omega)} \int_\Omega f\, dx \in \text{Int } R(\beta)
$$

implies the existence of a solution $u \in H^2(\Omega)$ of equation (5). Let $T = -\Delta$ and let $D(T) = \left\{ u \in H^2(\Omega) \,\middle|\, \dfrac{\partial u}{\partial n} = 0 \text{ on } \partial\Omega \right\}$. Then T is maximal monotone because $T = \partial\varphi$, where $\varphi(u) = \dfrac{1}{2} \displaystyle\int_\Omega \sum_{i=1}^N \left| \dfrac{\partial u}{\partial x_i} \right|^2 dx$. Let further $Su = \beta(u)$ with $D(S) = \{ u \in L^2(\Omega) | \beta(u) \in L^2(\Omega) \}$; then S is also maximal monotone because $S = \partial\psi$ with $\psi(u) = \displaystyle\int_\Omega j(u)\, dx$, where $j : \mathbb{R} \mapsto \overline{\mathbb{R}}$ is a proper convex l.s.c. function such that $\beta = \partial j$. Since

$$
(Tu, S_\lambda u) = -\int_\Omega \Delta u \beta_\lambda(u)\, dx = \int_\Omega \sum_{i=1}^N \left(\frac{\partial u}{\partial x_i} \right)^2 \beta_\lambda'(u)\, dx \geqslant 0,
$$

we get by the theorem that $R(T + S) \simeq R(T) + R(S)$.

On the other hand, any $g \in L^2(\Omega)$ can be expressed as $g_1 + g_2$, where

$$
g_1 = g - \frac{1}{\mu(\Omega)} \int_\Omega g\, dx \quad \text{and} \quad g_2 = \frac{1}{\mu(\Omega)} \int_\Omega f\, dx - \frac{1}{\mu(\Omega)} \int_\Omega (f - g)\, dx.
$$

We remark that $g_1 \in R(T)$. For $\| f - g \|_2$ small enough, we hawe $g_2 \in R(S)$ and $g \in R(T) + R(S)$. Hence $f \in \text{Int } [R(T) + R(S)] = \text{Int } [R(T + S)]$.

2) Consider now the Dirichlet problem

$$
\begin{cases}
-\Delta u - \gamma_1 u + \beta(u) \ni f & \text{in } \Omega, \\[1mm]
u = 0 & \text{on } \partial\Omega,
\end{cases}
\tag{6}
$$

where γ_1 is the first eigenvalue of $-\Delta$ with null boundary value on $\partial\Omega$, β and f are as above and $\beta(0)=0$. Let v_1 be the eigenfunction corresponding to γ_1.

A necessary condition for the existence of a solution of (6) in $H^2(\Omega)$ is

$$\frac{\displaystyle\int_\Omega fv_1\,dx}{\displaystyle\int_\Omega v_1\,dx} \in R(\beta).$$

In fact, if we multiply (6) by v_1 and integrate by parts, we obtain

$$\int_\Omega \left(\sum_{i=1}^N \frac{\partial u}{\partial x_i}\frac{\partial v_1}{\partial x_i} - \gamma_1 u\,v_1\right)dx = \int_\Omega (f\,v_1 - \beta(u)v_1)\,dx$$

or

$$0 = \int_\Omega (-\Delta v_1 - \gamma_1 v_1)u\,dx = \int_\Omega fv_1\,dx - c\int_\Omega v_1\,dx,$$

where $c = \beta(u(\xi))$ with a fixed $\xi \in \Omega$.

Let us prove now that a sufficient condition for the existence of a solution of (6) is

$$\frac{\displaystyle\int_\Omega fv_1\,dx}{\displaystyle\int_\Omega v_1\,dx} \in \text{Int } R(\beta). \tag{7}$$

Indeed, define the operators

$$Tu = -\Delta u - \gamma_1 u, \quad D(T) = H^2(\Omega) \cap H^1_0(\Omega),$$

$$Su = \beta(u), \quad D(S) = \{u \in L^2(\Omega)\,|\,\beta(u) \in L^2(\Omega)\}.$$

Both operators are maximal monotone because

$$T = \partial\varphi \quad \text{with} \quad \varphi(u) = \frac{1}{2}\int_\Omega\left[\sum_{i=1}^N\left(\frac{\partial u}{\partial x_i}\right)^2 - \gamma_1 u^2\right]dx, \quad D(\varphi)=H^1_0(\Omega)$$

and

$$S = \partial\psi \quad \text{with} \quad \psi(u) = \int_\Omega j(u)\,dx, \text{ where } \partial j = \beta.$$

The sum $T + S$ is also maximal monotone since

$$(Tu, S_\lambda u) = \int_\Omega (-\Delta u - \gamma_1 u)\beta_\lambda(u)\,dx = \int_\Omega\left[\sum_{i=1}^N\left(\frac{\partial u}{\partial x_i}\right)^2 - \gamma_1 u^2\right]\beta'_\lambda(u)\,dx \geq 0.$$

It follows from the theorem that $R(T + S) \simeq R(T) + R(S)$. On the other hand, for any $g \in L^2(\Omega)$ we can write $f + g = f_1 + f_2$, where

$$f_1 = (f + g) - \frac{\int_\Omega (f+g)v_1 \, dx}{\int_\Omega v_1 \, dx} \quad \text{and} \quad f_2 = \frac{\int_\Omega (f+g)v_1 \, dx}{\int_\Omega v_1 \, dx}.$$

As $\int_\Omega f_1 v_1 \, dx = 0$, it follows that $f_1 \in R(T)$. If $\|g\|_2$ is small, then so is $\int_\Omega g \, v_1 \, dx$ and by using assumption (7) we deduce that $f_2 \in R(S)$. Hence $f + g \in R(T) + R(S)$ and thus $f \in \text{Int } R(T) + R(S) = \text{Int } R(T + S)$.

3) Take the following problem with nonlinear boundary conditions: Given $f \in L^2(\partial\Omega)$ find $u \in H^2(\Omega) \cap H^1(\partial\Omega)$ such that

$$\begin{cases} \Delta u = 0 & \text{in } \Omega, \\ \dfrac{\partial u}{\partial n} + \beta(u) \ni f & \text{on } \partial\Omega. \end{cases} \tag{8}$$

Clearly if (8) has a solution, then integrating by parts we get

$$\frac{1}{\mu(\partial\Omega)} \int_{\partial\Omega} f \, ds \in R(\beta).$$

Let us add that, conversely, the condition

$$\frac{1}{\mu(\partial\Omega)} \int_{\partial\Omega} f \, ds \in \text{Int } R(\beta)$$

implies that (8) has at least one solution $u \in H^{\frac{3}{2}}(\Omega)$. In fact, given $u \in H^1(\partial\Omega)$, let $\tilde{u} \in H^{\frac{3}{2}}(\Omega)$ be the solution of the Dirichlet problem

$$\begin{cases} \Delta \tilde{u} = 0 & \text{in } \Omega, \\ \tilde{u} = u & \text{on } \partial\Omega. \end{cases}$$

Define $T : L^2(\partial\Omega) \mapsto L^2(\partial\Omega)$ by setting $Tu = \dfrac{\partial \tilde{u}}{\partial n}$. Then T is maximal monotone since

$$(Tu, u) = \int_{\partial\Omega} \frac{\partial \tilde{u}}{\partial n} \tilde{u} \, ds = \int_\Omega \sum_{i=1}^N \left(\frac{\partial \tilde{u}}{\partial x_i} \right)^2 dx > 0.$$

Let $Su = \beta(u)$ with $D(S) = \{u \in L^2(\partial\Omega) \mid \beta(u) \in L^2(\partial\Omega)\}$. It will suffice now to use the same arguments as in example 1).

4. Related topics and exercises

Throughout this section X is a reflexive Banach space with its dual space X^*, while H is a Hilbert space.

4.1. Let $\varphi: X \mapsto \overline{\mathbb{R}}$ be a continuous differentiable function such that $\varphi': X \mapsto X^*$ is a quasimonotone operator. Then φ is an l.s.c. function.

Proof: (By contradiction) Suppose that there exists a sequence $\{u_n\} \subset X$ such that $u_n \to u$ and $\lim \varphi(u_n) < \varphi(u)$. Then

$$\varphi(u_n) - \varphi(u) = \int_0^1 (\varphi'(u + t(u_n - u)), u_n - u) dt$$

and, by the mean value theorem, to each $n \in \mathbb{N}$ there corresponds a $t_n \in (0,1)$ such that

$$\varphi(u_n) - \varphi(u) = (\varphi'(u + t_n(u_n - u)), u_n - u).$$

Set $v_n = u + t_n(u_n - u)$. Then $v_n \to u$. By our assumption and the quasi-monotonicity of φ', we may choose n_0 such that

$$(\varphi'(v_n), u_n - u) \leqslant -\lambda < 0 \qquad (\forall)\ n \geqslant n_0. \tag{1}$$

As $\{t_n\} \subset [0,1]$, we may assume that $t_n \to t_0 \in [0,1]$. When $t_0 > 0$, then for a suitable $\varepsilon > 0$ we have $t_n \geqslant \varepsilon$ for all $n \geqslant n_1 \geqslant n_0$ and

$$(\varphi'(v_n), v_n - u) = t_n(\varphi'(v_n), u_n - u) \leqslant -\varepsilon\lambda < 0.$$

This contradicts the quasi-monotonicity of φ'. When $t_0 = 0$, then $v_n \to u$ and $\varphi'(v_n) \to \varphi'(u)$. It follows that $(\varphi'(v_n), u_n - u) \to 0$, which contradicts (1).

4.2. Let $P: X \mapsto X^*$ be a pseudo-monotone operator, such that

$$(Pu, u) \geqslant 0 \qquad (\forall)\ u \in X \text{ with } \|u\| \geqslant r,$$

for a number $r > 0$. Suppose that P^{-1} is locally bounded. Prove that $R(P) = X^*$.

SOLUTION For any $f \in X^*$, consider the homotopy

$$P_t u = Pu - (1 - t)f, \quad u \in X,\ t \in [0,1].$$

We claim that there exists an $R > 0$ such that $\|P_t u\| > \dfrac{1}{R}$ for all $\|u\| \geqslant R$ and all $t \in [0,1]$. Indeed, suppose that, on the contrary, there exist the sequences $\{t_n\} \subset [0,1]$ and $\{u_n\} \subset X$ with $\|u_n\| \geqslant n$ such that $\|P_{t_n} u_n\| < \dfrac{1}{n}$. We may assume that $t_n \to t_0$. Then $Pu_n \to (1 - t_0)f$ and $\|u_n\| \to \infty$ but this contradict the local boundedness of P^{-1} around $(1-t_0)f$. Hence, taking $p \geqslant \max (r, R)$, we get $(P_1 u, u) = (Pu, u) \geqslant 0$ for all $u \in \partial B(0, p)$. Now apply Proposition 1.3.

4.3. Let $X: T \mapsto X^*$ be a bounded demicontinuous operator and let $f \in X^*$. Define $A_t u = Tu - t\, T(-u)$ for all $t \in [0,1]$. Suppose that A_t is of type (S) and $A_t u - (1-t)f \neq 0$ for all $\|u\| = r > 0$ and all $t \in [0,1]$. Then there exists a solution in $B(0, r)$ for the equation $Tu = f$ (Nečas [5]).

4.4. Let $P_t u = P(u, t): X \times [0,1] \mapsto X^*$ be a mapping with the following properties:

    (k)   For fixed t, P_t is demicontinuous and quasi-monotone;

    (kk) $P(u, t)$ is continuous in t uniformly with respect to x on the bounded subsets of X;

    (kkk) There exists a continuous function $\varphi: \mathbb{R}_+ \mapsto \mathbb{R}_+$, such that for all $f \in X^*$ and $t \in [0,1]$, $P_t u = f$ implies that the a priori bound $\|u\| \leqslant \varphi(\|f\|)$ holds, (equivalent to the boundedness of P_t^{-1});

    (kv) P_1 is odd for $\|u\| \geqslant r$ and $P_0(\bar{B})$ is closed for each ball $\bar{B} \subset X$.

    Prove that $R(P) = X^*$.

SOLUTION For any $f \in X^*$, we can choose $R \geqslant r$ such that $\|P_t u\| > 1 + \|f\|$ for all $u \in X$ with $\|u\| = R$. Indeed, suppose that, on the contrary, there exist $\{u_n\}$ and $\{t_n\}$ such that $\|u_n\| \to \infty$ and $\|P_{t_n} u_n\| \leqslant 1 + \|f\|$. Set $g_n = P_{t_n} u_n$. By (kkk) we have $\|u_n\| \leqslant \varphi(\|g_n\|)$, which contradicts $\|u_n\| \to \infty$ because $\|g_n\| \leqslant 1 + \|f\|$ and φ is continuous.

$$\text{Now, for } 0 < \varepsilon < \frac{1}{R} \text{ set } A_t u = P_t u + \varepsilon\, Ju - tf. \text{ We have that } A_t u \neq 0 \text{ for } \|u\| \neq 0$$

since

$$\|A_t u\| \geqslant \|P_t u\| - \varepsilon \|Ju\| - t\|f\| > 0.$$

An application of Theorem 1.1 yields a zero of A_1. Thus there exists a $u_\varepsilon \in B(0, R)$ such that $(P_1 + \varepsilon J)u_\varepsilon = f$. As $\varepsilon \to 0$, $\|\varepsilon Ju_\varepsilon\| \leqslant \varepsilon R \to 0$. Thus $P_1 u_\varepsilon \to f$ and $f \in P_1(B(0, R))$, because $P_0(\bar{B})$ is closed.

4.5. Let $T: X \mapsto X^*$ be a bounded linear operator of type (S_+) and let $P: X \mapsto X^*$ be a completely continuous map, which is asymptotically zero. Suppose that $Tu = 0$ implies $u = 0$. Then $R(T + P) = X^*$.

HINT Let $f \in X^*$ and define the homotopy

$$A_t u = Tu + (1 - t)(Pu - f), \quad u \in X, \ t \in [0,1].$$

Then for fixed t, $A_t u$ is of type (S_+) and we can apply Theorem 1.1. Then we use the same technique as in the proof of Theorem 1.3.

4.6. Let T be a bounded demicontinuous mapping of type (S). Suppose that T admits a decomposition $T = L + Q$, with L a bounded linear operator of type (S) and Q an asymptotically zero mapping. Assume further that $R(Q) \subset N(L^*)^\perp$, where $N(L^*)$ is the null-space of the adjoint map of L. Then the equation $Tu = f$ admits solutions if and only if $f \in N(L^*)^\perp$, (this result was applied by Hess [4] to elliptic equations).

4.7. Consider in $L^2(\Omega)$ the Hammerstein equation

$$u(x) - \lambda \int_\Omega k(x, y)\, f(y, u(y))\ dy = w(x), \quad x \in \Omega, \tag{1}$$

where Ω is a bounded domain in $\mathbb{R}^N$, $\lambda \in \mathbb{R}$ and $w \in L^2(\Omega)$. Suppose that $f : \Omega \times \mathbb{R} \mapsto \mathbb{R}$ satisfies Caratheodory's conditions and the growth condition

$$\left| \frac{1}{t}\, f(y, tu) - a(y)u \right| \leqslant c(y, t) + d(t)|u| \quad \text{for } t \geqslant 1, \ a \in L^\infty(\Omega),$$

with $c(y, t) \to 0$ in $L^2(\Omega)$ and $d(t) \to 0$ as $t \to \infty$. If $k \in L^2(\Omega \times \Omega)$, then equation (1) has a solution provided that λ is not an eigenvalue for the asymptotic homogeneous equation

$$u(x) - \lambda \int_\Omega k(x, y)a(y)u(y)\ dy = 0. \tag{2}$$

Every solution $u \in L^2(\Omega)$ of equation (1) satisfies the a priori estimate

$$\|u\|_2 \leqslant C(1 + \|w\|_2). \tag{3}$$

Conversely, if (3) is satisfied for every solution of (1), then λ is not an eigenvalue of (2), (Nečas [4]).

4.8. Let $\varphi : H \mapsto \overline{\mathbb{R}}$ be a proper convex l.s.c. function so that $\varphi(0) = 0$ and $\varphi(u) > 0$ for all $u \in H$. Suppose that there exists a $K \subset H$ such that:

    (i) K is a convex cone with its vertex at the origin;
    (ii) The set $\{u \in H| \ \|u\| \leqslant c, \ \varphi(u) \leqslant c\}$ is compact for all $c \in \mathbb{R}$;
    (iii) $(\partial\varphi)^{-1}(0) \cap K$ is bounded in H;
    (iv) There exists a $v \in K - \{0\}$ such that $\varphi(v) < +\infty$;
    (v) $\varphi(P_K v) \leqslant \varphi(v)$ for all $v \in H$, where $P_K : H \mapsto K$ is the projection on K.
Then there exist $\lambda > 0$ and $u \in K - \{0\}$ such that $\lambda u \in Tu$.
    Conversely, if $\partial\varphi$ is homogeneous and there exist $\lambda > 0$ and $0 \neq u \, (\partial\varphi)^{-1}(0)$ such that $\lambda u \in \partial\varphi(u)$, then one can find a $K \subset H$ with the properties (i)–(v) (Abeasis, Dias and Lopes-Pinto [1]).

4.9. (A Sturm-Liouville problem). Let $\varphi : H \mapsto \overline{\mathbb{R}}$ be a proper convex l.s.c. function such that $\varphi(-u) = \varphi(u)$ for all $u \in H$ and $\partial\varphi(0) = \{0\}$. Then for every $r > 0$ there exists a sequence $\{u_n, \lambda_n\}$ $0 \neq u_n \in H$, $\lambda_n \in \mathbb{R}$ with the properties $u_n \neq u_m$ for $m \neq n$, $\lambda_n \geqslant 0$, $\lambda_n \to \infty$ and $\|u_n\| = r(1 + \lambda_n)^{-1} \to 0$ as $n \to \infty$ and $\lambda_n u_n \in Tu_n$ for all $n \in \mathbb{N}$ (Dias [1]).

4.10. Let $\varphi : X \mapsto \overline{\mathbb{R}}$ be a proper convex l.s.c function. Then $T = \partial\varphi$ satisfies a condition stronger than $(*)$: For any $z \in D(\varphi)$, $h \in D(\varphi^*)$, there exists a constant $k(z, h)$ such that

$$(f - h, x - z) \geqslant k(z, h) \qquad (\forall) \ [x, f] \in G(T).$$

SOLUTION Indeed for any $[x, f] \in G(T)$ we have

$$\varphi(x) - \varphi(z) \leqslant (f, x - z) = (f - h, x - z) + (h, x - z).$$

Then $(f - h, x - z) \geqslant \varphi(x) - \varphi(z) - (h, x) + (h, z) \geqslant k(z, h)$, where $k(z, h) = (h, z) - \varphi(z) - \varphi^*(h)$.

4.11. Let $T: X \mapsto X^*$ be a linear monotone operator with $G(T)$ closed in $X \times X^*$. Then T satisfies condition $(*)$ if and only if T satisfies one of the following equivalent inequalities:

(i) $(Tu, u - v) \geqslant - k(\|v\|^2 + \|Tv\|^2)$,

(ii) $(Tu - Tv, u) \geqslant - k(\|v\|^2 + \|Tv\|^2)$,

(iii) $|(Tu, v)| \leqslant 2\sqrt{k} \, (Tu, u)^{\frac{1}{2}} (\|v\|^2 + \|Tv\|^2)^{\frac{1}{2}}$,

(iv) $|(Tu, v)| \leqslant 2\sqrt{k} \, (Tu, u)^{\frac{1}{2}} (\|u\|^2 + \|Tu\|^2)^{\frac{1}{2}}$,

where k is a constant and $u, v \in D(T)$ (Brézis — Haraux [1]).

4.12. Let $T: H \mapsto H$ be a bounded monotone self-adjoint linear operator which satisfies the condition $(*)$. If $R(T)$ is closed, then T is angle-bounded.

SOLUTION $R(T) = N(T)$. The restriction $\tilde{T} = T|_{R(T)}$ is one-to-one and it has a continuous inverse map from $R(T)$ to $D(T)$. The condition $(*)$, i.e., $(Tu, u) \geqslant d\|Tu\|^2$, $d > 0$, implies that $(Tu, u) \geqslant d\|\tilde{T}u\|^2 \geqslant c\|u\|^2$, where $c > 0$. Write $x, y \in H$ as $x = x_1 + x_2$, $y = y_1 + y_2$ with $x_1, y_1 \in R(T)$ and $x_2, y_2 \in N(T)$. Then

$$(Tx, y) = (Tx_1, y_1) \leqslant \|Tx_1\| \, \|y_1\| \leqslant \frac{1}{\sqrt{d}} (Tx_1, x_1)^{\frac{1}{2}} \|y_1\| \leqslant \frac{1}{c\sqrt{d}} (Tx_1, x_1)^{\frac{1}{2}} (Ty_1, y_1)^{\frac{1}{2}} \leqslant$$

$$\leqslant \frac{1}{c\sqrt{d}} (Tx, x)^{\frac{1}{2}} (Ty, y)^{\frac{1}{2}},$$

i.e. T is angle-bounded.

4.13. Let T and S be two monotone mappings in H, which satisfy condition $(*)$. If $T + S$ is maximal monotone, then $R(T + S) \simeq R(T) + R(S)$.

HINT Apply Lemma 3.2 with $Q = T + S$ and $M = R(T) + R(S)$.

4.14. Let $\varphi: H \mapsto \overline{\mathbb{R}}$ be a proper convex l.s.c. function and let $T = \partial\varphi$. If $S: H \mapsto 2^H$ is a maximal monotone mapping such that $\varphi((I + \lambda S)^{-1}u) \leqslant \varphi(u)$ for all $u \in H$ and $\lambda > 0$, then $R(T + S) \simeq R(T) + R(S)$ (Brézis — Haraux [1]).

4.15. Consider the nonlinear periodic problem

$$\begin{cases} \dfrac{du}{dt} + \partial j(u) \ni f \text{ on } (0, t_0), \\ u(0) = u(t_0), \end{cases}$$

where $j: H \mapsto \overline{\mathbb{R}}$ is a proper convex l.s.c function and $f \in L^2(0, t_0; H)$.
Prove that

$$\frac{1}{t_0} \int_0^{t_0} f(t) \, dt \in R(\partial j) \quad \text{and} \quad \frac{1}{t_0} \int_0^{t_0} f(t) \, dt \in \text{Int } R(\partial j),$$

are a necessary condition and a sufficient condition for the existence of a solution of this problem.

HINT Apply Theorem 3.3 to the maximal monotone mappings

$$Tu = \frac{du}{dt} \qquad \text{with} \qquad D(T) = \{u \in H^1(0, t_0; H) \mid u(0) = u(t_0)\}$$

and

$$Su = \partial j(u) \qquad \text{with} \qquad D(S) = \{u \in L^2(0, t_0; H) \mid \partial j(u) \in L^2(0, t_0, H)\}.$$

Bibliographical comments

The task of homotopy arguments is to use the oddness of mappings instead of their coercivity conditions in fiding solutions to nonlinear equations. This idea can be found in Pohožaev's paper [1]. A complete discussion of nonlinear odd mappings defined on a bounded subset symmetric about the origin is given by Hess [5].

A direct method for bounded operators defined on balls is due to Calvet − Webb [1]. A refinement of this technique for unbounded operators was obtained by Hess [13] and it is exposed here in 1.1. Theorem 1.2 is due to Fitzpatrick [1]. The results concerning to perturbations homotopic to odd operators and their applications to Hammerstein equations are taken from Hess [5] and Pascali − Sburlan [2]. In this area, we also mention the works of Browder [10], [15], Petryšyn [3] and Skripnik [1, Chapter 3]. Some variants of these results are applied by Sburlan [2]−[3] to the boundary displacement problem of elastic equilibrium in nonlinear elasticity theory.

Since 1969. Nečas [2] was working on certain extensions of the Fredholm alternative for nonlinear bounded operators. These topics have been extensively developed in the monograph of Fučik−Nečas−Souček − Souček [1] and Rabinovitz [1].

The content of Sections 2.1−2.2 is based on the papers of Beirao-da-Vega − Dias [1] and Dias [2]. The discussion of the Fredholm alternative in Section 2.3 follows the reports of Nečas [6], Fučik − Nečas [1] and Pascali [3].

The study of range for sums of monotone mappings has been initiated by Schatzman [1], Brézis − Haraux [1] and Brézis [7]. The most important of these results can be found in Brézis − Nirenberg [1]−[2], Calvert − Gupta [1], Gupta [7] and Gupta −Hess [1].

Chapter VI

Variational problems and inequalities

Variational inequalities represent a powerful device for constructing in terms of convexity and duality mathematical models of various facts in physics. Their use allows to concentrate in a single inequality all the intrinsic features of a phenomenon: governing equations, boundary conditions, boundedness constraints and even jump conditions. Moreover, variational inequalities provide a natural means for numerical treatment of nonlinear problems.

1. Variational inequalities

1.1. Let K be a closed convex set in a reflexive Banach space X, and let T be a mapping from X into X^* such that $K \subset D(T)$ and $f \in X^*$. The element $x \in K$ is said to be a *solution of the variational inequality* $V(T, f, K)$ provided that

$$(Tx - f, y - x) \geqslant 0 \qquad (\forall)\, y \in K. \tag{1}$$

Obviously, if $K = X$ or x is an interior point of K then the y's range over a neighbourhood of x and thus inequality (1) actually reduces to the equality $(Tx - f, z) = 0$ for all $z \in X$, that is, $Tx = f$.

When X is a Hilbert space and $Tx = x$, inequality (1) becomes

$$(x - f, y - x) \geqslant 0 \qquad (\forall)\, y \in K$$

and this is just the problem of finding the projection of f on the convex set K, (see III.2.6).

We can consider the Debrunner-Flor Lemma III.2.8 as an existence result for solutions of variational inequalities in the case of compact operators.

In order to point out the connection between variational inequalities and boundary-value problems of partial differential equations we proceed with an example concerning the flow through semi-permeable walls.

Let Ω be a bounded domain in $\mathbb{R}^N$ with a sufficiently smooth boundary $\partial\Omega$ in the sense of Theorem II.2.7, and let $t_0>0$ be an arbitrarily fixed number. Consider the following problem:

Find a function $u(x, t)$ with $x \in \Omega$ and $t \in [0, t_0]$, such that u satisfies the equation

$$\frac{\partial u}{\partial t} - \Delta u = f(x, t) \quad in \; Q = \Omega \times [0, t_0], \tag{2}$$

the boundary initial conditions

$$u \geqslant 0, \; \frac{\partial u}{\partial n} \geqslant 0 \;\; and \;\; u\frac{\partial u}{\partial n} = 0 \;\; on \;\; \Sigma = \partial\Omega \times [0, t_0] \tag{3}$$

and the initial condition

$$u(x, 0) = u_0(x), \quad x \in \bar\Omega. \tag{4}$$

Here n is the outward normal to $\partial\Omega$, $u_0 \in L^2(\Omega)$ and $f(., t) \in L^2(\Omega)$ for fixed t.

Since u and $\dfrac{\partial u}{\partial n}$ are both non-negative on $\partial\Omega$, it follows by (3) that

$\partial\Omega$ consists of two parts Γ_0 and Γ_1 on which $u = 0$ and $\dfrac{\partial u}{\partial n} = 0$ respectively.

The parts Γ_0 and Γ_1 are not known a priori.

Let us pass to the variational formulation of this problem. Take the Sobolev space $H^1(\Omega)$, its dual space $(H^1(\Omega))^*$ and let $(.,.)$ be the pairing between $H^1(\Omega)$ and $(H^1(\Omega))^*$. For $p \geqslant 2$ and $\dfrac{1}{p} + \dfrac{1}{p'} = 1$ set $V = L^p(0, t_0; H^1(\Omega))$, $H = L^2(0, t_0; L^2(\Omega))$ and $V^* = L^{p'}(0, t_0; (H^1(\Omega))^*)$. From the chain of inclusions $H^1(\Omega) \subset L^2(\Omega) \subset (H^1(\Omega))^*$ it follows that $V \subset H \subset V^*$.

Define in $H^1(\Omega)$ the bilinear form

$$a(u, v) = \int_\Omega \nabla u \cdot \nabla v \, dx$$

and consider the convex cone with the vertex in the origin

$$K = \{v \in V \mid v \geqslant 0 \text{ on } \Sigma\}.$$

We shall prove the equivalence of problem (2) – (4) with the following variational inequality:

Let $u_0 \in L^2(\Omega)$ and $f \in H$ be given. Find a $u \in K$ such that $u(0) = u_0$, $\dfrac{du}{dt} \in V^*$

and

$$\int_0^{t_0} \left(\frac{du}{dt} - f, v - u\right) dt + \int_0^{t_0} a(u, v - u) \, dt \geqslant 0, \quad (\forall) \; v \in K. \tag{5}$$

We can readily see that any solution of problem (2) — (4) satisfies the variational inequality (5).

Conversely, if $v = u \pm z$, where $z \in C_0^\infty(Q)$ in (5), then

$$\left(\frac{du}{dt} - f, z\right) + a(u, z) = 0 \qquad (\forall) \, z \in C_0^\infty(Q), \tag{6}$$

that is, u also satisfies equation (2) in the sense of distributions. It remains to prove that this solution u satisfies boundary conditions (3).

We assert that (5) is equivalent to the system

$$\begin{cases} \int_0^{t_0} \left(\frac{du}{dt} - f, u\right) dt + \int_0^{t_0} a(u, u) \, dt = 0, \\[2mm] \int_0^{t_0} \left(\frac{du}{dt} - f, v\right) dt + \int_0^{t_0} a(u, v) \, dt \geqslant 0 \qquad (\forall) \, v \in K. \end{cases} \tag{7}$$

In fact, for any $w \in K$ take $v = u + w$ in (5); then we have

$$\int_0^{t_0} \left(\frac{du}{dt} - f, w\right) dt + \int_0^{t_0} a(u, w) \, dt \geqslant 0 \qquad (\forall) \, w \in K,$$

i.e., the second inequality in (7). Setting $v = 0$ in (5) we get

$$\int_0^{t_0} \left(\frac{du}{dt} - f, u\right) dt + \int_0^{t_0} a(u, u) \, dt \leqslant 0.$$

Inequality (5) follows now from (7) by subtraction.

Let us prove that the boundary conditions (3) hold. For any fixed $t \in [0, t_0]$, the condition $u(t) \in H^1(\Omega)$ implies that $u(t) \in H^{\frac{1}{2}}(\Gamma)$ and $\dfrac{\partial u}{\partial n} \in H^{-\frac{1}{2}}(\Gamma)$. Thus the product $u \dfrac{\partial u}{\partial n}$ is well defined (see II.2.7). By Green's formula from (3) we obtain

$$\int_\Sigma v \frac{\partial u}{\partial n} \, d\sigma = \int_0^{t_0} \left[\left(\frac{du}{dt} - f, v\right) + a(u, v) \right] dt \qquad (\forall) \, v \in K.$$

In virtue of (6) and (7) we deduce that

$$\int_\Sigma u \frac{\partial u}{\partial n} \, d\sigma = 0 \quad \text{and} \int_\Sigma v \frac{\partial u}{\partial n} \, d\sigma \geqslant 0 \qquad (\forall) \, v \in K,$$

that is, all boundary conditions (3) are verified.

Define on $H^1(\Omega)$ the proper convex l.s.c. function

$$\Phi(v) = \frac{1}{2} \int_\Omega |\nabla v|^2 \, dx.$$

The function Φ is Gâteaux differentiable and $(D\Phi(u), v) = a(u, v)$. Thus in virtue of the subgradient inequality

$$a(u, v - u) = (D\Phi(u), v - u) \leqslant \Phi(v) - \Phi(u) \quad (\forall)\, v \in H^1(\Omega),$$

the variational inequality (5) can be put in the form

$$\int_0^{t_0} \left(\frac{du}{dt} - f, u - v \right) dt \leqslant \int_0^{t_0} [\Phi(v) - \Phi(u)] dt \quad (\forall)\, v \in H^1(\Omega).$$

Finally, we remark that for an evolution equation of the form

$$\frac{\partial u}{\partial t} + A(t, u(t)) \ni f(x, t),$$

where the mapping $A(t,.)$ is of subdifferential type in $H^1(\Omega)$, the corresponding function Φ is directly dependent on t.

Equivalent formulations of minimum problems

1.2. Variational inequalities may serve to define more accurately the monotonicity property of subdifferentials.

We proceed now to give equivalent formulations of the minimum problem

$$\inf \{\varphi(x),\, |\, x \in K\}, \tag{1}$$

where φ is a proper convex l.s.c. function defined on a closed convex subset K of a Banach space X. The existence of a minimizing point of φ is assured by the Weierstrass theorem.

(*I*) *Direct formulation:* Find an element $x \in K$ such that

$$\varphi(x) \leqslant \varphi(y) \quad \text{for all} \ \ y \in K. \tag{2}$$

If we introduce the level set $\varphi^<(\varphi(y)) = \{z \in K \mid \varphi(z) \leqslant \varphi(y)\}$ for any fixed $y \in K$, then (2) is equivalent to

$$x \in \cap \{\varphi^<(\varphi(y)) | y \in K\},$$

that is, the set of solutions of problem (1) is convex and closed.

By our hypotheses φ is subdifferentiable (see I.2.6). Denote by $D\varphi$ a single-valued section of its subdifferential $\partial\varphi : K \mapsto 2^{X^*}$. If φ is Gâteaux-differentiable on K then $D\varphi : K \mapsto X^*$ coincides with its differential. Using the subgradient inequality

$$\varphi(y) - \varphi(x) \geqslant (D\varphi(x), y - x) \quad (\forall)\, y \in K,$$

we arrive at the

(*II*) *Gradient formulation:* Find an element $x \in K$ such that

$$(D\varphi(x), y - x) \geqslant 0 \quad (\forall) \, y \in K. \tag{3}$$

These two formulations are nonlinear inequalities in x. On account of the monotonicity of the subdifferential, we can write that

$$(D\varphi(y), y - x) \geqslant (D\varphi(x), y - x) \quad (\forall) \, x, y \in K,$$

which leads to the

(*III*) *Linearized formulation.* Find an element $x \in K$ such that

$$(D\varphi(y), y - x) \geqslant 0 \quad (\forall) \, y \in K.$$

THEOREM *Let $\varphi : K \mapsto \overline{\mathbb{R}}$ be a proper subdifferentiable function such that $D\varphi : K \mapsto X^*$ is a hemicontinuous map. Then the formulations* (*I*), (*II*) *and* (*III*) *are mutually equivalent.*

Proof: (*I*) $\Rightarrow$ (*III*). By property (iii) of the subdifferential, $0 \in \partial\varphi(x)$. Since $\partial\varphi$ is a monotone map, we obtain that

$$(D\varphi(y), y - x) \geqslant 0 \quad (\forall) \, y \in K.$$

(*III*) $\Rightarrow$ (*II*): Take in (*III*) $y_t = (1 - t)x + ty, t \in (0, 1)$ instead of y; by the hemicontinuity of $D\varphi$, we get (*II*) as $t \to 0$.

(*II*) $\Rightarrow$ (*I*) follows by the subgradient inequality.

When φ is not subdifferentiable, we characterize the minimizing element by means of the epigraph of φ.

Consider in $\tilde{X} = X \times \mathbb{R}$ the epigraph $\mathrm{eg}\,\varphi = \{[x, t] \in \tilde{X} \,|\, \varphi(x) \leqslant t\}$ and the subset $\tilde{K} = \{[x, t] \in \tilde{X} \,|\, x \in K, \varphi(x) \leqslant t\}$, which is the closed convex subset obtained by the intersection of $\mathrm{eg}\,\varphi$ and the "cylinder" $K \times \mathbb{R}$ in $\tilde{X}$. Clearly, x minimizes φ on K if and only if $[x, \varphi(x)]$ minimizes the function $\tilde{\varphi}([x, t]) = t$ on $\tilde{K}$. The function $\tilde{\varphi}$ is obviously Gâteaux differentiable with a constant gradient $D\tilde{\varphi} = [0, 1]$ on $\tilde{K}$.

(*IV*) *Epigraph formulation:* Find an element $u = [x, \varphi(x)]$ in $\tilde{K}$ such that

$$(D\tilde{\varphi}(u), v - u) \geqslant 0, \quad (\forall) \, v \in \tilde{K}.$$

A glance at the form of $D\tilde{\varphi}$ shows the equivalence of the formulations (*I*) and (*IV*).

Finally, let H be a Hilbert space, identified with its dual space, let K be a closed convex subset of H and let $P_K: H \mapsto K$ be the projection on K. Then the minimizing points of a proper subdifferentiable function $\varphi : K \mapsto \overline{\mathbb{R}}$ can be characterized by the

(*V*) *Fixed point formulation:* Find $x \in K$ such that

$$P_K(I + rD\varphi)\, x = x,$$

where I is the identity map in H and $r > 0$ is a real fixed number.

Clearly $D\varphi(x) = 0$ if and only if $P_K x = x$. Furthermore, we have the

PROPOSITION *If* $\varphi : K \mapsto \overline{\mathbb{R}}$ *is a proper subdifferentiable function, then the formulations* (II) *and* (V) *are equivalent.*

Proof: By the properties of the projection, $x = P_K z$ with $z = x - rD\varphi(x)$ is a solution of (V) if and only if

$$(x - (x - rD\varphi(x)), y - x) \geqslant 0 \text{ for all } y \in K,$$

which is equivalent to (II) because $r > 0$.

We note that the fixed point formulation suggests an iterative algorithm

$$x_{n+1} = P_K(x_n - r \, D\varphi(x_n)), \quad n \in \mathbb{N},$$

yielding the solution x of problem (V). This algorithm, which is called the *gradient projection method*, converges whenever $P_K(I - r \, D\varphi)$ is a contraction for a suitable $r > 0$.

1.3. We extend the equivalence of formulations (I) and (II) to the problem of the minimization of a sum of two proper functions. This allows the expression of the constraint $x \in K$ by means of the indicator function $I_K : X \mapsto \mathbb{R}$ (see I.2.1). Clearly, the minimization of $\varphi + I_K$ on X is equivalent to the minimization of φ on $K = D(I_K)$.

THEOREM *Let* $\varphi : X \mapsto \mathbb{R}$ *be a proper convex function and let* $\psi : X \mapsto \overline{\mathbb{R}}$ *be a proper subdifferentiable function with* $D\psi$ *hemicontinuous. Then the inequalities*

(j) $(\varphi + \psi)(x) \leqslant (\varphi + \psi)(y) \quad (\forall) \, y \in X;$

(jj) $\varphi(x) - \varphi(y) \leqslant (D\psi(x), y - x) \quad (\forall) \, y \in X,$

are mutually equivalent.

Proof: (j) $\Rightarrow$ (jj). Since φ is convex we have for any $y \in X$ and $t \in (0, 1)$

$$\varphi(x) + \psi(x) \leqslant \varphi((1 - t)x + ty) + \psi((1 - t)x + ty) \leqslant$$

$$\leqslant \psi((1 - t)x + ty) + (1 - t) \, \varphi(x) + t\varphi(y)$$

and since $\psi(x) < \infty$,

$$\varphi(x) - \varphi(y) \leqslant \frac{1}{t} [\psi(x + t(y - x)) - \psi(x)] \leqslant (D\psi(x + t(y - x)), y - x).$$

Using the hemicontinuity of $D\psi$ and taking the limit as $t \to 0$, we deduce (jj).

Conversely, the implication (jj) $\Rightarrow$ (j) follows by the subgradient inequality $\psi(y) - \psi(x) \geqslant (D\psi(x), y - x)$.

Problem (jj) is the general (mixed) form of a variational inequality. More generally, for given $f \in X^*$ let us find an element $x \in X$ such that

$$(Tx - f, y - x) \geqslant \varphi(x) - \varphi(y) \quad (\forall) \, y \in X, \tag{k}$$

where $T : X \mapsto X^*$ and $\varphi : X \mapsto \overline{\mathbb{R}}$ is a proper function. For that purpose let us take $K = \text{eg } \varphi = \{[x, t] \in \tilde{X} \mid \varphi(x) \leqslant t\}$ and $\tilde{f} = [f, 0] \in \tilde{X}^*$ and let us extend the epigraph formulation to the product space $\tilde{X} = X \times \mathbb{R}$: Find $u \in K$ as the solution of the inequality

$$(\tilde{T}u - \tilde{f}, v - u) \geqslant 0 \quad (\forall) \, v \in K, \tag{kk}$$

where $\tilde{T}$ maps $\tilde{X}$ into $\tilde{X}^* = X^* \times \mathbb{R}$ with $\tilde{T}([x, t]) = [Tx, 1]$, for all $x \in X$ and $t \in \mathbb{R}$.

PROPOSITION *The element $x \in X$ is a solution of the variational inequality* (k) *if and only if the pair $[x, \varphi(x)] = u$ is a solution of the inequality* (kk).

Proof: Suppose that x verifies (k). Then for any $r \geqslant \varphi(y)$ we have that

$$0 \leqslant (Tx - f, y - x) + \varphi(y) - \varphi(x) \leqslant (Tx - f, y - x) + 1(r - \varphi(x)) =$$

$$= (\tilde{T}u - \tilde{f}, y - u),$$

where $y = [y, r]$. Hence $u = [x, \varphi(x)]$ fulfils (kk).

Conversely, if $u = [x, s]$ satisfies (kk), then for $\varphi(x) \leqslant s$ and any $v = [y, r]$ with $\varphi(y) \leqslant r$ we have,

$$0 \leqslant (\tilde{T}u - \tilde{f}, y - u) = (Tx - f, y - x) + r - s.$$

In particular, take $x = y$ and $r = \varphi(x)$. Then $s \leqslant r \leqslant \varphi(x)$, whence $s = \varphi(x)$. Thus for $r \leqslant \varphi(y)$ we deduce that

$$0 \leqslant (Tx - f, y - x) + r - \varphi(x),$$

which means that x verifies (k) when $r = \varphi(y)$.

The penalty method

1.4. We shall give presently some sufficient conditions for the existence of solutions of the variational inequality

$$x \in K, \quad (Tx - f, y - x) \geqslant 0 \quad (\forall) \, y \in K, \tag{1}$$

where K is a closed convex set in the reflexive Banach space and T maps X into the dual space X^*.

As we have seen in Theorem 1.2, if T is monotone and hemicontinuous, then the inequality (1) is equivalent to the linearized variational inequality

$$x \in K \qquad (Ty - f, y - x) \geqslant 0 \qquad (\forall) \ y \in K. \tag{2}$$

The existence of solutions of the variational inequality (1) can be established by the same technique as those employed in the proof of Debrunner-Flor's lemma. However, we prefer to use the penalty method because it can be more easily handled in various concrete situations.

A bounded demicontinuous monotone operator $P : X \mapsto X^*$ with the property that

$$K = \ker P = N(P) = \{x \in X \mid Px = 0\}$$

is said to be the *penalty operator* of K. Its structure is described by the following

PROPOSITION *Let* $J : X \mapsto X^*$ *be the duality map and let* $P_K : X \mapsto K$ *be the projection on K. Then the mapping*

$$P = J(I - P_K),$$

(I denotes the identity map on X) is a penalty operator of K.

Proof: As we have seen in III.2.6, for any $x \in X$ the element $u = P_K x$ is characterized by

$$(J(x - u), u - v) \geqslant 0 \qquad (\forall) \ v \in K.$$

In particular, for any two elements $x, y \in X$ we get that

$$(J(x - P_K x), P_K x - P_K y) \geqslant 0 \text{ and } (J(y - P_K y), P_K y - P_K x) \geqslant 0.$$

By the definition of P and the above inequalities, we get

$$(Px - Py, x - y) = (J(x - P_K x) - J(y - P_K y), (x - P_K x) - (y - P_K y)) +$$
$$+ (J(x - P_K x), P_K x - P_K y) + (J(y - P_K y), P_K y - P_K x) \geqslant 0,$$

that is, P is monotone. It is also bounded, demicontinuous and $Px = 0$ means that $(I - P_K)x = 0$, whence $P_K x = x \in K$.

Examples. (a) If $K = \{u \in H_0^1(\Omega) \mid u \geqslant 0 \text{ a.e. in } \Omega\}$, then $P_K u = u^+$ and $Pu = u - u^+ = -u^-$. Here u^+ and u^- denote the positive and negative part of $u \in H_0^1(\Omega)$ and Ω is a domain in $\mathbb{R}^N$.

(b) If $K = \{u \in W_0^{1,p}(\Omega) \mid u > 0 \text{ a.e. in } \Omega\}$, $p > 2$, then $Ju = \|u\|^{p-2}u$ and $Pu = -\|u\|^{p-2}u^-$.

The main existence result is given by the following

THEOREM *Let* T *be a bounded pseudo-monotone mapping from* X *into* X^* *and let* K *be a closed convex subset of* $D(T)$. *If* T *is coercive with respect to* $x_0 \in K$, *i.e.,*

$$\frac{(Tx, y - x_0)}{\|x\|} \to \infty \text{ as } \|x\| \to \infty, \tag{3}$$

then there exists at least one solution of the variational inequality (1).

Proof: Let P be the penalty operator of K. We prove first that for each $\varepsilon > 0$ there exists at least one solution of the equation

$$Tx + \frac{1}{\varepsilon} Px = f. \tag{4}$$

Indeed, by Proposition III.2.5, the sum $T + \frac{1}{\varepsilon} P$ is pseudo-monotone. Since x_0 lies in K we have that

$$(Tx, x - x_0) + \frac{1}{\varepsilon} (Px, x - x_0) = (Tx, x - x_0) + \frac{1}{\varepsilon} (Px - Px_0, x - x_0) \geqslant$$

$$\geqslant (Tx, x - x_0)$$

by the monotonicity of P, and therefore

$$\frac{1}{\|x\|} \left[(Tx, x - x_0) + \frac{1}{\varepsilon} (Px, x - x_0) \right] \to \infty \quad \text{as} \quad \|x\| \to \infty.$$

Hence, by Theorem III.2.9, for each $\varepsilon > 0$ there exists an $x_\varepsilon \in X$ which is a solution of equation (4).

Moreover, by the coercivity condition (3), we deduce that the set $\{x_\varepsilon | \varepsilon > 0\}$ remains bounded as $\varepsilon \to 0$. The boundedness of T and the fact that $Px_\varepsilon = \varepsilon(f - Tx)$ implies that $\| Px_\varepsilon \| \leqslant C$, where $C > 0$ is a constant. By the reflexivity of X, we can choose a sequence $\{x_n\}$, $x_n = x_{\varepsilon_n}$, such that $x_n \to x$ in X as $\varepsilon_n \to 0$. Using the fact that $\| Px_n \| \to 0$ as $n \to \infty$, and taking the limit in the monotonicity relation, we get that

$$0 \leqslant \lim (Py - Px_n, y - x_n) = (Py, y - x) \qquad (\forall) \; y \in X.$$

Set $y = x + tz$ with $t > 0$ and let z be arbitrarily chosen in X. Hence $(P(x + tz), z) \geqslant 0$ and, by the hemicontinuity of P, we get that $(Px, z) \geqslant 0$ for all $z \in X$ as $t \to 0$. Hence $Px = 0$, i.e., $x \in K$.

Now, by (4) we have that

$$(Tx_n - f, x - x_n) = \frac{1}{\varepsilon_n} (Px - Px_n, x - x_n) \geqslant 0,$$

because of the monotonicity of P. Hence

$$\limsup (Tx_n, x_n - x) \leqslant \limsup(f, x_n - x) = 0.$$

By the pseudo-monotonicity of T, we get that

$$(Tx, x - y) \leqslant \liminf (Tx_n, x_n - y) \leqslant (f, x - y) \qquad (\forall) \; y \in K,$$

therefore $(Tx - f, y - x) \geqslant 0$ for all $y \in K$, as claimed.

COROLLARY *Let $S: X \mapsto X^*$ be a bounded pseudo-monotone operator and let $\varphi: X \mapsto \overline{\mathbb{R}}$ be a proper convex l.s.c. function. Suppose S verifies the modified coercivity condition: There is a $u_0 \in X$ such that $\varphi(u_0) < \infty$ and*

$$\frac{(Su, u - u_0) + \varphi(u)}{\|u\|} \to \infty \text{ as } \|u\| \to \infty.$$

Then for each given $g \in X^*$ there exists at least one solution of the variational inequality

$$(Su - g, v - u) \geqslant \varphi(u) - \varphi(v) \qquad (\forall) \; v \in X. \tag{5}$$

Proof: In virtue of Proposition 1.3, inequality (5) is equivalent to inequality (1) where $T = [S, 1]$ maps $\tilde{X} = X \times \mathbb{R}$ into $\tilde{X}^* = X^* \times \mathbb{R}$, $f = [g, 0] \in \tilde{X}^*$ and $K = \text{eg } \varphi$. The modified coercivity condition implies the coercivity of T with respect to $x_0 = [u_0, 0] \in K$ and the corollary follows by the theorem.

Convergence of solutions of variational inequalities

1.5. The convexity property of variational inequalities is singled out by the fact that it is stable with respect to perturbations of the convex sets which arise in structure.

We shall give an outline concerning the approximate methods for variational inequalities without numerical details.

Let K be a closed convex set in a reflexive Banach space X and let T be a mapping from X into X^* such that $K \subseteq D(T)$. Consider the variational inequality

$$x \in K, \qquad (Tx, y - x) \geqslant 0 \qquad (\forall) \; y \in K. \tag{I}$$

For the sake of simplicity, suppose that X is separable and denote by X_n the finite-dimensional subspace of X spanned by the first n elements of a linear independent dense subset $\{e_n\}$ of X. Thus, to each element $x_n = \sum_{j=1}^{n} x_j^{(n)} e_j$ in X_n there corresponds the vector $x^n = (x_1^{(n)}, \ldots, x_n^{(n)})$ in $\mathbb{R}^n$.

We choose a convex subset K_n of X_n, such that the sequence $\{K_n\}$ realizes a suitable approximation of K, and associate with the initial problem (I) the following approximate variational inequality

$$x_n \in K_n, \qquad (T_n x_n, y_n - x_n) \geqslant 0 \qquad (\forall) \; y_n \in K_n, \tag{I_n}$$

where T_n denotes the restriction of T to K_n.

We consider the convex set in $\mathbb{R}^n$

$$C^n = \left\{ x_n \in \mathbb{R}^n \;\middle|\; x_n = \sum_{j=1}^{n} x_j^{(n)} e_j \in K_n \right\}$$

and the operator $T^n : C^n \mapsto C^n$ whose components are defined by

$$T_j^n x^n = \left(T_n \left(\sum_{i=1}^{n} x_i^{(n)} e_i \right), e_j \right), \qquad 1 \leqslant j \leqslant n.$$

Then the approximate problem (I_n) takes on the discrete form in $\mathbb{R}^n$:

$$x^n \in C^n, \quad \sum_{i,\, j=1}^{n} T_j^n x^{(n)} (y_j^{(n)} - x_j^{(n)}) \geqslant 0 \qquad (\forall)\; y^n \in C^n. \tag{I_n'}$$

When T is linear, then the system (I_n') reduces to the system of linear inequalities

$$x^n \in C^n, \quad \sum_{i,\, j=1}^{n} T_{ij}^n x_i^{(n)} (y_j^{(n)} - x_j^{(n)}) \geqslant 0 \qquad (\forall)\; y^n \in C^n,$$

with the $n \times n$ matrix (T_{ij}^n) given by

$$T_{ij}^n = (T_n e_i, e_j), \qquad 1 \leqslant i, j \leqslant n.$$

The form of problem (I_n) depends on the choice of the linear independent dense subset $\{e_n\}$ of X and on the specification of the convex subset K_n of X_n. Such choice is suitable provided that the corresponding discrete system is easily solvable and the solution x_n of (I_n) converges to a solution x of (I) as K_n approaches K.

The most natural candidate for K_n, namely,

$$K_n = K \cap X_n$$

is sometimes a bad choice. Indeed, if K is the one-dimensional subspace of $\mathbb{R}^n$ spanned by a vector v_0 which has non-zero components with respect to a given basis, then all the subspaces $\mathbb{R}^m$, $1 \leqslant m \leqslant n - 1$, of $\mathbb{R}^n$ intersect K only in the origin. This situation remains valid in a general separable space X, when K is the one-dimensional subspace spanned by a vector v_0 of X with infinitely many non-zero components, where, as above, we have $K_n = K \cap X_n = \{0\}$.

Further, we define a topology on the family of all closed convex subsets of a Banach space, weak enough to allow a family of finite-dimensional K_n to converge to a possible infinite-dimensional set K.

Let $\{K_n\}$ be a sequence of closed convex sets of X. We define

$s - \mathrm{Lim\ inf}\; K_n = \{x \in X \mid \text{ there exists a sequence } \{x_n\} \text{ with}$

$$x_n \in K_n \text{ such that } x_n \to x \text{ in } X\}$$

and

$w - \mathrm{Lim\ sup}\; K_n = \{x \in X \mid \text{ there exists a sequence } \{x_j\} \text{ with } x_j \in K_{n_j} \text{ for a}$

$$\text{subsequence } \{K_{n_j}\} \text{ of } \{K_n\} \text{ such that } x_j \to x \text{ in } X\}.$$

We say that the sequence $\{K_n\}$ *converges* to a (closed convex) set K of X and write

$$K = \text{Lim } K_n$$

provided that

$$w - \text{Lim sup } K_n \subseteq K \subseteq s - \text{Lim inf } K_n.$$

If $K_n \subseteq K$ for every $n \in \mathbb{N}$ then $K = \text{Lim } K_n$ is equivalent to the single inclusion $K \subseteq s - \text{Lim inf } K_n$. If $K \subseteq K_n$ for every $n \in \mathbb{N}$ then $K = \text{Lim } K_n$ reduces to the condition $w - \text{Lim sup } K_n \subseteq K$.

In this set-up, the stability of the variational inequality (I) with respect to the perturbation of the closed convex set K is rendered by the following

THEOREM *Let X be a reflexive separable Banach space, let $T: X \mapsto X^*$ be a bounded hemicontinuous monotone operator and let $\{K_n\}$ be a sequence of closed convex sets of $D(T)$ such that $\text{Lim } K_n = K$. Suppose that the approximate problems*

$$x_n \in K_n, \ (Tx_n, z - x_n) \geqslant 0 \qquad (\forall) z \in K_n, \tag{1}$$

admit a solution $x_n \in K_n$ for all $n \in \mathbb{N}$ and that the sequence $\{x_n\}$ is bounded in X. Then the variational inequality (1) has at least one solution $x \in K$. Moreover, if this solution is unique, then $x_n \rightharpoonup x$ in X and $(Tx_n - Tx, x_n - x) \to 0$.

Proof: First, we show that the weak limit of a subsequence of approximate solutions (1) is a solution of the variational inequality (I). Indeed let $\{x_j\}$ be a subsequence of $\{x_n\}$ such that $x_j \rightharpoonup x$ in X. Since $w - \text{Lim sup } K_n \subseteq K$ we have that $x \in K$. On the other hand, $K \subseteq s - \text{Lim inf } K_n$ implies that to each $y \in K$ there corresponds a sequence $\{y_n\}$ with $y_n \in K_n$ such that $y_n \to y$ in X. We take $z = y_n$ in (1) and obtain

$$(Tx_n, y - x_n) \geqslant (Tx_n, y - y_n).$$

By the monotonicity of T at the left-hand term, we have

$$(Ty, y - x_n) \geqslant (Tx_n, y - x_n).$$

Since $\{x_n\}$ and $\{Tx_n\}$ are bounded sequences, we deduce that x is a solution of the linearized problem

$$x \in K, \qquad (Ty, y - x) \geqslant 0 \qquad (\forall) \ y \in K.$$

The latter is equivalent to (I) as it has been stated in Section 1.4.

Under the uniqueness hypothesis for the solution of problem (1), the boundedness of $\{x_n\}$ and the reflexivity of X imply that $x_n \rightharpoonup x$ in X.

Finally, since $K \subseteq s - \text{Lim inf } K_n$, there exists a sequence $\{u_n\}$ with $u_n \in K_n$ such that $u_n \to u$ in X. Replacing z by u_n in (1), we get

$$(Tx_n, u_n - x) + (Tx_n, x - x_n) = (Tx_n, u_n - x_n) \geqslant 0,$$

whence lim sup $(Tx_n, x_n - x) \leqslant 0$ or equivalently

   lim sup $(Tx_n - Tx, x_n - x) \leqslant 0$.

Thus the monotonicity of T implies that

   $(Tx_n - Tx, x_n - x) \to 0,$

as claimed.

   By the last part of the proof, it follows that if T is moreover of type (S), then $x_n \to x$ in X.

   This result points towards some necessary conditions which could be used in the numerical study of solutions of variational inequalities.

   For a more exact description of the above topology, we refer the reader to Mosco's lectures [1], Kemnochi [1] and Stroescu [1].

2. Variational boundary-value problems

The variational problems play an important role in the existence theory of weak solutions for a wide class of boundary-value problems. The operators of monotone type are useful in the investigation of variational problems for quasilinear equations.

Generalized divergence equations

2.1. Let Ω be a domain in $\mathbb{R}^N, (N \geqslant 2)$ with the boundary $\partial\Omega$ for which Green's formula holds. Denote by $x = (x_1,...,x_N)$ its generic point and denote, as usually, the elementary differential operators by $D^\alpha = D_1^{\alpha_1}...D_N^{\alpha_N}$, where $D_j = \dfrac{\partial}{\partial x_j}$ and $\alpha = (\alpha_1, ..., \alpha_N)$ are any N-tuples of non-negative integers with $|\alpha| = = \alpha_1 + ... + \alpha_N$. To write nonlinear partially differential operators in a more convenient form, we introduce the vector space $\mathbb{R}^{Nm}$ whose elements are of the form $\xi = \{\xi_\alpha \,|\, |\alpha| \leqslant m\}$. Within this framework, a nonlinear divergence operator of order $2m$ has the form

   $$Au = \sum_{|\alpha| \leqslant m} (-1)^{|\alpha|} D^\alpha A_\alpha(x, \xi(u(x))),$$

where $\xi(u) = \{D^\alpha u |\, |\alpha| \leqslant m\}$ and the function A_α maps $\Omega \times \mathbb{R}^{Nm}$ into $\mathbb{R}$, for each α. Every linear differential operator can be written in divergence form provided that its coefficients are sufficiently smooth. The nonlinear divergence operators appear for instance in the variational calculus.

For a given function f, which we shall specify more exactly later, we consider the boundary-value problem

$$\begin{cases} Au = f & \text{in } \Omega, \\ B_j u = 0 & \text{on } \partial\Omega, \ 0 \leqslant j \leqslant m - 1, \end{cases} \tag{1}$$

where B_j is a nonlinear differential operator of order $m - j - 1$.

When $A_\alpha(.\,,\,\xi(u(\,.\,))) \in L^{p'}(\Omega)$ for every $u \in W^{m,p}(\Omega)$ with $\dfrac{1}{p} + \dfrac{1}{p'} = 1$, we assign to the operator A its Dirichlet form

$$a(u, v) = \sum_{|\alpha| \leqslant m} \int_\Omega A_\alpha(x, \xi(u(x)))\, D^\alpha v(x)\, dx \qquad (\forall)\, u, v \in W^{m,p}(\Omega),$$

and a closed space V such that $W_0^{m,p}(\Omega) \subseteq V \subseteq W^{m,p}(\Omega)$.

Problem (1) is said to be of *variational type* provided that its boundary conditions are implicitly verified by the constraints

$$u \in V \quad \text{and} \quad (Au, v) = a(u, v) \quad \text{for all } v \in V. \tag{2}$$

Here $(.\,,.)$ denotes the duality pairing between V and its dual space V^*. By repeated integration of (Au, v), we see that the condition $(Au, v) = a(u, v)$ implies that certain integrals over $\partial\Omega$ vanish.

Of course, different variational boundary conditions are determined by the different choices of V. In the case when $V = W_0^{m,p}(\Omega)$, the condition $(Au, v) = a(u, v)$ is automatically satisfied. The constraint $u \in V$ implies that $B_j = \dfrac{\partial^j}{\partial n^j}$ in virtue of Theorem II.2.7, where $\dfrac{\partial}{\partial n}$ denotes the normal derivative. These are the *generalized Dirichlet boundary conditions*. In the oposite extreme case where $V = W^{m,p}(\Omega)$, the first constraint in (2) contains no information, while the second, yields

$$\sum_{j=0}^{m=1} \int_{\partial\Omega} C_j(u)\, \frac{\partial^j v}{\partial n^j}\, d\sigma = 0 \quad \text{for all } v \in V,$$

and the boundary conditions in (1) are given by $B_j = C_j$. These are the *generalized Neumann boundary conditions*. Finally, for the intermediate choice $W_0^{m,p}(\Omega) \subset V \subset W^{m,p}(\Omega)$, we must deal with boundary conditions of mixed type. Notice that all variational boundary conditions except for the Dirichlet ones are nonlinear.

With the variational problem (1) we associate the problem (A, V): Given $f \in V^*$, find $u \in V$ such that

$$a(u, v) = (f, v) \quad \text{for all} \quad v \in V. \tag{3}$$

A solution of the problem (A, V) is said to be a *variational solution* of problem (1).

If there exists a continuous function $\varphi: \mathbb{R}_+ \to \mathbb{R}_+$ such that

$$|a(u, v)| \leqslant \varphi(\|u\|_{m, p})\|v\|_{m, p},$$

then the function $a(u,.): V \to \mathbb{R}$ is linear and bounded for all $u \in V$, and it defines an operator (generally nonlinear) $T: V \to V^*$ for which

$$a(u, v) = (Tu, v) \quad \text{for all} \quad v \in V.$$

Hence a variational solution of problem (1) is also a solution of the functional equation $Tu = f$ in V.

A variational solution of problem (1) is a solution in the weak sense. We have to find the conditions under which the variational solution can exist and be unique. In order to be in agreement with the physical laws modelled by boundary-value problems, there arises the second problem of investigating the regularity of variational solutions, as dependent on the smoothness of inhomogeneous terms. In the sequel let us restrict our investigation to the first problem.

The existence of variational solutions

2.2. Let Ω be a domain in $\mathbb{R}^N$, let $f \in V^*$ and assume that the functions A_α in the definition of a nonlinear divergence operator of order $2m$ satisfy the following set of hypotheses (I):

(I_1) Each $A_\alpha: \Omega \times \mathbb{R}^{Nm} \to \mathbb{R}$ satisfies the Caratheodory conditions and there exists a p with $1 < p < \infty$ such that

$$|A_\alpha(x, \xi)| \leqslant c|\xi|^{p-1} + g(x),$$

where $c > 0$ is a constant and $g \in L^{p'}(\Omega)$ with $\dfrac{1}{p} + \dfrac{1}{p'} = 1$;

(I_2) $\displaystyle\sum_{|\alpha| \leqslant m} [A_\alpha(x, \xi) - A_\alpha(x, \xi')] (\xi_\alpha - \xi'_\alpha) \geqslant 0,$

for every x in Ω and every pair ξ, ξ' in $\mathbb{R}^{Nm}$;

(I_3) There exists an $r > 0$ so that

$$\sum_{|\alpha| \leqslant m} [A_\alpha(x, \xi) - f]\xi_\alpha > 0 \quad \text{whenever} \quad \Big(\sum_{|\alpha| \leqslant m} |\xi_\alpha|^p\Big)^{\frac{1}{p}} \geqslant r.$$

THEOREM *Under the hypotheses* (I), *the problem* (A, V) *has solutions and the set of its solutions is for each given* $f \in V^*$ *a bounded convex closed subset of* V.

Proof: By hypothesis (I_1) and Theorem IV.1.1, the functions $A_\alpha(., \xi(u(.)))$ lie in $L^{p'}(\Omega)$ for any $u \in W^{m,p}(\Omega)$. The Hölder and Minkowski inequalities yield that

$$\left| \int_\Omega A_\alpha(x, \xi(u)) \, D^\alpha v \, dx \right| \leq \left(\int_\Omega |A_\alpha(x, \xi(u))|^{p'} dx \right)^{\frac{1}{p'}} \left(\int_\Omega |D^\alpha v|^p \, dx \right)^{\frac{1}{p}} \leq$$

$$\leq \left[c \sum_{|\beta| \leq m} \left(\int_\Omega |D^\beta u|^{p'(p-1)} dx \right)^{\frac{1}{p'}} + \left(\int_\Omega |g(x)|^{p'} dx \right)^{\frac{1}{p'}} \right] \left(\int_\Omega |D^\alpha v|^p dx \right)^{\frac{1}{p}} \leq$$

$$\leq (C\|u\|_{m,p}^{p-1} + \|g\|_{p'}) \left(\int_\Omega |D^\alpha v|^p dx \right)^{\frac{1}{p}},$$

with $C > 0$, for all $u, v \in W^{m,p}(\Omega)$. Adding, we obtain

$$|a(u, v)| \leq \varphi(\|u\|_{m,p})\|v\|_{m,p},$$

where $\varphi(r) = Cr^{p-1} + \|g\|_{p'}$. Hence the operator $T: V \mapsto V^*$ is well defined and the problem (A, V) is equivalent to the equation $Tu = f$ in V.

In order to obtain a solution of this equation it is sufficient to show that T verifies the conditions established in Section III.2.9:

T is continuous. Indeed, by Theorem IV.1.1, condition (I_1) implies that the Nemitskyi operators $F_\alpha u = A_\alpha(., \xi(u))$ are continuous from $W^{m,p}(\Omega)$ into $L^{p'}(\Omega)$. Now let $\{u_n\}$ be a sequence strongly convergent to u in V. Then

$$\|Tu_n - Tu\| = \sup_{\|v\|_{m,p} \leq 1} |(Tu_n - Tu, v)| \leq$$

$$\leq \sup_{\|v\|_{m,p} \leq 1} \sum_{|\alpha| \leq m} \int_\Omega |A_\alpha(x, \xi(u_n)) - A_\alpha(x, \xi(u))| \, |D^\alpha v| \, dx \to 0$$

as $n \to \infty$, and hence T is continuous on V.

T is monotone. By (I_2) we have

$$(Tu - Tv, u - v) = a(u, u - v) - a(v, u - v) =$$

$$= \int_\Omega \sum_{|\alpha| \leq m} [A_\alpha(x, \xi(u)) - A_\alpha(x, \xi(v))] [\xi_\alpha(u) - \xi_\alpha(v)] \, dx \geq 0,$$

for all u and v in V.

T is coercive. By (I_3) we get

$$a(u, u) - (f, u) > 0 \quad \text{whenever} \quad \|u\|_{m,p} \geq r.$$

Hence by Corollary III.2.9, the equation $Tu = f$ has solutions.

The closedness, convexity and boundedness of $T^{-1}(f)$ in V follow by the properties of maximal monotone operators. This completes the proof.

To assume the existence of a solution of the variational problem (A, V) for every $f \in V^*$ we need a hypothesis stronger than (I_3):

(I'_3) There exist a positive constant c_0 and a function $h \in L^1(\Omega)$ such that

$$\sum_{|\alpha| \leqslant m} A_\alpha(x, \xi) \, \xi_\alpha \geqslant c_0 |\xi|^p - h(x),$$

for all x in Ω and all ξ in $\mathbb{R}^{N_m}$.

In that case we can easily see that T is coercive and thus surjective.

We remark that conditions (I_2) and (I'_3), which imply the coerciveness of T, replace the strong ellipticity condition in the theory of linear elliptic problems. In addition, hypotheses (I) bear on the dependence of the functions A_α on all the derivatives of order less than or equal to m, but not on the highest order derivatives. These assumptions impose no restriction on the boundedness of Ω.

Other assumptions

2.3. In this section, we shall apply the Sobolev's imbedding theorem to the study of variational solutions of divergence equations. As the assumptions seem to be more sophisticated, they may be omitted in the first reading.

Let Ω be a bounded domain in $\mathbb{R}^N$ for which the Sobolev imbedding theorem holds. We formulate a new set of hypotheses (II):

(II_1). The functions $A_\alpha: \Omega \times \mathbb{R}^{N_m} \mapsto \mathbb{R}$ satisfy the Caratheodory conditions. Let $b = b_{m, N, p}$ be the greatest integer less than $m - \dfrac{N}{p}$ and let $\xi_b = \{\xi_\alpha | \, |\alpha| \leqslant b\}$ denote a vector in $\mathbb{R}^{N_b}$. Suppose that there exist continuous functions $c_\alpha: \mathbb{R}^{N_b} \mapsto L^{s_\alpha}(\Omega)$ and $c_1: \mathbb{R}^{N_b} \mapsto \mathbb{R}$ such that the inequalities

$$|A_\alpha(x, \xi)| \leqslant c_\alpha(\xi_b) + c_1(\xi_b) \sum_{m - \frac{N}{p} \leqslant |\beta| \leqslant m} |\xi_\beta|^{s_{\alpha\beta}} \tag{1}$$

hold, where the exponents s_α and $s_{\alpha\beta}$ satisfy

$$s_\alpha = p' \quad \text{for} \quad |\alpha| < m,$$

$$s_\alpha > p' \quad \text{for} \quad m - \frac{N}{p} \leqslant |\alpha| < m, \quad \frac{1}{p_\alpha} = \frac{1}{p} - \frac{m - |\alpha|}{N}, \quad \frac{1}{p_\alpha} + \frac{1}{p'_\alpha} = 1;$$

$$s_\alpha = 1 \quad \text{for} \quad |\alpha| < m - \frac{N}{p};$$

and

$$S_{\alpha\beta} \leqslant p - 1 \quad \text{for} \quad |\alpha| = |\beta| = m,$$

$$S_{\alpha\beta} \leqslant \frac{p_\beta}{p_\alpha'} \quad \text{for } m - \frac{N}{p} \leqslant |\alpha| \leqslant m \text{ and } |\beta| < m;$$

$$S_{\alpha\beta} \leqslant p_\beta \quad \text{for } |\alpha| < m - \frac{N}{p}.$$

(II$_2$). Take $\xi = (\eta, \zeta)$ such that $\eta = \{\xi_\alpha \,\big|\, |\alpha| < m\}$ and $\zeta = \{\xi_\alpha \,\big|\, |\alpha| = m\}$. Suppose that $A_\alpha(x, \eta, \zeta)$ verifies the inequality

$$\sum_{|\alpha|=m} [A_\alpha(x, \eta, \zeta) - A_\alpha(x, \eta, \zeta')] (\zeta_\alpha - \zeta_\alpha') > 0,$$

for all $\zeta \neq \zeta'$, $x \in \Omega$ and $\eta \in \mathbb{R}^{N_{m-1}}$.

(II$_3$). There exist constants $c_0 > 0$ and c_2 such that

$$\sum_{|\alpha|\leqslant m} A_\alpha(x, \xi) \xi_\alpha \geqslant c_0 |\xi|^p - c_2$$

for all x in Ω and all $\xi \in \mathbb{R}^{N_m}$.

THEOREM *Let Ω be a bounded domain in $\mathbb{R}^N$ for which the Sobolev imbedding theorem holds. Then, under assumptions (II), the problem (A, V) admits a solution.*

Proof: First, we show that the Dirichlet form $a(u, v)$ in the problem (A, V) is well defined for all $u, v \in W^{m,p}(\Omega)$.

In fact, the map $D^\alpha : W^{m,p}(\Omega) \mapsto W^{m-|\alpha|, p}(\Omega)$ is continuous and by the Sobolev imbedding theorem so is the map

$$D^\alpha : W^{m,p}(\Omega) \mapsto L^{p_\alpha}(\Omega) \quad \text{for} \quad \frac{1}{p_\alpha} = \frac{1}{p} - \frac{m - |\alpha|}{N}.$$

To lend meaning to $a(u, v)$, we must show that $A_\alpha(x, \xi(u)) \in L^{p_\alpha'}(\Omega)$. Let us consider the different values of α in the assumptions (II):

(a) Set $|\alpha| = m$. We prove that $A_\alpha(., \xi(u(.))) \in L^{p_\alpha'}(\Omega)$. These functions satisfy (1), where c_α maps $\mathbb{R}^{N_b}$ into $L^{p'}(\Omega)$. For $u \in W^{m,p}(\Omega)$, by Theorem II.2.6, the components ξ_b are continuous in Ω and bounded by $\|u\|_{m,p}$. Also, the functions c_α and c_1 have their absolute values bounded by the continuous functions of $\|u\|_{m,p}$. Moreover, $D^\beta u \in L^{p_\beta}(\Omega)$ and $|\xi_\beta(u)|^{s_{\alpha\beta}} \in L^{\frac{p_\beta}{p_{\alpha\beta}}}(\Omega)$. Since $\frac{p_\beta}{s_{\alpha\beta}} \geqslant p_\alpha' = p'$, where $n - \frac{N}{p} \leqslant |\beta| \leqslant m$, we can deduce that

$$\|A_\alpha(x, \xi(u))\|_{p'} \leqslant \varphi_1(\|u\|_{m,p}),$$

where $\varphi_1: \mathbb{R}_+ \mapsto \mathbb{R}_+$ is a continuous function. Hence for $|\alpha| = m$ we have

$$|(A_\alpha(x, \xi(u)), D^\alpha v)| \leqslant \varphi\left(\|u\|_{m,p}\right)\|v\|_{m,p}. \tag{2}$$

(b) If $m - \dfrac{N}{p} \leqslant |\alpha| < m$, then $D_\alpha v \in L^{p_\alpha}(\Omega)$ and $c_\alpha(\xi_b) \in L^{p_\alpha'}(\Omega)$. Since

$D^\beta u \in L^{p_\beta}(\Omega)$, $|\xi_\beta(u)|^{s_{\alpha\beta}} \in L^{\frac{p_\beta}{s_{\alpha\beta}}}(\Omega)$ and $\dfrac{p_\beta}{s_{\alpha\beta}} \geqslant p_\alpha'$, it follows by (1) that

$A_\alpha(., \xi(u(.))) \in L^{p_\alpha'}(\Omega)$ for every $u \in W^{m,p}(\Omega)$. Hence an inequality of type (2) holds.

(c) For $|\alpha| < m - \dfrac{N}{p}$, by the Sobolev imbedding theorem, D^α carries

$W^{m,p}(\Omega)$ into $C(\bar{\Omega})$ and $c_\alpha(\xi_b) \in L^1(\Omega)$, $s_{\alpha\beta} \leqslant p_\beta$. Thus $|\xi_b(u)|^{s_{\alpha\beta}} \in L^1(\Omega)$ and thereby, $A_\alpha(., \xi(u(.))) \in L^1(\Omega)$. We obtain again an inequality of type (2).

Therefore, the operator $T: V \mapsto V^*$ is well defined and the problem (A, V) is equivalent to the equation $Tu = f$ in V.

Inequality (2) is equivalent to $\|Tu\| \leqslant \varphi(\|u\|_{m,p})$. This implies the boundedness of T. Hypothesis (II_1) and Theorem IV.1.1 assure the demicontinuity of T.

Let us show now that T is of type (S_+). For this purpose, let $\{u_n\}$ be a sequence in V converging weakly to u for which

$$\lim\sup\,(Tu_n - Tu, u_n - u) = \lim\sup\,[a(u_n, u_n - u) - a(u, u_n - u)] = \tag{3}$$

$$= \lim\sup \int_\Omega \sum_{|\alpha| \leqslant m} (A_\alpha(., \eta(u_n), \zeta(u_n)) - A_\alpha(., \eta(u), \zeta(u))) D^\alpha(u_n - u)\, dx \leqslant 0$$

Since $W^{m,p}(\Omega)$ is compactly imbedded in $W^{m-1,p}(\Omega)$, we can select a subsequence $\{u_k\}$ such that $D^\alpha u_k \to D^\alpha u$ in $L^p(\Omega)$ for $|\alpha| \leqslant m-1$ and inequality (3) becomes

$$\lim\sup \int_\Omega \sum_{|\alpha|=m} (A_\alpha(., \eta(u), \zeta(u_k)) - A_\alpha(., \eta(u), \zeta(u))) D^\alpha(u_k - u)\, dx \leqslant 0.$$

Since hypothesis (II_3) implies that

$$\sum_{|\alpha|=m} A(x, \eta, \zeta)\, \zeta_\alpha \geqslant c_0\,|\zeta|^p - g_2, \tag{4}$$

where

$$g_2 = c_0 \sum_{i=1}^{p} (-1)^i C_p^i |\zeta|^{p-i}|\eta|^i + \sum_{|\alpha|<m} A_\alpha(x, \eta, \zeta)\, \eta_\alpha$$

lies in $L^1(\Omega)$, we can apply Proposition IV.1.1. Then condition (4) and hypothesis (II_2) imply that $D^\alpha u_k \to D^\alpha u$ in $L^p(\Omega)$ also for $|\alpha| = m$. Hence T is of type (S_+).

Hypothesis (II_3) guarantees the coerciveness of T and the surjectivity of T follows by Corollary V.1.1.

3. Strongly nonlinear problems

The study of the boundary-value problems for a nonlinear equation of the form

$$Au(x) + g(x, u(x)) = f(x)$$

is of foremost importance as concerns applications. Here A denotes a general divergence operator and g is a function satisfying certain conditions in the second variable. The above equation is *strongly nonlinear* in the sense that no growth restriction upon g is imposed.

The investigation of these problems leads to the introduction of new classes of mappings of monotone type in Banach spaces. The induced non-linear operators are generally neither everywhere defined nor bounded.

Mappings of type (M) with respect to two Banach spaces

3.1. Let W and V be two real reflexive separable Banach spaces with the norms $\|\cdot\|_W$, $\|\cdot\|_V$ and the dual spaces W^* and V^* respectively. Suppose that $W \subset V$ with a continuous natural injection $i: W \mapsto V$. Then $V^* \subset W^*$ in the following sense: If $f|_W = i^*f$ denotes the restriction of $f \in V^*$ to W, then $i^*f \in W^*$, where $i^*: V^* \mapsto W^*$ is the adjoint map of i. We do not assume that W is dense in V. (f, u) will denote the pairing either between $f \in W^*$ and $u \in W$ or between $f \in V^*$ and $u \in V$.

Let T be a nonlinear operator from V to W^* such that $W \subseteq D(T) \subseteq V$. Then T is said to be *of type (M) with respect to (W, V)*, provided that

(M_1) T is continuous from finite-dimensional subspaces of W to the weak topology on W^*;

(M_2) Whenever $\{w_n\} \subset W$ and $h \in V^*$ are such that $w_n \to v$ in V, $Tw_n \to i^*h$ in W^* and $\limsup (Tw_n, w_n) \leqslant (h, v)$, it follows that $v \in D(T)$ and $Tv = i^*h$.

We observe that when $W = V$ then we obtain the operators of type (M) investigated in III.5.1.

Within this framework, T is *quasi-bounded* if for any $\{w_n\}$ in W with $w_n \to v$ in V and $(Tw_n, w_n) \leqslant c \|w_n\|_V$ with some constant $c > 0$, the sequence $\{Tw_n\}$ remains bounded in W^*. This concept extends the one in Section III.1.3 and it is trivially satisfied if T carries sets of W, bounded in the V-norm, into bounded sets of W^*. The difference between the quasi-boundedness and the boundedness of T will be emphasised in Lemma 3.2.

An abstract existence result for the equations defined by these operators is given by the following

THEOREM *Let T be a quasi-bounded mapping of type (M) with respect to (W, V).
Suppose that T is coercive in f ∈ V*, i.e.,*

$$(Tw - i^*f, w) > 0 \text{ for all } w \in W \text{ with } \|w\|_V = r$$

for some r > 0. Then there exists a v ∈ D(T) with $\|v\|_V < r$ such that Tv=i*f.

Proof: Let $\{W_n\}$ be an increasing sequence of finite-dimensional subspaces of W with dim $W_n = n$, the union of which is dense in W. For every $n \in \mathbb{N}$, consider the injection $j_n: W_n \mapsto W$ and let $j_n^*: W^* \mapsto W_n^*$ be its adjoint map. Then the mapping $T_n = j_n^* T j_n: W_n \mapsto W_n^*$ is continuous and

$$(T_n w - i^*f, w) = (Tw - i^*f, w) > 0 \text{ for all } w \in W_n \text{ with } \|w\|_V = r.$$

Since the W- and the V-norms are equivalent on the finite-dimensional space W_n, the open balls in W_n with respect to the V-norm remain bounded and open in the W-norm. By Proposition III.2.8, there exists a $v_n \in W_n$ with $\|v_n\|_V < r$ such that $T_n v_n = j_n^* i^* f$.

By the reflexivity of V, passing if necessary to a subsequence, we conclude that $v_n \rightharpoonup v$ in V. Since

$$(Tv_n, v_n) = (T_n v_n, v_n) = (j_n^* i^* f, v_n) = (i^* f, v_n) \leqslant \|f\|_V \|v_n\|_V,$$

the quasi-boundedness of T implies that the sequence $\{Tv_n\}$ is bounded in W^*.

For each $w \in \bigcup \{W_m \mid m \in \mathbb{N}\}$ there exists an $n \in \mathbb{N}$ such that $w \in W_n$ and therefore

$$(Tv_n - i^*f, w) = (T_n v_n - j_n^* i^* f, w) = (0, w) = 0.$$

Since $\bigcup \{W_n \mid n \in \mathbb{N}\}$ is dense in W and $\{\|Tv_n\|_{W^*}\}$ is bounded, we conclud that $Tv_n \rightharpoonup i^*f$ in W^*. As T is of type (M) with respect to (W, V), it follows that $v \in D(T)$ and $Tv = i^*f$.

COROLLARY *Let T be a quasi-bounded mapping of type (M) with respect to (W, V). If*

$$\frac{(Tw, w)}{\|w\|_V} \rightarrow \infty \text{ as } w \in W, \|w\|_V \rightarrow \infty,$$

then R(T) is all of W*.

Note that the above results generalize those of Section III.5.4.

The application of the above results requires some prerequisites on truncated functions.

Let Ω be a bounded domain in $\mathbb{R}^N$ and let $V = W^{m,p}(\Omega)$ with $1 < p < \infty$. For a real number $k > 0$ we denote by v^k the function obtained from $v \in V$ by truncation

$$v^k(x) = \begin{cases} -k & \text{if } v(x) \leqslant -k, \\ v(x) & \text{if } |v(x)| \leqslant k, \\ k & \text{if } v(x) \geqslant k. \end{cases}$$

Clearly $v^k \in V$ and $v^k(x) \to v(x)$ for a.a. $x \in \Omega$ as $k \to \infty$.

PROPOSITION *There exists a sequence $\{\varphi_j\}$ in $C^\infty(\overline{\Omega})$ with $|\varphi_j(x)| \leqslant k$ for all $j \in \mathbf{N}$ and $x \in \Omega$ such that $\varphi_j \to v^k$ in V as $j \to \infty$.*

Proof: Indeed, since $v \in V = W^{m,p}(\Omega)$, there exists a sequence $\{\psi_n\} \subset C^\infty(\overline{\Omega})$ such that $\psi_n \to v$ in V as well as pointwise almost everywhere in Ω (see Theorem II.2.5). Let ψ_n^k be the truncated function of ψ_n. Since $\|\psi_n^k\| = \|\psi_n\| \leqslant \text{const.}$, passing if necessary to a subsequence, we may assume that
$$\psi_n^k \to u \text{ in } V \text{ and } \psi_n^k(x) \to u(x) \text{ for a.a. } x \in \Omega, \text{ as } n \to \infty.$$
But we know that $\psi_n^k(x) \to v^k(x)$ for a.a $x \in \Omega$ as $n \to \infty$. Hence $u = v^k$ and therefore $\psi_n^k \to v^k$ in V as $n \to \infty$.

For every ψ_n^k we construct the mollified functions Φ_{jn} with $\Phi_{jn} \to \psi_n^k$ in V as $j \to \infty$. Then $\Phi_{jn} \in C^\infty(\overline{\Omega})$ at least for sufficiently large j and $|\Phi_{jn}(x)| \leqslant k$ for $x \in \Omega$. Choosing $\varphi_j = \Phi_{jn(j)}$ as a suitable diagonal subsequence we obtain the desired sequence.

Strongly nonlinear variational problems

3.2. Let Ω be a bounded domain in $\mathbf{R}^N$ such that the Sobolev imbedding theorem holds. We shall apply Theorem 3.1 to the variational boundary-value problems for the equation

$$Au(x) + g(x, u(x)) = f(x), \qquad x \in \Omega, \tag{1}$$

under the following assumptions upon the strongly nonlinear term:
  (III$_1$) the function $g: \Omega \times \mathbf{R} \to \mathbf{R}$ satisfies the Caratheodory conditions;
  (III$_2$) $g(., t)$ is esentially bounded for t bounded;
  (III$_3$) $g(x, t)$ satisfies the sign condition

$$g(x, t) t \geqslant 0 \text{ for a.a. } x \in \Omega \text{ and } t \in \mathbf{R}.$$

Suppose that the boundary conditions determine a reflexive Banach space V such that $W_0^{m,p}(\Omega) \subseteq V \subseteq W^{m,p}(\Omega)$.

Concerning the divergence operator

$$Au = \sum_{|\alpha| \leqslant m} (-1)^{|\alpha|} D^\alpha A_\alpha(., \xi(u)) \text{ with } \xi(u) = \{D^\alpha u \mid |\alpha| \leqslant m\},$$

we shall make the following assumptions:
  (III$_4$) Each $A_\alpha: \Omega \times \mathbf{R}^{N_m} \to \mathbf{R}$ satisfies the Caratheodory conditions and there exists a $p \in (1, \infty)$ such that

$$|A_\alpha(x, \xi)| \leqslant c(x) |\xi|^{p-1} + h(x),$$

where $h \in L^{p'}(\Omega)$ and c is a non-negative $L^\infty(\Omega)$-function.

As we have seenin Section 2.2, hypothesis (III$_4$) implies that the Dirichlet form

$$a(u, v) = \sum_{|\alpha| \leqslant m} \int_\Omega A_\alpha(x, \xi(u(x)))\, D^\alpha v(x)\, \mathrm{d}x \qquad (\forall)\, u, v \in V$$

is well-defined and determines a bounded continuous operator $T_1 \colon V \mapsto V^*$ such that

$$(T_1 u, v) = a(u, v) \text{ for all } u, v \in V.$$

(III$_5$) The mapping T_1 defined above is of type (M).

Observe that condition (III$_5$) holds in particular if A is linear of if A is nonlinear satisfiyng hypothesis (I$_2$) in 2.2 or hypothesis (II$_2$) in 2.3. Indeed, if A is linear, i.e.,

$$Au = \sum_{|\alpha|,\, |\beta| \leqslant m} (-1)^{|\alpha|} D^\alpha[a_{\alpha\beta}(.)\, D^\beta u],$$

with $a_{\alpha\beta} \in L^\infty(\Omega)$, then T_1 is linear bounded and thus of type (M) by Proposition III.5.1; in this case, we take $p = 2$. Otherwise, if A is nonlinear and satisfies (I$_2$), then T_1 is maximal monotone and thus a fortiori of type (M) (see Proposition III.5.2.). Finally, if A is nonlinear and satisfying (II$_2$), then T_1 is demicontinuous of type (S_+) and hence of type (M).

Let $W = W^{s,p}(\Omega) \cap V$ with $s > 1 + \left[\dfrac{N}{p}\right]$. Then $W \subset C(\bar{\Omega})$ by the

Sobolev imbedding theorem. Set

$$V_1 = \{u \in V \mid g(., u) \in L^1(\Omega),\ g(., u)\, u \in L^1(\Omega)\}.$$

Clearly, $W \subset V_1 \subset V$. Given $u \in V_1$, the form

$$b(u, w) = \int_\Omega g(x, u(x))\, w(x)\, \mathrm{d}x$$

is linear in $w \in W$. It is bounded because

$$|b(u, w)| \leqslant \|g(., u)\|_\infty \|w\|_{C(\bar{\Omega})} \qquad (\forall)\, w \in W,$$

and it determines a nonlinear operator $T_2 \colon V_1 \mapsto W^*$ such that

$$(T_2 u, w) = b(u, w) \qquad (\forall)\, u \in V_1,\ w \in W.$$

For a given $f \in V^*$, a function $u \in V_1$ is said to be a *variational solution* of equation (1) with respect to the boundary conditions imposed by V provided that

$$a(u, w) + b(u, w) = (f, w) \qquad (\forall)\, w \in W \tag{2}$$

holds.

LEMMA *Under the assumptions* (III), $T = i*T_1 + T_2: V_1 \mapsto W*$ *is a quasi-bounded operator of type* (M) *with respect to* (W,V).

Proof: The condition (M_1) concerning the finite continuity of T follows easily. Let us verify (M_2). Let $\{w_n\}$ be a sequence in W and let $h \in V*$ be such that $w_n \rightharpoonup v$ in V, $Tw_n \rightharpoonup i*h$ and $\lim \sup (Tw_n, w_n) \leqslant (h, v)$. By the Sobolev theorem, $w_n \to v$ in $L^p(\Omega)$ and $w_n(x) \to v(x)$ for a.a. $x \in \Omega$, at least for fome subsequence. In virtue of $(III_2) - (III_3)$, the sequence of non-negative functions $\{g(x,w_n(x))w_n(x)\}$ converges to $g(x, v(x))v(x)$ a.e. in Ω. Since $i*T_1$ is bounded, we have

$$\lim \sup \int_\Omega g(x, w_n(x))w_n(x) \, dx = \lim \sup (T_2 w_n, w_n) \leqslant C,$$

where C is a positive constant and thus by Fatou's lemma

$$\int_\Omega g(x, v(x)) v(x) \, dx = (T_2 v, v) \leqslant \lim \inf (T_2 w_n, w_n) \leqslant C,$$

and hence $g(., v)v \in L^1(\Omega)$.
Now write

$$|g(x, t)| = |t|^{-1} g(x, t)t \text{ for } t \neq 0$$

and set $h_0(x) = \text{ess sup } \{|g(x, t)| \,|\, t \in [0, 1]\}$. By (III_2), h_0 is an l.s.c. function and thus a summable one. Therefore

$$|g(x, t)| \leqslant g(x, t)t + h_0(x) \qquad (\forall) \, t \in \mathbb{R}.$$

This implies that $g(., v) \in L^1(\Omega)$. Consequently, $v \in V_1$.

Next, for a given $\delta > 0$ we set $K = \text{ess sup} \left\{ g(x, t) \,\middle|\, |t| < \dfrac{1}{\delta} \right\}$. By (III_2) it follows that $K < \infty$. For each fixed $n \in \mathbb{N}$ take

$$\Omega_1 = \left\{ x \in \Omega \,\middle|\, |w_n(x)| \leqslant \frac{1}{\delta} \right\} \text{ and } \Omega_2 = \left\{ x \in \Omega \,\middle|\, |w_n(x)| > \frac{1}{\delta} \right\}.$$

Then $\Omega = \Omega_1 \cup \Omega_2$ and $g(x, w_n(x)) \leqslant \delta g(x, w_n(x))w_n(x)$ for any $x \in \Omega_2$. Moreover, for any measurable subset $A \in \Omega$, we have the estimate

$$\int_A g(x, w_n(x)) \, dx \leqslant \int_{A \cap \Omega_1} g(x, w_n(x)) \, dx +$$

$$+ \int_{A \cap \Omega_2} g(x, w_n(x)) \, w_n(x) \, dx \leqslant K\mu(A) + \delta C,$$

where K depends only on δ. By Vitali's theorem. $g(.\,, w_n) \to g(.\,, v)$ in $L^1(\Omega)$ and thus

$$(T_2 w_n, u) = \int_\Omega g(x, w_n(x))\, u(x)\, dx \to \int_\Omega g(x, v(x))\, u(x)\, dx = (T_2 v, u)$$

for all $u \in W$, that is, $T_2 w_n \rightharpoonup T_2 v$ in W^*.

As T_1 is a bounded operator and the sequence $\{T_2 w_n\}$ is bounded, we deduce that T is quasi-bounded. The reflexivity of V implies that (at least on a subsequence) $T_1 w_n \rightharpoonup \chi$ in V^* and $i^*\chi + T_2 v = i^*h$. The equality $\chi = T_1 v$ follows from the fact that T_1 is of type (M) provided that we prove that

$$\limsup\, (T_1 w_n, w_n - v) \leqslant 0. \tag{3}$$

For any $z \in W$ we write

$$\limsup\, (T_1 w_n, w_n - v) = \limsup\, (T_1 w_n, w_n - z) + (\chi, z - v) =$$

$$= \limsup\, [(T w_n, w_n - z) - (T_2 w_n, w_n - z)] + (\chi, z - v) \leqslant$$

$$\leqslant (h - \chi, v - z) - (T_2 v, v - z).$$

For any $k > 0$, let $v^k \in V$ be the truncation of v and let $\{\varphi_j\} \subset C^\infty(\bar\Omega)$ be a sequence with $|\varphi_j(x)| \leqslant k$ and $\varphi_j \to v^k$ in V. We set $z = \varphi_j$ in the above inequality and pass to the limit as $j \to \infty$. Then we have

$$\limsup\, (T_1 w_n, w_n - v) \leqslant (h - \chi, v - v^k) - (T_2 v, v - v^k).$$

Finally we let $k \to \infty$ above and thus we obtain (3). This proves the lemma.

Note that T is a quasi-bounded operator but not necessarily bounded.
Furthermore, *the restriction of T_2 to W is demicontinuous.* Indeed, if $w_n \to w$ in W, then $w_n \to w$ in V and there exists a constant $C_1 > 0$ such that $|w_n(x)| \leqslant C_1$ for all $x \in \Omega$ and $n \in \mathbb{N}$. Hence by $(III_2) - (III_3)$, $g(x, w_n(x))w_n(x) \leqslant C_2$ for some constant $C_2 > 0$ and therefore

$$(T_2 w_n, w_n) = \int_\Omega g(x, w_n(x))\, w_n(x)\, dx \leqslant C_2 \mu(\Omega).$$

The demicontinuity of T_2 follows by its quasi-boundedness.
From theorem 3.1 we may conclude the validity of the following

THEOREM *Let hypotheses* (III) *be verified. If there exists an* $r > 0$ *such that*

$$a(w, w) + b(w, w) - (f, w) > 0, \qquad (\forall)\, w \in W \text{ with } \|w\|_V = r, \tag{4}$$

then problem (2) *has a solution.*

As (III_3) implies that $b(w, w) \geqslant 0$, condition (4) is satisfied for any $f \in V^*$ if there exists a constant $c > 0$ such that

$$a(w, w) \geqslant c\|w\|_V^p \qquad (\forall) \, w \in V.$$

We consider now the case of second order operators

$$Au = - \sum_{i=1}^{N} D_i A_i(x, u, \nabla u), \quad \nabla u = (\zeta_i)_{1 \leqslant i \leqslant N}, \quad \zeta_i = D_i u \tag{5}$$

for which there exists a constant $c_0 > 0$ such that

$$\sum_{i=1}^{N} A_i(x, t, \zeta) \, \zeta_i \geqslant c_0 \sum_{i=1}^{N} |\zeta_i|^p, \tag{6}$$

for almost all $x \in \Omega$ and all $(t, \zeta) \in \mathbb{R} \times \mathbb{R}^N$.

We define

$$g_+ (x) = \liminf_{t \to +\infty} g(x, t) \quad \text{and} \quad g_-(x) = \limsup_{t \to -\infty} g(x, t).$$

(g_+ and g_- may attain the values $+\infty$ and $-\infty$).

PROPOSITION *Suppose that for a given $f \in L^{p'}(\Omega)$ the inequalities*

$$\int_\Omega g_- \, dx < \int_\Omega f \, dx < \int_\Omega g_+ \, dx \tag{7}$$

hold. Then under the hypotheses (III), *problem* (1) *associated to the operator* (5) *has a variational solution.*

Proof: We shall show the existence of a constant $r > 0$ such that inequality (4) holds.

Let us assume the contrary. Then there exists a sequence $\{w_n\}$ in W with $\|w_n\|_V \to \infty$ and

$$a(w_n, w_n) + b(w_n, w_n) - (f, w_n) \leqslant 0. \tag{8}$$

Set $v_n = \|w_n\|_V^{-1} w_n$. Passing, if necessary, to a subsequence, we may assume that $v_n \to v$ in V and $v_n \to v$ in $L^p(\Omega)$ as well as $v_n(x) \to v(x)$ a.a. $x \in \Omega$. From (6) and (8) we infer that

$$c_0 \int_\Omega \sum_{i=1}^{N} |D_i v_n|^p \, dx + \frac{1}{\|w_n\|_V^{p-1}} \left[\int_\Omega g(x, w_n) \, w_n \, dx - (f, w_n) \right] \leqslant 0, \tag{9}$$

for all $n \in \mathbb{N}$. Since

$$g(x, w_n) \, v_n = \frac{1}{\|w_n\|_V} g(x, w_n) \, w_n \geq 0$$

and

$$\frac{1}{\|w_n\|^{p-1}} (f, v_n) \to 0,$$

we deduce that

$$\int_\Omega \sum_{i=1}^N |D_i v_n|^p \, dx \to 0 \text{ as } n \to \infty.$$

Thus $v_n \to v$ in V with $v = \alpha = \text{const.} \neq 0$, because $\|v_n\|_V = 1$. Thus, inequality (9) yields

$$\limsup \int_\Omega g(x, w_n) \, v_n \, dx \leq (f, v) = \int_\Omega f v \, dx.$$

Suppose first that $\alpha > 0$. As $w_n = \|w_n\|_V \, v_n \to \infty$ a.e. in Ω, by Fatou's lemma we have

$$\int_\Omega g_+ \, v \, dx \leq \liminf \int_\Omega g(x, w_n) \, v_n \, dx.$$

Hence

$$\int_\Omega g_+ \, v \, dx \leq \int_\Omega f v \, dx$$

and thus

$$\int_\Omega g_+ \, dx \leq \int_\Omega f \, dx.$$

When $\alpha < 0$, we similarly derive

$$\int_\Omega g_- \, dx \geq \int_\Omega f \, dx.$$

Both cases contradict (8). The proposition is thus proved.

Strongly nonlinear inequalities

3.3. The mappings of monotone type with respect to two Banach spaces can be adapted to the investigation of the existence of bounded solutions in variational inequalities.

Let Ω be a bounded domain in $\mathbb{R}^N$ for which the Sobolev inbedding theorem holds. We shall consider the equations of the form

$$Au + g(., u) = f \text{ in } \Omega, \tag{1}$$

under the hypotheses $(III_1)-(III_3)$ upon the strongly nonlinear term g. Here A is a linear differential operator of second order

$$Au(x) = - \sum_{i,j=1}^{N} D_i[a_{ij}(x) D_j u(x)].$$

The boundary conditions are determined by a Hilbert space V which satisfies $H_0^1(\Omega) \subseteq V \subseteq H^1(\Omega)$. Let $W = V \cap L^\infty(\Omega)$. Unlike in the cases studied earlier, W is not a reflexive space.

Suppose that the coefficients a_{ij} are in $L^\infty(\Omega)$ and that they satisfy the condition of uniform ellipticity

$$\sum_{i,j=1}^{N} a_{ij}(x) \xi_i \xi_j \geq c_0 |\xi|^2 \quad (\forall)x \in \Omega, \quad \xi \in \mathbb{R}^N$$

for some constant $c_0 > 0$.

As previously, let $T_1 : V \to V^*$ and $T_2 : V_1 \to W^*$ be the operators generated by

$$a(u, v) = \sum_{i,j=1}^{N} \int_\Omega a_{ij} D_i u D_j v \, dx \quad \text{and} \quad b(u, v) = \int_\Omega g(., u) v \, dx,$$

where $V_1 = \{u \in V \mid g(., u) \in L^1(\Omega), g(., u)u \in L^1(\Omega)\}$. As above, we have now also that $W \subset V_1 \subset V$ and $T = i^* T_1 + T_2$.

LEMMA *The operator* $T : V_1 \to W^*$ *satisfies the following conditions:*

(A_1) *T is continuous from the finite-dimensional subspaces of W to the weak* topology on W^*;*

(A_2) *If $\{u_n\}$ is a sequence in W with $u_n \rightharpoonup u$ in V and $(Tu_n, u_n) \leq const.$ for all $n \in \mathbb{N}$, then $u \in D(T)$ and $Tu_n \to Tu$ in W^* (endowed with the weak* topology) and there exists a number $\lambda(u)$ such that*

$$\lambda(u) \leq \liminf (Tu_n, u_n).$$

Proof: Condition (A_1) is easily verified. For every $\{u_n\} \subset W$ such that $u_n \to u$ in V and $(Tu_n, u_n) \leqslant$ const., we get

$$\int_\Omega g(\cdot, u_n) u_n \, dx \leqslant \text{const.} \qquad (\forall) \, n \in \mathbb{N}.$$

Passing, if necessary, to a subsequence, we may assume that $u_n \to u$ in $L^2(\Omega)$ and $u_n(x) \to u(x)$ a.a. $x \in \Omega$. Using the same arguments as in the proof of Lemma 3.2, we deduce that $u \in V_1$ and $T_2 u_n \to T_2 u$ in W^*. On the other hand, the boundedness and linearity of T_1 imply that $T_1 u_n \to T_1 u$ in V^* and thus $T_1 u_n \to T_1 u$ in W^*. Hence, $Tu_n \to Tu$ in W^*.

In addition, by the monotonicity of T_1 and by Fatou's lemma,

$$\lambda(u) = a(u, u) + b(u, u) \leqslant \lim \inf \left[a(u_n, u_n) + b(u_n, u_n) \right].$$

The lemma is proved.

Now, we give a formulation of the above problem in operator terms. Let V be a reflexive Banach space with dual V^* and let W be a normed space which is continuously imbedded in V. Then $V^* \subset W^*$. Let T be a mapping with the domain $D(T)$ such that $W \subseteq D(T) \subseteq V$ and the range contained in W^*, and conditions $(A_1) - (A_2)$ hold.

The number $\lambda(u)$ replaces (Tu, u) for a general mapping $T : V_1 \to W^*$, which for $Tu \in W^*$ cannot be defined for all $u \in D(T)$.

THEOREM *Let* $T : D(T) \to W^*$ *be a mapping satisfying the hypotheses* $(A_1)-(A_2)$ *and let K be a closed convex set of V containing the origin. If for a given $f \in V^*$ and some $r > 0$, the coercitivity condition*

$$(Tw - f, w) > 0, \qquad (\forall) w \in K \cap W \text{ with } \|w\|_V = r,$$

holds, then the variational inequality

$$(Tu, w) - \lambda(u) - (f, w - u) \geqslant 0 \qquad (\forall) \, w \in K \cap W, \qquad (2)$$

admits a solution $u \in K \cap D(T)$ with $\|u\|_V < r$.

Proof: We follow the same procedure as in the proof of Proposition III.2.9. Let Λ be the set of all finite-dimensional subspaces F of W, provided with the norms $\| \cdot \|_F = \| \cdot \|_V$. For each $F \in \Lambda$, let $j_F : F \to W$ be the inclusion and let $j_F^* : W^* \to F^*$ be its adjoint map. As a consequence of (A_1), the mapping $T_F = =j_F^* T j_F : F \to F^*$ is continuous. Applying Theorem III.2.8 to the operator T_F and the sets $C = K \cap B(0, r)$ and $G = K \times \{0\}$, we see that there exists a $u_F \in F \cap C$ such that

$$(Tu_F - f, w - u_F) \geqslant 0, \qquad (\forall) \, w \in K \cap W. \qquad (3)$$

For each F' we put $U_{F'} = \{u_F \mid F \in \Lambda, \ F \supseteq F'\}$ and $u \in \bigcap \{\tilde{U}_{F'} \mid F' \in \Lambda\}$, where $\tilde{U}_{F'}$ is the weak closure of $U_{F'}$ in V. Then $u \in C$ and for fixed $F_0 \in \Lambda$, there exists a sequence $\{u_n\} \subset U_{F_0}$ with $u_n \to u$ in V. Hence

$$(Tu_n - f, w - u_n) \geqslant 0 \text{ for all } n \in \mathbb{N}. \qquad (4)$$

Taking $w = 0$, we obtain $(Tu_n, u_n) \leqslant (f, u_n) \leqslant$ const. By (A_2), we have $u \in D(T)$, $Tu_n \to Tu$ in W^* and thus $\lambda(u) \leqslant \lim \inf (Tu_n, u_n)$. Passing to the limit in (3), we may conclude that

$$(Tu, w) + \lambda(u) - (f, w - u) \geqslant 0 \qquad (\forall) \, w \in K \cap W.$$

The theorem is thus proved.

In particular, (2) is solvable for all $f \in V^*$ if T is coercive on K, i.e.,

$$\frac{(Tw, w)}{\|w\|_V} \to \infty \;\; \text{as} \;\; \|w\|_V \to \infty \qquad w \in K \cap W.$$

REMARK Let $z \in W$ be an element with the property that $w + z \in K \cap W$ for all $w \in K \cap W$. Choose $F_0 \in \Lambda$ such that $z \in F_0$ and set $w_n = u_n + z$ in (4). Then we have

$$(Tu - f, z) \geqslant 0.$$

If, for example, K is a cone with vertex at the origin and X is a linear manifold contained in $K \cap W$, then $(Tu - f, z) = 0$ for all $z \in X$.

Let us revert to the initial problem. An immediate consequence of the lemma is the following

PROPOSITION *Let $K \subset V$ be a closed convex set containing the origin and suppose that for a given $f \in V^*$ and some $r > 0$*

$$a(w, w) + b(w, w) - (f, w) > 0 \qquad (\forall) \, w \in K \cap L^\infty(\Omega) \;\; \text{with} \;\; \|w\|_V = r.$$

Then there exists a $u \in K \cap V_1$ with $\|u\|_V \leqslant r$ such that

$$a(u, w - u) + b(u, w - u) \geqslant (f, w - u) \qquad (\forall) \, w \in K \cap L^\infty(\Omega). \qquad (5)$$

In particular, inequality (5) is solvable for all $f \in V^*$ provided that

$$\frac{a(w, w) + b(w, w)}{\|w\|_V} \to \infty \;\; \text{as} \;\; \|w\|_V \to \infty, \, w \in K \cap L^\infty(\Omega).$$

When K is a cone with vertex at the origin and $C_0^\infty(\Omega)$ is contained in K, the above remark implies that a solution of variational inequality (5) is a solution in the distribution sense of problem (1) with some boundedness constraints.

4. Other methods

In the case of Dirichlet boundary-value problems for elliptic equations of second order we can prove the existence of variational solutions in the absence of the coercitivity condition.

Odd perturbations

4.1. We shall apply a homotopy argument in the discussion of solution of the Dirichlet problem for the elliptic strongly nonlinear equations

$$\begin{cases} Au(x) + g(x, u(x)) = f(x) \text{ in } \quad \Omega, \\ \qquad\qquad u(x) = 0 \quad \text{ on } \partial\Omega. \end{cases}$$

Here Ω is a bounded domain in $\mathbb{R}^N$ and A is a linear differential operator of second order

$$Au(x) = - \sum_{i,j=1}^{N} [a_{ij}(x) \, D^j u(x)] + \sum_{i=1}^{N} b_i(x) \, D^i u(x) + c(x) \, u(x).$$

We consider the following set of hypotheses (IV):
(IV$_1$) a_{ij}, b_i and c are $L^\infty(\Omega)$-functions;
(IV$_2$) The uniform ellipticity condition

$$\sum_{i,j=1}^{N} a_{ij}(x) \, \xi_i \xi_j \geqslant c \, |\xi|^2 \text{ a.e. on } \Omega \qquad (\forall) \xi \in \mathbb{R}^N,$$

holds for some constant $c > 0$;
(IV$_3$) The function $g : \Omega \times \mathbb{R} \mapsto \mathbb{R}$ satisfies the Caratheodory conditions;
(IV$_4$) $g(.,t)$ is essentially bounded for t bounded;
(IV$_5$) $g(,.)$ is odd.
With the notation used in Section 3.2, a variational solution of the above problem satisfies

(DP$_0$) $u \in V_1$, $a(u, v \, + \, b(u, v) = (f, v)$ $\qquad (\forall) w \in W,$

where $V = H_0^1(\Omega)$, $V_1 = \{v \in V \mid g(., v) \in L^1(\Omega), \ g(., v)v \in L^1(\Omega)\}$ and $W = H_0^s(\Omega)$ with $s > 1 + \left[\dfrac{N}{2}\right]$. Here f is an element in $V^* = H^{-1}(\Omega)$.

In this case we can prove the existence of a variational solution under a condition of Leray-Schauder type, weaker than coerciveness.

We need the following convergence criterion which extends the technique of Lemma 3.2.

PROPOSITION *Let* $\{u_n\}$ *be a sequence of strongly measurable functions from* Ω *into a Banach space* X *and let* $\{g_n\}$ *be a sequence of functions from* $\Omega \times X$ *into* $\mathbb{R}$ *such that*
(i) $\{g_n\}$ *is uniformly bounded on* $\Omega \times B$ *for any bounded set* $B \subset X$;
(ii) $|g_n(., u_n(.))|$ *is measurable and there is a constant* $c > 0$ *such that*

$$\int_\Omega |g_n(x, u_n(x))| \, \|u_n(x)\|_X \, dx \leqslant c \, ;$$

(iii) $g_n(x, u_n(x)) \to v(x)$ *a.e. in* Ω.
Then $v \in L^1(\Omega)$ *and* $g_n(., u_n(.)) \to v$ *in* $L^1(\Omega)$.

Proof: Put $v_n = g_n(\cdot, u_n(\cdot))$. On the set where $\|u_n(x)\|_X \geq 1$ we have $|v_n(x)| \leq |v_n| \|u_n(x)\|_X$. On its complement $\{u_n(x)\}$ is uniformly bounded in X. Thus $\{v_n\}$ is bounded in $L^1(\Omega)$ and, by Fatou's lemma, $v \in L^1(\Omega)$.

For each $\varepsilon > 0$, by Egorov's theorem there exists a measurable subset Ω_ε of Ω such that $\mu(\Omega_\varepsilon) < \varepsilon$ and $v_n \to v$ uniformly on $\Omega - \Omega_\varepsilon$. For a given $\delta > 0$ it is sufficient to prove that the integral of v_n over Ω_ε is less than δ for all $n \in \mathbb{N}$.

By (i), we can find an $M > 0$ with the property that if $|g_n(x, t)| > M$ for some n and some $x \in \Omega$, then $\|t\|_X > \dfrac{2c}{\delta}$. Let Ω' be the set of points in Ω_ε with $|g_n(x, u_n(x))| > M$ and let Ω'' be its complement relative to Ω. If $\varepsilon < \dfrac{\delta}{2M}$, then for any $n \in \mathbb{N}$ we can deduce that

$$\int_{\Omega''} |v_n(x)| \, dx \leq M\varepsilon < \frac{\delta}{2}$$

and

$$\int_{\Omega'} |v_n(x)| \, dx \leq \frac{\delta}{2c} \int_{\Omega} |v_n(x)| \, \|u_n(x)\|_X \, dx < \frac{\delta}{2}.$$

Hence

$$\int_{\Omega_\varepsilon} |v_n(x)| \, dx \leq \delta \quad (\forall) \, n \in \mathbb{N}$$

and Vitali's theorem assures that $v_n \to v$ in $L^1(\Omega)$, as claimed.

Now we consider the Dirichlet problem

$$(DP_t) \quad u_t \in V_1, \quad a(u_t, w) + b(u_t, w) = (1 - t)(f, w) \quad (\forall) \, w \in W$$

for $t \in [0, 1]$.

Let us prove that the coercivity assumption

$$\frac{1}{\|w\|_{1,2}} \{a(w, w) + b(w, w)\} \to \infty \quad \text{as} \quad w \in V_1 \, \|w\|_{1,2} \to \infty, \tag{1}$$

implies the uniform boundedness in the V-norm of the solutions $\{u_t\}$ of problems (DP_t) for $t \in [0, 1]$. Indeed by setting $w = u_t$ in (DP_t), we obtain that

$$a(u_t, u_t) + b(u_t, u_t) = (1 - t)(f, u_t) \leq \|f\|_{V*} \|u_t\|_V$$

and condition (1) implies that $\|u_t\|_V < M, 0 \leq t \leq 1$, for some constant $M > 0$. This remark leads to the

THEOREM *Under the hypotheses* (IV), *if there exists a constant $M > 0$ such that $\|u_t\|_V < M$ for any solution of (DP_t) with $0 \leq t \leq 1$, then problem (DP_0) admits a variational solution.*

Proof: For each $n \in \mathbf{N}$ take the truncated function

$$g_n(x, t) = \begin{cases} -n & \text{if} \quad g(x, t) \leqslant -n, \\ g(x, t) & \text{if} \quad |g(x, t)| \leqslant n, \\ n & \text{if} \quad g(x, t) \geqslant n, \end{cases}$$

and set

$$b_n(u, w) = \int_\Omega g_n(x, u(x))\, w(x)\, \mathrm{d}x, \quad u, w \in V.$$

Consider the associated problem

$(DP_t)_n \quad u_{tn} \in V. \quad a(u_{tn}, w) + b_n(u_{tn}, w) = (1 - t)(f, w) \quad (\forall)\, w \in V.$

We shall prove first that there exists an n_0 such that $\|u_{tn}\|_V < M$ for all $n \geqslant n_0$ and $t \in [0, 1]$.

Suppose that, on the contrary, for each n we can choose elements $v_n \in V$ with $\|v_n\|_V = M$ and $t_n \in [0, 1]$ such that

$$a(v_n, w) + b_n(v_n, w) = (1 - t_n)(f, w) \quad (\forall)\, w \in V. \tag{2}$$

We may assume that $v_n \rightharpoonup v$ in V and $t_n \to t_0$ as $n \to \infty$. Setting $w = v_n$ in (2) we deduce that

$$b_n(v_n, v_n) = \int_\Omega g_n(\cdot, v_n) v_n \, \mathrm{d}x \leqslant \text{const.} \quad (\forall)\, n \in \mathbf{N}.$$

The proposition implies that $v \in V_1$ and $g_n(\cdot, v_n) \to g(\cdot, v)$ holds in $L^1(\Omega)$. Letting $n \to \infty$ in (2) we get

$$a(v, w) + b(v, w) = (1 - t_0)(f, w) \quad (\forall)\, w \in W,$$

i.e., v is a solution of problem (DP_{t_0}).

On the other hand, for fixed $k > 0$ let v^k be the truncation of v and take $w = v_n - v^k$ in (2). We get

$$a(v_n, v_n - v^k) = -b_n(v_n, v_n - v^k) + (1 - t_n)(f, v_n - v^k).$$

Thus

$$\limsup a(v_n, v_n - v) = \limsup a(v_n, v_n - v^k) + a(v, v^k - v) \leqslant \tag{3}$$
$$\leqslant -b(v, v - v^k) + (1 - t)(f, v - v^k) + a(v, v^k - v).$$

Since the left side of (3) is independent of k, passing to limit as $k \to \infty$, we obtain

$$\limsup a(v_n - v, v_n - v) = \limsup a(v_n, v_n - v) \leqslant 0.$$

By the condition of uniform ellipticity we can deduce Gårding's inequality

$$a(v_n - v, v_n - v) \geqslant c_1 \|v_n - v\|_V^2 - c_2 \|v_n - v\|_2^2 \text{ with } c_1 > 0.$$

As $v_n \rightharpoonup v$ in V, it follows that $v_n \to v$ in $L^2(\Omega)$ and by the above inequality, $v_n \to v$ in V; therefore $\|v\|_V = M$. This contradicts the *a priori* estimate of the solution for (DP_t).

Moreover, the last argument implies that the operator $T_1 : V \mapsto V^*$ defined by

$$(T_1 u, v) = a(u, v) \quad (\forall) \, v \in V$$

is of type (S_+).

Next, for any fixed $n > n_0$, define the bounded operator $T_{2n} : V_1 \mapsto V^*$ by

$$(T_{2n} u, v) = b_n(u, v) \quad (\forall) \, v \in V.$$

By Theorem IV.1.1, the Nemitskyi operator T_{2n} is continuous from $L^2(\Omega)$ into $L^2(\Omega)$. Moreover, T_{2n} is compact because $V = H_0^1(\Omega)$ is compactly imbedded in $L^2(\Omega)$.

The problem $(DP_t)_n$ can be written as

$$(T_1 + T_{2n})u = (1 - t)f.$$

Since the mapping $A_t u = (T_1 + T_{2n})u - (1 - t)f$ verifies in $B(0, M)$ the conditions of Theorem V.1.1, there exists a solution $u_{0n} \in V$ with $\|u_{0n}\|_V < M$ of the equation

$$(DP_0)_n \quad (T_1 + T_{2n})u_{0n} = f.$$

Finally, the existence of a solution for problem (DP_0) follows by a limit argument as $n \to \infty$ similar to that used in the first part of the proof.

A problem with upper and lower solutions

4.2. Let us describe a criterion for the solvability of a simple variational problem involving upper and lower solutions.

Let Ω be a bounded domain in $\mathbb{R}^N$ for which the Sobolev imbedding theorem holds. Let $f \in H^{-1}(\Omega)$ and $h \in H^{\frac{1}{2}}(\partial \Omega)$ be given. We shall be concerned with the nonlinear Dirichlet problem

$$\begin{cases} Au + g(.,u) = f & \text{in} \quad \Omega, \\ u = h & \text{on} \quad \partial \Omega, \end{cases} \tag{1}$$

under the hypotheses $(IV_1) - (IV_3)$, where we shall assume the existence of a lower solution φ and an upper solution ψ of (1) with $\varphi \leqslant \psi$ in Ω.

A function $u \in H^1(\Omega)$ is said to be a *variational solution* of problem (1) provided that $u|_{\partial \Omega} = h$ in $H^{\frac{1}{2}}(\partial \Omega)$, $g(.,u) \in L^2(\Omega)$ and

$$a(u, v) + \int_\Omega g(., u)v \, dx = (f, v) \quad (\forall) \, v \in H_0^1(\Omega).$$

A natural extension of the classical concept of upper solution is the *variational upper solution* of problem (1), which is defined by

$\psi \in H^1(\Omega)$, $\psi \mid_{\partial\Omega} \geqslant h$ in $H^{\frac{1}{2}}(\partial\Omega)$, $g(.,\psi) \in L^2(\Omega)$ and

$$a(\psi, v) + \int_\Omega g(.,\psi) \, v \, dx \leqslant (f, v), \quad (\forall)v \in H_0^1(\Omega) \text{ with } v \geqslant 0 \text{ a.e. in } \Omega.$$

Similarly, for a *variational lower solution* $\varphi \in H^1(\Omega)$ the inequality signs above are reversed.

THEOREM *Assume that φ and ψ are variational lower and upper solutions of problem (1), respectively, with $\varphi \leqslant \psi$ a.e. in Ω. If*

$$\int \sup_{\varphi(x) \leqslant t \leqslant \psi(x)} |g(x, t)|^2 \, dx < \infty, \tag{2}$$

then problem (1) admits a variational solution with $\varphi \leqslant u \leqslant \psi$ in Ω.

Proof: Let $\hat{h} \in H^1(\Omega)$ be an extension of h to $\bar{\Omega}$, (i.e., $\hat{h}\mid_{\partial\Omega} = h$) such that $\varphi \leqslant \hat{h} \leqslant \psi$ in $\bar{\Omega}$. Replacing u by $u - \hat{h}$, we have $\hat{h} = 0$ and $\varphi \leqslant 0 \leqslant \psi$ in Ω. We shall now search for a variational solution of problem (1) in $H_0^1(\Omega)$.

As a consequence of the uniform ellipticity of A, we have

$$a(v, v) \geqslant c_0\|v\|_{1,2}^2 - c_1\|v\|_2^2 \quad (\forall) \ v \in H_0^1(\Omega)$$

with some constant $c_1 \geqslant 0$. Moreover, $a(u, v)$ defines a linear bounded operator of type (S_+).

Introduce the modified function

$$\hat{g}(x, t) = \begin{cases} g(x, \varphi(x)) & \text{if} \quad t \leqslant \varphi(x) \\ g(x, t) & \text{if} \quad \varphi(x) \leqslant t \leqslant \psi(x) \\ g(x, \psi(x)) & \text{if} \quad t \geqslant \psi(x). \end{cases}$$

By assumption (2), there exists a constant $c_2 > 0$ such that

$$\left| \int_\Omega \hat{g}(.,u) \, v \, dx \right| \leqslant c_2\|v\|_2, \quad (\forall)u, \ v \in L^2(\Omega).$$

This function defines a compact Nemitskyi operator on $H_0^1(\Omega)$.

Denote by $v^+ = \max\{v, 0\}$ and $v^- = \min\{v, 0\}$ the positive and negative part of any function v. For $C > c_1$, let us consider the perturbed problem

$$b(u, v) \equiv a(u, v) + \int_\Omega \hat{g}(.,u)v \, dx + C \int_\Omega [(u - \varphi)^- + (u - \psi)^+]v \, dx =$$

$$= (f, v) \quad (\forall) \ v \in H_0^1(\Omega), \tag{3}$$

and prove that its solution $u \in H_0^1(\Omega)$ turns out to be a variational solution for problem (1).

To show the existence of a solution of problem (3), we remark that for fixed $x \in \Omega$, we have

$$[(t - \varphi(x))^- + (t - \psi(x))^+]t \geq |t|^2 - \max \{|\varphi(x)|, |\psi(x)|\}|t| \quad (\forall) \, t \in \mathbb{R},$$

whence

$$\int_\Omega [(v - \varphi)^- + (v - \psi)^+] \, v \, dx \geq \|v\|_2^2 - c_3 \|v\|_2$$

with some constant $c_3 > 0$. Hence this term defines a bounded operator also of type (S_+). We conclude that the sum of all the above operators is of type (S_+) and coercive,

$$b(v, v) \geq c_0 \|v\|_{1,2}^2 - (c_2 + Cc_3) \|v\|_2 \quad (\forall) \, v \in H_0^1(\Omega).$$

Thus, in virtue of Corollary V.1.1, there exists a $u \in H_0^1(\Omega)$ such that

$$b(u, v) = (f, v) \quad (\forall) \, v \in H_0^1(\Omega).$$

In the remaining part of this section we are going to show that u is a variational solution of problem (1).

Note that $(u - \varphi)^-$ lies in $H_0^1(\Omega)$. Setting $v = (u - \varphi)^-$ in (3), we obtain

$$a(u, (u - \varphi)^-) + \int_\Omega \hat{g}(., u) (u - \varphi)^- dx + C \|(u - \varphi)^-\|_2^2 = (f, (u - \varphi)^-),$$
$$(4)$$

because $\varphi \leq \psi$ implies that

$$\int_\Omega (u - \psi)^+ (u - \varphi)^- dx = 0.$$

Since φ is a variational lower solution, we have

$$a(\varphi, (u - \varphi)^-) + \int_\Omega g(., \varphi)(u - \varphi)^- dx \geq (f, (u - \varphi)^-)$$

and by subtraction from (4) we get

$$a(u - \varphi, (u - \varphi)^-) + \int_\Omega (\hat{g}(., u) - g(., u))(u - \varphi)^- dx + C \|(u - \varphi)^-\|_2^2 \leq 0.$$

Recalling the definition of $\hat{g}$ we readily see that

$$\int_\Omega (\hat{g}(., u) - g(., \varphi)) (u - \varphi)^- dx = 0.$$

Observing that $a(u - \varphi, (u - \varphi)^-) = a((u - \varphi)^-, (u - \varphi)^-)$ and using (IV$_2$), we obtain that $c_0 \|(u - \varphi)^-\|_{1,2}^2 \leq 0$. This implies that

$$(u - \varphi)^- = 0 \qquad (5)$$

and thus $u \geq \varphi$ a.e. in Ω.

Similarly, one can prove that

$$(u - \psi)^+ = 0 \tag{6}$$

and thus $u \leqslant \psi$ a.e. in Ω.

By (5) and (6), $\hat{g}(.,u) = g(.,u)$ and equation (3) is reduced to

$$a(u, v) + \int_\Omega g(.,u) \, v \, dx = (f, v) \quad (\forall) \, v \in H_0^1(\Omega),$$

i.e., the function $u \in H_0^1(\Omega)$ is a variational solution of problem (1).

The above treatment can be applied to the solution of the Neumann problem (Hess [16]).

Landesman-Lazer theorem

4.3. Let us discuss now another necessary and sufficient condition for the solvability of linear divergence equations with nonlinear perturbations.

Let Ω be a bounded domain in $\mathbb{R}^N$ for which the Sobolev imbedding theorem holds and let

$$Au = \sum_{|\alpha|, |\beta| \leqslant 1} (-1)^\alpha D^\alpha(a_{\alpha\beta}(x) \, D^\beta u)$$

be a uniformly elliptic, formally selfadjoint linear operator whose coefficients $a_{\alpha\beta} = a_{\beta\alpha} \in L^\infty(\Omega)$. Let $g \in \mathbb{R} \mapsto \mathbb{R}$ be a bounded continuous function such that

$$g(-\infty) = \lim_{t \to -\infty} g(t) \quad \text{and} \quad g(\infty) = \lim_{t \to \infty} g(t)$$

are finite and non-vanishing, and suppose that

$$g(-\infty) < g(t) < g(\infty) \quad \text{for all } t \in \mathbb{R}. \tag{1}$$

Consider the Dirichlet problem

$$\begin{cases} Au + g(u) = f & \text{in} \quad \Omega, \\ u = 0 & \text{on} \quad \partial\Omega, \end{cases} \tag{2}$$

for a given function $f \in L^2(\Omega)$. Here $g(u)$ is the realization of g in $L^2(\Omega)$.

Associate to A the symmetric bounded bilinear form

$$a(u, w) = \sum_{\alpha, \beta \leqslant 1} \int_\Omega a_{\alpha\beta} D^\alpha u \, D^\beta w \, dx \quad (\forall) \, u, w \in H_0^1(\Omega),$$

and, as usually, call $u \in H_0^1(\Omega)$ a *variational solution* of problem (1) provided that

$$a(u, w) + \int_\Omega (g(u) - f) \, w \, dx = 0 \quad (\forall) \, w \in H_0^1(\Omega).$$

Under the assumptions made on A, it is well-known (Agmon [1, p. 102]) that the generalized null-space of $A - \mu I$ is finite-dimensional and that the Fredholm alternative holds. In particular, the linear equation $(A - \mu I)u = h$ admits a variational solution in $H_0^1(\Omega)$ for any $h \in L^2(\Omega)$ provided that $Au = \mu u$ has only the solution $u = 0$, i.e., μ is not an eigenvalue of A.

For the sake of simplicity, we assume that $\mu = 0$ is an eigenvalue of A and the corresponding eigenspace is spanned by the system $\{\varphi_1, \ldots, \varphi_k\}$, orthonormal in $L^2(\Omega)$. Then every solution $\Phi \in H_0^1(\Omega)$ of the equation

$$a(\Phi, v) = 0 \quad \text{for all} \quad v \in H_0^1(\Omega)$$

is of the form $\Phi = \sum\limits_{i=1}^{k} \lambda_i \varphi_i$ with $\lambda_i \in \mathbb{R}$, $1 \leqslant i \leqslant k$.

THEOREM *Problem* (1) *with* $f \in L^2(\Omega)$ *admits a variational solution if and only if*

$$g(-\infty) \int_{\Omega_+} |\varphi| \, dx - g(\infty) \int_{\Omega_-} |\varphi| \, dx < \int_{\Omega} f\varphi \, dx <$$

$$< g(\infty) \int_{\Omega_+} |\varphi| \, dx - g(-\infty) \int_{\Omega_-} |\varphi| \, dx, \tag{3}$$

where $\Omega_{\pm} = \{x \in \Omega \mid \varphi(x) \gtrless 0\}$ *and* $\varphi = \lambda_1 \varphi_1 + \ldots + \lambda_k \varphi_k$ *for all* $\lambda_i \in \mathbb{R}$ *with* $\lambda_1^2 + \ldots + \lambda_k^2 = 1$.

Proof: The "if" part follows from the orthogonality condition

$$\int_{\Omega} (f - g(u))\varphi \, dx = 0.$$

Thus

$$\int_{\Omega} f\varphi \, dx = \int_{\Omega} g(u) \, \varphi \, dx = \int_{\Omega_+} g(u) \, |\varphi| \, dx - \int_{\Omega_-} g(u) \, |\varphi| \, dx \leqslant$$

$$\leqslant g(\infty) \int_{\Omega_+} |\varphi| \, dx - g(-\infty) \int_{\Omega_-} |\varphi| \, dx.$$

Similarly, the first inequality follows easily by (1).

The "only if" part. Define the operators.

$$P_n = Au + \frac{1}{n} u \quad \text{and} \quad Qu = g(u).$$

We can easily verify that P_n is a bounded linear operator of type (S_+) for every $n \in \mathbb{N}$, while Q is compact and asymptotically zero. Since the eigenvalues of A in $H_0^1(\Omega)$ are discrete, there exists a small enough neighbourhood of $\mu = 0$ which does not contain other eigenvalue of A. Hence $P_n u = 0$ implies that $u = 0$ for a sufficiently large n. Then, for any $f \in L^2(\Omega)$, by Theorem V.1.3 there exists $u_n \in H_0^1(\Omega)$ such that

$$a(u_n, w) = \frac{1}{n} \int_{\Omega} u_n w \, dx + \int_{\Omega} (g(u_n) - f)w \, dx \qquad (\forall) \, w \in H_0^1(\Omega). \tag{4}$$

We claim that under condition (3), the sequence $\{u_n\}$ remains bounded in $H_0^1(\Omega)$. Suppose that on the contrary $\|u_n\| \to \infty$ as $n \to \infty$ and denote $v_n = \|u_n\|_{1,2}^{-1} u_n$. Then

$$a(v_n, w) - \frac{1}{n}\int_\Omega v_n w \, dx = \frac{1}{\|u_n\|_{1,2}}\int_\Omega (g(u_n) - f) w \, dx \qquad (\forall) \, w \in H_0^1(\Omega). \qquad (5)$$

By the reflexivity of $H_0^1(\Omega)$, $v_n \rightharpoonup v$ in $H_0^1(\Omega)$ and from (5) we have

$$a(v, w) = \lim a(v_n, w) = 0 \qquad (\forall) \, w \in H_0^1(\Omega),$$

because of the boundedness of g. Setting $w = v_n - v$ in (5), we get

$$a(v_n - v, v_n - v) = a(v_n, v_n - v) \to 0 \text{ as } n \to \infty.$$

By the Gårding inequality we have

$$a(v_n - v, v_n - v) \geqslant c_0 \|v_n - v\|_{1,2}^2 - c_1 \|v_n - v\|_2^2,$$

and by the compact imbedding of $H_0^1(\Omega)$ in $L^2(\Omega)$, we deduce that $v_n \to v$ in $H_0^1(\Omega)$. Hence, $\|v\|_{1,2} = 1$ and so $v = \lambda \varphi$ with $\lambda \neq 0$.

We substitute $w = v$ in (5) and observe that $a(v_n, v) = a(v, v_n) = 0$ for all n. Thus

$$\int_\Omega v_n v \, dx \to \int_\Omega v^2 \, dx > 0.$$

Therefore

$$\int_\Omega (g(u_n) - f) v \, dx < 0 \text{ for large } n,$$

and

$$\limsup \int_\Omega g(u_n) v \, dx \leqslant \int_\Omega f v \, dx.$$

Since $u_n = \|u_n\|_{1,2} v_n$ with $\|u_n\|_{1,2} \to \infty$ and since $v_n(x) \to v(x)$ for a.a $x \in \Omega$, by Lebesgue's dominated convergence theorem, we obtain for $\lambda > 0$ that at least on some subsequence

$$g(\infty)\int_{\Omega_+} |\varphi| \, dx - g(-\infty)\int_{\Omega_-} |\varphi| \, dx = \lim \int_\Omega g(u_n) \varphi \, dx \leqslant \int_\Omega f\varphi \, dx.$$

Similarly, for $\lambda < 0$ we obtain

$$g(-\infty)\int_{\Omega_+} |\varphi| \, dx - g(\infty)\int_{\Omega_-} |\varphi| \, dx = \lim \int_\Omega g(u_n) \varphi \, dx =$$

$$= \frac{1}{\lambda}\lim \int_\Omega g(u_n) v \, dx \geqslant \frac{1}{\lambda}\int_\Omega f v \, dx = \int_\Omega f\varphi \, dx.$$

As both results above contradict (3), the sequence $\{u_n\}$ remains bounded in $H_0^1(\Omega)$. We may assume, at least for a subsequence, that $u_n \rightharpoonup u$ in $H_0^1(\Omega)$. Passage to the limit in (4) yields

$$a(u, v) = \int_\Omega (g(u) - f)w \, dx \quad (\forall) \, w \in H_0^1(\Omega),$$

which proves the sufficiency of condition (3).

5. Boundary-value problems in unbounded domains

The problem of extension of the existence results concerning variational solutions for elliptic nonlinear equations to unbounded domains meets with some nontrivial difficulties. The main drawback lies in the fact that the compactness in the Sobolev imbedding theorem is no longer valid for general unbounded domains in $\mathbb{R}^N$. In order to circumvent this difficulty we shall use either an approximation of unbounded domains by a suitable increasing sequence of bounded domains or an elliptic super-regularization.

An approach with approximate domains

5.1. Let Ω be a domain of σ-finite measure in $\mathbb{R}^N$, which can be represented in the form

$$\Omega = \bigcup \{\Omega_n \mid n \in \mathbb{N}\},$$

where $\{\Omega_n\}$ is an increasing sequence of bounded measurable domains, with the property that there exists a $p \in (1, \infty)$ such that $W^{m, p}(\Omega_n)$ is compactly imbedded in $L^p(\Omega_n)$, for every $n \in \mathbb{N}$.

These assumptions are realized when the boundary $\partial\Omega$ is piecewise sufficiently smooth.

We consider, as previously, the divergence equation

$$\sum_{|\alpha| \leqslant m} (-1)^{|\alpha|} D^\alpha A_\alpha(\,.\,, \xi(u)) = f \quad \text{in } \Omega, \tag{1}$$

whose boundary conditions are determined by a closed space V such that $W_0^{m, p}(\Omega) \subseteq V \subseteq W^{m, p}(\Omega)$.

We impose the following variant of hypotheses (I):

(I$_1'$) Each $A_\alpha: \Omega \times \mathbb{R}^{Nm} \mapsto \mathbb{R}$ verifies the Caratheodory conditions and satisfies the growth condition

$$|A_\alpha(x, \xi)| \leqslant c|\xi|^{p-1} + g(x), \quad (\forall) \; \xi \in \mathbb{R}^{Nm}, \quad \text{for a.a. } x \in \Omega,$$

where $c > 0$ and $g \in L^{p'}(\Omega)$ with $\dfrac{1}{p} + \dfrac{1}{p'} = 1$;

Write $\xi = (\eta, \zeta)$ with $\eta = \{\xi_\alpha \mid |\alpha| < m\}$ and $\zeta = \{\xi_\alpha \mid |\alpha| = m\}$.

(I$_2'$) $\displaystyle\sum_{|\alpha| = m} [A_\alpha(x, \eta, \zeta) - A_\alpha(x, \eta, \zeta')] (\zeta_\alpha - \zeta'_\alpha) > 0$, for all $\zeta \neq \zeta'$, $\eta \in \mathbb{R}^{Nm-1}$ and a.a. $x \in \Omega$;

(I$_3'$) There exists a constant $c > 0$ such that

$$\sum_{|\alpha| = m} A_\alpha(x, \eta, \zeta) \, \zeta_\alpha \geqslant c|\zeta|^p$$

for all $(\eta, \zeta) \in \mathbb{R}^{Nm}$ and a.a. $x \in \Omega$.

Note that hypothesis (I$_2'$) is weaker than (I$_2$) since it involves only the highest order derivatives.

As in Section 2.1, by hypothesis (I$_1'$), the form

$$a(u, v) = \sum_{|\alpha| \leqslant m} \int_\Omega A_\alpha(\, . \, , \xi(u)) \, D^\alpha v \, dx$$

is well-defined on $W^{m, p}(\Omega) \times W^{m, p}(\Omega)$ and it determines a bounded continuous operator $T : V \mapsto V^*$ defined by $a(u, v) = (Tu, v)$ for all $v \in V$.

As usually, for given $f \in V^*$, a solution $u \in V$ of the equation

$$a(u, v) = (f, v) \quad (\forall) \; v \in V,$$

is called a *variational solution* of problem (1) with boundary conditions given by V. A variational solution of problem (1) is also a solution of operational equation $Tu = f$ in V.

Let χ_n be the characteristic function of Ω_n. Let $V_n = V \cap W^{m, p}(\Omega_n)$ for every $n \in \mathbb{N}$. We break up the form $a(u, v)$ on Ω_n into its constituent parts

$$a_{1, n}(u, v) = \sum_{|\alpha| = m} \int_\Omega A_\alpha(\, . \, , \chi_n\eta(u), \zeta(u)) \, D^\alpha v \, dx,$$

$$a_{2, n}(u, v) = \sum_{|\alpha| \leqslant m-1} \int_\Omega \chi_n A_\alpha(\, . \, , \eta(u), \zeta(u)) \, D^\alpha v \, dx$$

and define a bounded continuous operator $T_n: V_n \mapsto V_n^*$ by

$$(T_n u, v) = a_{1, n}(u, v) + a_{2, n}(u, v) \quad (\forall) \; u, v \in V_n.$$

Under hypotheses (I'), we show that T_n *is an operator of type* (M), i.e., if $u_j \rightharpoonup u$ in V_n, $T_n u_j \rightharpoonup h$ in V_n^* and

$$\lim \sup (T_n u_j - T_n u, u_j - u) \leqslant 0,$$

then $T_n u = h$. Clearly,

$$\lim \sup (T_n u_j - T_n u, u_j - u) = \lim \sup [a_{1,n}(u_j, u_j - u) - a_{1,n}(u, u_j - u)] +$$

$$+ \lim \sup [a_{2,n}(u_j, u_j - u) - a_{2,n}(u, u_j - u)] \leqslant 0.$$

By the Sobolev imbedding theorem, $u_j \to u$ in $W^{m-1,p}(\Omega_n)$ and thus

$$a_{2,n}(u_j, u_j - u) - a_{2,n}(u, u_j - u) \to 0 \quad \text{as} \quad j \to \infty.$$

Then,

$$\lim \sup [a_{1,n}(u_j, u_j - u) - a_{1,n}(u, u_j - u)] \leqslant 0.$$

Since $\chi_n \eta(u_j) \to \chi_n \eta(u)$, this inquality becomes

$$\lim \sup \int_\Omega \sum_{|\alpha|=m} [A_\alpha(., \chi_n \eta(u), \zeta(u_j)) - A_\alpha(., \chi_n \eta(u), \zeta(u))] \, D^\alpha(u_j - u) \, dx \leqslant 0.$$

Hypothesis (I_2') asserts that the above integrand is strictly monotone in $\zeta(u)$. By virtue of inequality (I_3') and by Proposition IV.1.1, we deduce that $D^\alpha u_j \to D^\alpha u$ in $L^p(\Omega)$ also for $|\alpha| = m$. Therefore, $u_j \to u$ in V_n, $T_n u_j \to T_n u$ in V_n^* and hence $T_n u = h$.

In particular, T_n is of type (S_+).

The coerciveness of T_n is compensated by the addition of the duality map $J \colon W^{m-1,p}(\Omega) \mapsto (W^{m-1,p}(\Omega))^*$. For $\varepsilon > 0$ there exists, by Corollary V.1.1, an element $u \in V_n$ such that

$$\varepsilon (J u_n, v) + (T_n u_n, v) = (f, v) \qquad (\forall) \; n \in \mathbb{N}. \tag{2}$$

The coerciveness of $\varepsilon J + T$ implies the boundedness of $\{u_n\}$. Passing to a subsequence $\{u_k\} \subset \{u_n\}$, we can assume that $u_k \rightharpoonup u$ in V.

Set $v = u_k - u$ in (2). Since $\varepsilon J + T_n$ is of type (S_+), we deduce that $u_k \to u$ in V_n and thus $D^\alpha u_k \to D^\alpha u$ a.e. in Ω_n, with $|\alpha| \leqslant m$. When $n \to \infty$, we choose the diagonal subsequence and then

$$D^\alpha u_k \to D^\alpha u \quad \text{a.e. in } \Omega, \quad |\alpha| \leqslant m.$$

By the continuity of $A_\alpha(x, .)$,

$$A_\alpha(., \eta(u_k), \zeta(u_k)) \to A_\alpha(., \eta(u), \zeta(u)) \quad \text{a.e. in } \Omega \quad (\forall) \; |\alpha| \leqslant m.$$

We can pass now to limit in (2) and infer that $u = u_\varepsilon$ is a solution of the equation

$$\varepsilon (J u, v) + (T u, v) = (f, v) \quad (\forall) \; v \in V. \tag{3}$$

Let $\{u_j\}$ be the sequence with $u_j = u_{\varepsilon_j}$ and $\varepsilon_j \to 0$. The coerciveness of $\varepsilon_j J + T$ implies that $\|u_j\|_V \leqslant$ const. By the reflexivity of V and the boundedness of T we may assume that

$$u_j \to u \text{ in } V \text{ and } Tu_j \to g \text{ in } V^*.$$

Since

$$\varepsilon_j \|u_j\|_V^2 + (Tu_j, u_j) = (f, u_j),$$

it follows that $\varepsilon_j \|u_j\|_V^2 \leqslant$ const. and thus $\varepsilon_j u_j \to 0$ in V. Set $u = u_j$ in (3) and let $j \to \infty$; we have $(g, v) = (f, v)$ for all $v \in V$, whence $g = f$. We may conclude now that $Tu = f$, because the weak limit and pointwise limit coincide in $L^p(\Omega)$ almost everywhere.

Elliptic super-regularization

5.2. The second approach used in the existence theory of variational elliptic problems is a regularization method of singular perturbation type. This makes use of the compact imbedding property for separable Banach spaces and of a suitable truncation for the elements of a Sobolev space.

THEOREM *Let X be a separable reflexive Banach space and let S be a countable subset of X. Then there exists a separable Hilbert space H and a compact one-to-one linear operator $Q : H \mapsto X$ such that $S \subset Q(H)$ and $Q(H)$ is dense in X.*

Proof: Let $\|.\|$ be the norm in X. Without loss of generality, we may assume that the elements $\{e_1, \ldots, e_n, \ldots\}$ of S are normalized and linearly independent in X. Let H_0 be the closed subspace of X spanned by S. Each element in H_0 can be represented as a linear combination of the $\{e_j\}$ in which only a finite number of coefficients are non-vanishing. Define in H_0 the inner product

$$\langle u, v \rangle = \frac{\pi^2}{6} \sum_{k=1}^{\infty} k^2 a_k b_k \text{ for all } u = \sum_{k=1}^{\infty} a_k e_k \text{ and } v = \sum_{k=1}^{\infty} b_k e_k.$$

The corresponding norm $|u| = \langle u, u \rangle^{\frac{1}{2}}$ satisfies the inequality

$$\|u\| \leqslant \sum_{k=1}^{\infty} |a_k| \leqslant \left(\sum_{k=1}^{\infty} k^{-2} \right)^{\frac{1}{2}} \left(\sum_{k=1}^{\infty} |ka_k|^2 \right)^{\frac{1}{2}} = |u| \quad (\forall) \ u \in H_0.$$

Let H_1 be the completion of H_0 with respect to this inner product. Then H_1 is a separable Hilbert space and the natural injection $Q_1 : H_0 \mapsto X$ can be extended by continuity to a bounded linear map of H_1 into X such that $S \subset Q_1(H_1)$. Let $N(Q_1)$ be the null-space of Q_1 and set $H_2 = H_1/N(Q_1)$. Then $Q_1 : H_2 \mapsto X$ is one-to-one and $\|Q_1\|_{\mathcal{L}(H_2, X)} \leqslant 1$. Denote $Q_1^{-1}(S)$ by S_1.

To complete the proof, it is sufficient to show the existence of a separable Hilbert space H and a one-to-one compact linear map $Q_2: H \mapsto H_2$ such that $S_1 \subset Q_2(H)$. In fact, the composite $Q = Q_1 Q_2$ is a one-to-one compact linear map of H into X. Moreover $S \subset Q(H)$ and $Q(H)$ is dense in X.

Let $\{h_i\}$ be the linearly independent elements spanning S_1. Without loss of generality, we may assume, by multiplication with suitable constants, that $\|h_i\| \leqslant \dfrac{1}{2^i}$ for all $i \in \mathbb{N}$. Let H be the same Hilbert space H_2 in which we choose an orthonormal basis $\{g_i \mid u \in \mathbb{N}\}$. We define the linear map $Q_2: H \mapsto H_2$ by setting $Q_2 g_i = h_i$. This map is compact as the strong limit of its restrictions to finite-dimensional subspaces spanned by $\{g_i \mid 1 \leqslant i \leqslant n\}$, (see Theorem II.1.1). Since the elements of S_1 are linearly independent, the map Q_2 is one-to-one and $S_1 \subset Q_2(H)$. The theorem is proved.

We were interested to exhibit the general conditions under which H and Q can exist, and not in an explicit construction of them.

In the case of unbounded domains a special form of truncation is required, which differs from the previous one and allows some simplification of the limiting process.

Let k be a positive integer and define the function $g_k : \mathbb{R} \mapsto \mathbb{R}$ with $g_k(-s) = -g_k(s)$ as follows

$$
g_k(s) = \begin{cases}
s & \text{for } 0 \leqslant s \leqslant k-1, \\[2mm]
(k-1) + \dfrac{2}{\pi}\sin\dfrac{\pi}{2}(s-k+1), & \text{for } k-1 \leqslant s \leqslant k, \\[2mm]
k-1+\dfrac{2}{\pi}, & \text{for } s \geqslant k.
\end{cases}
$$

We can readily see that g_k and g_k' are both continuous and bounded with $|g_k(s)| \leqslant |s| \leqslant k$ and $|g_k'(s)| \leqslant 1$. Moreover g_k'' exists almost everywhere and $|g_k''(s)| \leqslant \dfrac{\pi}{2}$.

The truncation at the level k of a real function u defined on a general domain $\Omega \subset \mathbb{R}^N$ is given by

$$
g_k(u) = g_k(u(x)) \quad \text{for } x \in \Omega.
$$

LEMMA *If $u \in W_0^{1,p}(\Omega)$ with $1 < p < \infty$, then $g_k(u) \in W_0^{1,p}(\Omega)$ and*

$$
D_i(g_k(u)) = g_k'(u)\, D_i u
$$

in the sense of distributions. Moreover $g_k(u) \to u$ in $W_0^{1,p}(\Omega)$ as $k \to \infty$.

Proof: Let $u \in W_0^{1, p}(\Omega)$ and let $\{u_n\}$ be a sequence in $C_0^\infty(\Omega)$ such that $u_n \to u$ in $W_0^{1, p}(\Omega)$. For each $n \in \mathbb{N}$, the derivative $D_i(g_k(u_n)) = g_k'(u_n) D_i u_n$ exists in the classical sense. As $g_k(u_n) = 0$ outside the support of u_n, it follows that $g_k(u_n) \in W_0^{1, p}(\Omega)$. Since

$$|g_k(u_n)| \leqslant |u_n| \quad \text{and} \quad |D_i g_k(u_n)| \leqslant |D_i u_n|,$$

the sequence $\{g_k(u_n)\}$ is bounded in $W_0^{1, p}(\Omega)$. By the reflexivity of $W_0^{1, p}(\Omega)$, there exists a subsequence $\{g_k(u_j)\}$ weakly convergent to an element $w \in W_0^{1, p}(\Omega)$. But the boundedness of g_k' yields

$$\|g_k(u_n) - g_k(u)\|_p \leqslant \|u_n - u\|_p,$$

whence $w = g_k(u)$. Now, for any $\varphi \in C_0^\infty(\Omega)$, we get

$$\int_\Omega \varphi D_i g_k(u) \, dx = -\int_\Omega g_k(u) D_i \varphi \, dx = \int_\Omega \varphi g_k'(u) D_i u \, dx,$$

which proves the first part of the lemma.

We show that $g_k(u) \to u$ in $W_0^{1, p}(\Omega)$ as $k \to \infty$. By the definition, $g_k(u) = u$ and $D_i g_k(u) = D_i u$ on the set $\{x \in \Omega \mid |u(x)| \leqslant k - 1 \text{ a.e. } \Omega\}$. Letting $k \to \infty$ we get $g_k(u(x)) \to u(x)$ and $D_i g_k(u(x)) \to D_i u(x)$ for a.a. $x \in \Omega$.

Let A be a measurable subset of Ω. We have

$$\int_A |g_k(u)|^p \, dx \leqslant \int_A |u|^p \, dx \quad \text{and} \quad \int_A |D_i g_k(u)|^p \, dx \leqslant \int_A |D_i u|^p \, dx, \quad 1 \leqslant i \leqslant N.$$

Hence $g_k(u) \in L^p(\Omega)$ and $D_i g_k(u) \in L^p(\Omega)$. By an analogous argument we can find for $\varepsilon > 0$ a subdomain $\Omega_\varepsilon \subset \Omega$ of finite measure such that

$$\int_{\Omega - \Omega_\varepsilon} |g_k(u)|^p \, dx < \varepsilon \quad \text{and} \quad \int_{\Omega - \Omega_\varepsilon} |D_i g_k(u)|^p \, dx < \varepsilon.$$

Vitali's theorem guarantees that $g_k(u) \to u$ and $D_i g_k(u) \to D_i u$ in $L^p(\Omega)$ as $k \to \infty$. Therefore $g_k(u) \to u$ in $W_0^{1, p}(\Omega)$, as required.

We take now a closer look at the countable set S in the above theorem.

PROPOSITION *Let* $V = W_0^{1, p}(\Omega)$ *and* $W = W_0^{1, s}(\Omega) \cap V$ *with* $s > N$. *Then there exists a countable set S contained in W such that for each $v \in V$ and each $k \in \mathbb{N}$, there exists a sequence $\{\varphi_n\} \subset S$ such that $|\varphi_n(x)| \leqslant k$ for all $x \in \Omega$ and $\varphi_n \to g_k(v)$ in V as $n \to \infty$.*

Proof: Since V is separable, there exists a countable set S_1 contained in $C_0^\infty(\Omega)$ and dense in V. Consider the set

$$S = \bigcup_{k=1}^\infty \{g_k(f) \mid f \in S_1\}$$

wnich is countable, as a countable union of countable sets. Moreover, as each $f \in S_1 \subset W$, by the lemma, $g_k(f) \in W$ and $g_k(f) \to f$ in W Hence, $S \subset W$.

In order to see that S has the required properties, let $v \in V$ and $k \geqslant 1$ be given. As S_1 is dense in V, there exists a sequence $\{f_n\} \subset S_1$ such that $f_n \to v$ in V. This implies that

$$\| f_n - v \|_p \to 0 \quad \text{and} \quad \| D_i f_n - D_i v \|_p \to 0, \ 1 \leqslant i \leqslant N.$$

Thus there exists a subsequence $\{f_j\}$ such that $f_j(x) \to v(x)$ and $D_i f_j(x) \to D_i v(x)$ for a.a. $x \in \Omega$. Since g_k and g_k' are continuous we see that $g_k(f_j(x)) \to g_k(v(x))$ and

$$D_i g_k(f_j(x)) = g_k'(f_j)\, D_i f_j(x) \to g_k'(v)\, D_i v, \quad \text{a.e. in } \Omega.$$

By the properties of g_k and the same arguments as in the proof of the lemma, we deduce, *via* Vitali's convergence theorem that $g_k(f_j) \to v$ in V as $j \to \infty$, Set $\varphi_j \equiv g_k(f_j)$. The proof is complete.

5.3. We shall apply now the above results to the study of the Dirichlet problem for a strongly nonlinear elliptic equation in unbounded domains.

Consider an equation of the form

$$Au(x) + q(x)\, g(u(x)) = f(x), \quad x \in \Omega, \tag{1}$$

where A is a divergence operator of second order

$$Au = \sum_{|\alpha| \leqslant 1} (-1)^{|\alpha|}\, D^\alpha A_\alpha(x, u, \nabla u).$$

The Dirichlet problem in the unbounded domain Ω requires a solution $u(x)$ of equation (1) such that

$$u(x) = 0 \text{ on } \partial\Omega \text{ and } u(x) \to 0 \text{ as } |x| \to \infty. \tag{2}$$

We assume that the functions $A_\alpha : \Omega \times \mathbb{R}^{N+1} \mapsto \mathbb{R}$ satisfy the hypotheses (I) listed in Section 2.2 for the case $m = 1$. The strongly nonlinear term $g : \mathbb{R} \mapsto \mathbb{R}$ is continuous and it satisfies the sign condition

$$g(t)\, t > 0 \quad \text{for all } t \in \mathbb{R}.$$

In order to adopt the procedure used in Section 3.1, it is necessary to multiply g by a non-negative function $q \in L^1(\Omega)$.

As above, let $V = W_0^{1,\,p}(\Omega)$, $1 < p < \infty$, and $W = W_0^{1,\,s}(\Omega) \cap V$ with $s > N$. By Proposition II.2.9, $W \subset C(\Omega) \cap L^\infty(\Omega)$.

As we have seen in Theorem 2.1, the form

$$a(u, v) = \sum_{|\alpha| \leqslant 1} \int_\Omega A_\alpha(x, u, \nabla u)\, D^\alpha v \, dx \quad (\forall)\ u, v \in V,$$

defines a continuous monotone coercive operator $T_1 : V \mapsto V^*$.

Les us introduce the subspace

$$V_1 = \{v \in V \mid q\, g(v) \in L^1(\Omega), \quad q\, g(v)\, v \in L^1(\Omega)\}$$

and remark that $W \subset V_1 \subset V$. Then, given any u in V_1, the form

$$b(u, w) = \int_\Omega q\, g(u)\, w\, dx \qquad (\forall)\ w \in W$$

is bounded because

$$|b(u, w)| \leqslant \|q\, g(u)\|_1\, \|w\|_\infty .$$

Hence the linear functional $b(u, .)$ determines an operator $T_2 \colon V_1 \mapsto W^*$ by the formula

$$b(u, w) = (T_2 u, w) \qquad (\forall)\ w \in W.$$

Given $f \in V^*$, we say that $u \in V_1$ is a *variational solution* of the Dirichlet problem (1)–(2) provided that

$$a(u, w) + b(u, w) = (f, w) \qquad (\forall)\ w \in W. \tag{3}$$

Clearly, if $i \colon W \mapsto V$ denotes the natural injection, $i^* \colon V^* \mapsto W^*$ its adjoint map and

$$T = i^* T_1 + T_2 \colon V_1 \mapsto W^*, \tag{4}$$

then in operatorial terms the variational solution satisfies the functional equation $Tu = f$.

Our aim is now to establish some properties of T_2. The key-step in this direction is the following

LEMMA *Let* $\{u_n\}$ *be a sequence in* W *such that* $u_n \rightharpoonup v$ *in* V *and*

$$\limsup (T_2 u_n, u_n) \leqslant C < \infty.$$

Then $v \in V_1$ *and* $T_2 u_n \rightharpoonup T_2 v$ *in* W^*.

Proof: By Corollary II.2.10, passing, if necessary, to a subsequence, we get $u_n(x) \to v(x)$ a.a. $x \in \Omega$, and by Fatou's lemma we deduce that

$$\int_\Omega q(x)\, g(v(x))\, v(x)\, dx \leqslant C.$$

Moreover, there is a constant $C_1 > 0$ such that

$$\int_\Omega q(x)\, g(v(x))\, dx < \infty.$$

In fact, $K = \sup \{|g(t)| \mid |t| \leq 1\} < \infty$, since g is a continuous function. Let us write $\Omega = \Omega_1 \cup \Omega_2$, where

$$\Omega_1 = \{x \in \Omega \mid |v(x)| \leq 1\} \quad \text{and} \quad \Omega_2 = \{x \in \Omega \mid |v(x)| > 1\}.$$

Since $q(x) g(v(x)) \leq q(x) g(v(x)) v(x)$ for $x \in \Omega_2$, we get

$$\int_\Omega q(x) |g(v(x))| \, dx \leq K \int_{\Omega_1} q(x) \, dx + \int_{\Omega_2} q(x) g(v(x)) v(x) \, dx \leq$$

$$\leq K \int_\Omega q(x) \, dx + C = C_1.$$

Hence $v \in V_1$.

We show now the weak convergence in W of the sequence $\{T_2 u_n\}$. Let w be fixed in W and consider

$$(T_2 u_n, w) = \int_\Omega q(x) g(u_n(x)) w(x) \, dx.$$

Given $\delta > 0$, $n \in \mathbb{N}$ and $x \in \Omega$, we have either $|u_n(x)| < \dfrac{1}{\delta}$ or $|g(u_n(x))| \leq \delta u_n(x) g(u_n(x))$. Clearly $K(\delta) = \sup \left\{ |g(t)| \, \middle| \, |t| \leq \dfrac{1}{\delta} \right\} < \infty$. If E is a measurable subset of Ω, then

$$I(E) = \int_E q(x) \mid g(u_n(x)) \mid w(x) \, dx \leq K(\delta) \int_E q(x) \mid w(x) \mid dx +$$

$$+ \delta \int_E q(x) u_n(x) g(u_n(x)) \mid w(x) \mid dx \leq K(\delta) \|w\|_\infty \int_E q(x) \, dx + \delta \|w\|_\infty C.$$

For any $\varepsilon > 0$ we choose $\delta > 0$ such that $\delta \|w\|_\infty C < \dfrac{1}{2} \varepsilon$ and we obtain that $I(E) \to 0$ as $\mu(E) \to 0$.

On the other hand, since $q \in L^1(\Omega)$, there exists a measurable subset $E_\varepsilon \subset \Omega$ with $\mu(E_\varepsilon) < \infty$ such that

$$\int_{\complement E_\varepsilon} q(x) \, dx < \frac{\varepsilon}{2K(\delta) \|w\|_\infty},$$

where $\complement E_\varepsilon$ denotes the complementary set of E_ε relative to Ω. It follows that

$$\int_{\complement E_\varepsilon} q(x) |g(u_n(x))| \mid w(x) \mid dx < \varepsilon \quad \text{for all } n \in \mathbb{N}.$$

Thus the sequence $\{v_n\} \subset L^1(\Omega)$, where $v_n(x) = q(x)\,g(u_n(x))\,w(x)$, satisfies the conditions of Vitali's theorem. Therefore

$$\int_\Omega q(x)\,w(x)\,[g(u_n(x)) - g(v(x))]\,dx \to 0 \quad \text{as } n \to \infty,$$

that is, $T_2 u_n \rightharpoonup T_2 v$ in W^*.

We note that *the restriction of T_2 to W is demicontinuous*. In fact, let $u_n \to u$ in W and let C_2 be a positive constant such that $\sup\{u_n(x) \mid x \in \Omega\} \leqslant C_2$. Obviously $u_n \to u$ in V and by the continuity of g we have $g(u_n(x))\,u_n(x) \leqslant C_3$ for all $n \in \mathbf{N}$, and

$$(T_2 u_n,\, u_n) = \int_\Omega q(x)\,g(u_n(x))\,u_n(x)\,dx \leqslant C_3 \|q\|_1.$$

By the lemma, we deduce that $T_2 u_n \rightharpoonup T_2 u$ in W^* as claimed.

In particular, *T_2 is also quasi-bounded*.

A simple argument shows that T is quasi-bounded and of type (M) with respect to (W, V). Since T_1 is coercive and $(T_2 v, v) \geqslant 0$, Corollary 3.1 implies that for any given $f \in V^*$ there exists a $u \in V_1$ such that $Tu = i^* f$.

Next, let us apply to problem (1)–(2) a method of singular perturbation.

Let S be a countable subset of W with the properties set out in proposition 5.2. Let H be the separable Hilbert space, with the inner product $\langle .\,,\,. \rangle$, and let $Q : H \mapsto W$ be the one-to-one linear compact operator with $S \subset Q(H)$ and $Q(H)$ dense in W which are given by Theorem 5.2. Denote by $Q^* : W^* \mapsto H$ the adjoint map of Q and let $T : V_1 \mapsto W^*$ be defined by (4).

PROPOSITION *Suppose that the above hypotheses hold. Then, for every $\varepsilon > 0$ and any $f \in V^*$. there exists at least one $u \in H$ such that*

$$\varepsilon u + Q^* TQu = Q^* f. \tag{5}$$

Proof: Since T_1 and T_2 are both demicontinuous, so is T and the map $Q^* TQ : H \mapsto H$ is compact. By the Leray-Schauder principle, it is sufficient to show that there exists a constant $r > 0$ such that $u \in H$ and $|u| = r$ imply

$$\varepsilon \lambda u + Q^*(TQu - f) \neq 0, \quad \text{for all } \lambda > 1.$$

In fact, in the opposite case, let $|u| = r$ and

$$\varepsilon \lambda |u|^2 + (TQu, Qu) = (f, Qu) \quad \text{for any } \lambda > 1. \tag{6}$$

Then the inequality $(TQu, Qu) \leqslant \|f\|_{V^*} \|Qu\|_V$ and the coerciveness of T yield $\|Qu\|_V \leqslant C$. As T_1 is a bounded map of V to V^*, we have $|(T_1 Qu, Qu)| \leqslant C_1$. Since $(T_2 Qu, Qu) \geqslant 0$, we get $(TQu, Qu) \geqslant -C_1$. We substitute this into (6), to obtain

$$\varepsilon \lambda |u|^2 = (f, Qu) - (TQu, Qu) \leqslant C\|f\|_{V^*} + C_1 \equiv C_2.$$

Thus $\|u\|^2 \leqslant \dfrac{C_2}{\varepsilon\lambda} < \dfrac{C_2}{\varepsilon}$ since $\lambda > 1$, that is, $\|u\| < \left(\dfrac{C_2}{\varepsilon}\right)^{\frac{1}{2}} \equiv C_3$. If we choose $r > C_3$, we get a contradiction. Now apply the Leray-Schauder principle and the conclusion of the proposition follows easily.

THEOREM *Let u_ε be a solution in H of equation (5). Then $\|Qu_\varepsilon\|_V \leqslant C$ and there exists a sequence $\varepsilon_j \to 0$ such that the sequence $\{Qu_{\varepsilon_j}\}$ converges weakly in V to an element $v \in V_1$ which satisfies $Tv = f$.*

Proof: We have seen in the proof of the proposition that $\|Qu_\varepsilon\|_V \leqslant C$. As V is a reflexive Banach space. there exists a sequence $\varepsilon_j \to 0$ such that $\{Qu_{\varepsilon_j}\}$ converges weakly to an element v in V. We write $\{u_j\}$ instead of $\{u_{\varepsilon_j}\}$. By the above proposition. $\varepsilon_j\|u_j\|^2 \leqslant C_2$. Moreover,

$$\varepsilon_j\|u_j\|^2 + (TQu_j, Qu_j) = (f, Qu_j)$$

and the boundedness of T_1 imply that $(T_2Qu_j, Qu_j) \leqslant C_4$.

By the lemma. it follows that $v \in V_1$ and $T_2Qu_j \to T_2v$ in W^*. Since T_1 is bounded, we may assume, passing if need be to a subsequence, that $\{T_1Qu_j\}$ converges weakly to z in V^*.

Multiplying (5) by an arbitrary $h \in H$, we obtain

$$\varepsilon_j\langle u_j, h\rangle + (TQu_j, Qh) = (f, Qh).$$

Since $|\varepsilon_j\langle u_j, h\rangle| \leqslant \varepsilon_j\|u_j\|\,\|h\|$, passing to the limit as $j \to \infty$, we find that $\varepsilon_j\langle u_j, h\rangle \to 0$ and therefore

$$(z, Qh) + (T_2v, Qh) = (f, Qh) \qquad (\forall)\ h \in H.$$

As $Q(H)$ is dense in W, Theorem 5.2, implies that

$$z + T_2v = f.$$

To complete the proof we must show that $z = T_1v$. This follows from the fact that the monotone mapping T_1 is of type (M), provided we establish that

$$\lim \sup (T_1Qu_j, Qu_j - v) \leqslant 0.$$

We write $(T_1Qu_j, Qu_j - v) = \langle Q^*T Qu_j, u_j - h_1\rangle + (T_1Qu_j, Qh_1 - v)$ for a suitably chosen $h_1 \in H$. Then

$$\lim \sup (T_1Qu_j, Qu_j - v) \leqslant \lim \sup \langle Q^*T_1Qu_j, u_j - h_1\rangle + (z, Qh_1 - v) =$$

$$= F_j + (z, Qh_1 - v).$$

Taking the inner product of (5) by $u_j - h_1$, we get

$$\varepsilon_j\langle u_j, u_j - h_1\rangle + \langle Q^*TQu_j, u_j - h_1\rangle = \langle Q^*f, u_j - h_1\rangle;$$

whence we obtain

$$\varepsilon_j |u_j|^2 - \varepsilon_j \langle u_j, h_1 \rangle + F_j + (T_2 Q u_j, Q u_j - Q h_1) = (f, Q u_j - Q h_1).$$

By Fatou's lemma

$$\int_\Omega q\, g(v)\, v\, \mathrm{d}x \leqslant \lim \inf (T_2 Q u_j, Q u_j),$$

it follows that

$$\lim \sup F_j \leqslant (T_2 v, Q h_1) + (f, v - Q h_1) - \int_\Omega q\, g(v)\, v\, \mathrm{d}x$$

and

$$\lim \sup (T_1 Q u_j, Q u_j - v) \leqslant (f - z, v - Q h_1) +$$

$$+ \int_\Omega q\, g(v)\, Q h_1\, \mathrm{d}x - \int_\Omega q\, g(v)\, v\, \mathrm{d}x. \tag{7}$$

Recall that $Q(H)$ contains S and take the elements φ_n instead of $Q h_1$. By Proposition 5.2, for each $k \in \mathbf{N}$, there exists a sequence $\{\varphi_n\} \subset S$ such that $|\varphi_n(x)| < k$ for all $n \in \mathbf{N}$ and $\varphi_n \to g_k(v)$ in V. Thus

$$(f - z, v - \varphi_n) \to (f - z, v - g_k(v))$$

and by virtue of Lebesgue's dominated convergence theorem,

$$\int_\Omega q\, g(v)\, \varphi_n\, \mathrm{d}x \to \int_\Omega q\, g(v)\, g_k(v)\, \mathrm{d}x.$$

Let $k \to \infty$; then, by Lemma 5.2,

$$(f - z, v - g_k(v)) \to 0,$$

and by Lebesgue's theorem again, since $q\, g(v)\, g_k(v)$ is dominated by $q\, g(v)\, v$ in $L^1(\Omega)$,

$$\int_\Omega q\, g(v)\, g_k(v)\, \mathrm{d}x \to \int_\Omega q\, g(v)\, v\, \mathrm{d}x.$$

Since the left-hand side of the inequality (7) is independent of the choice of h, we have that

$$\lim \sup (T_1 Q u_j, Q u_j - v) \leqslant 0$$

and thus $T_1 v = z$. This completes the proof.

The above theory may be applied to establish the existence of a variational solution of the Dirichlet problem for

$$-\Delta u + q u e^u = f \quad \text{in } \Omega,$$

where q is any non-negative $L^1(\Omega)$-function.

6. Related topics and exercises

For the sake of simplicity, we use in this section a unitary system of numbering equations.

6.1. (Signorini's problem). Define on $H^1(\Omega)$ the bilinear form

$$a(u, v) = \sum_{i,j=1}^{N} \int_{\Omega} a_{ij} D_i u \, D_j v \, dx + \int_{\Omega} a_0 u \, v \, dx, \quad a_{ij}, \, a_0 \in L^\infty(\Omega)$$

and the linear function

$$(f, v) = \int_{\Omega} f_0 v \, dx + \int_{\partial\Omega} g \, v \, d\sigma, \quad f_0 \in L^2(\Omega), \, g \in H^{-1/2}(\partial\Omega).$$

Set $K = \{v \in H^1(\Omega) \mid v \geqslant 0 \text{ a.e. on } \partial\Omega\}$ and consider the variational inequality

$$u \in K, \quad a(u, v - u) \geqslant (f, v - u) \qquad (\forall) \, v \in K. \tag{1}$$

Putting

$$Au = - \sum_{i,j=1}^{N} D_i(a_{ij} D_j u) + a_0 u,$$

prove that

(i) The inequality (1) is equivalent to the system

$$\begin{cases} Au = f & \text{in } \Omega, \\ u \geqslant 0, \quad \dfrac{\partial u}{\partial v} \geqslant g \text{ and } u\left(\dfrac{\partial u}{\partial v} - g\right) = 0 \text{ on } \partial\Omega, \end{cases}$$

where $\dfrac{\partial u}{\partial v} = \sum_{i,j=1}^{N} a_{ij} D_j u \cos(n, x_i)$ and n is the outward normal to $\partial\Omega$;

(ii) The solution of inequality (1) exists provided that

$$\sum_{i,j=1}^{N} a_{ij} \xi_i \xi_j \geqslant c \sum_{i=1}^{N} \xi_i^2, \quad a_0 \geqslant \alpha \text{ and } c, \, \alpha > 0,$$

(see Duvaut and Lions [3]).

6.2. (A free boundary problem). Let $a: H_0^1(\Omega) \times H_0^1(\Omega) \mapsto \mathbb{R}$ be defined by

$$a(u, v) = \int_{\Omega} \nabla u \, \nabla v \, dx.$$

For a given function $\theta \in C(\bar{\Omega})$ such that $\theta \leqslant 0$ on $\partial\Omega$, take

$$K = \{v \in H_0^1(\Omega) \,|\, v \geqslant \theta \text{ a.e. in } \Omega\}.$$

Prove that inequality (1) with $f \in L^2(\Omega)$ is equivalent to the following conditions

$$\begin{cases} -\Delta u - f \geqslant 0, \, u \geqslant \theta \text{ and } (u - \theta)(-\Delta u - f) = 0 \text{ in } \Omega \\ \qquad\qquad u = 0 \qquad\qquad\qquad\qquad \text{on } \Omega \end{cases}$$

(see Lions [3]).

6.3. (Elasto-plastic torsion). Take $a(u, v)$ and f as in 6.2 and set $K = \{v \in H_0^1(\Omega) \,|\, |\nabla v(x)| \leqslant 1$ a.a. $x \in \Omega\}$. Prove that the variational inequality (1) with this K is equivalent to the following conditions

$$\begin{cases} -\Delta u = f \text{ in } \Omega_e, \quad |\nabla u| = 1 \text{ in } \Omega_p, \\ u|_{\Sigma_e} = u|_{\Sigma_p} \text{ and } \nabla u|_{\Sigma_e} = \nabla u|_{\Sigma_p}, \end{cases}$$

where $\Omega_e = \{x \in \Omega \,|\, |\nabla u(x)| < 1\}$, $\Omega_p = \{x \in \Omega \,|\, |\nabla u(x)| = 1\}$ and Σ is the separating surface between Ω_e and Ω_p (see Lions [3] and Mazilu-Sburlan [1, p. 108]).

6.4. Let $\theta \in C(\bar{\Omega})$ be such that $\theta \leqslant 0$ on Ω. We set

$$K_\theta = \{v \in H_0^1(\Omega) \,|\, v \geqslant \theta \text{ a.e. in } \Omega\}$$

and, for fixed $t_0 > 0$, we put $X = L^2(0, t_0; H_0^1(\Omega))$, $Q = \Omega \times [0, t_0]$, $\Sigma = \partial Q$ and $K = \{v \in X \,|\, v(t) \in K_\theta \text{ a.a. } t \in [0, t_0]\}$. Consider the variational inequality

$$u \in K, \qquad \int_0^{t_0} \left(\frac{\partial u}{\partial t} - f, v - u\right) dt + \int_0^{t_0} a(u, v - u) \, dt \geqslant 0 \qquad (\forall) v \in K, \qquad (2)$$

where $a(u, v)$ is defined in 6.2 and $f \in L^2(0, t_0; H^{-1}(\Omega))$. Prove that there exists a unique function $u(t, x)$ such that

$$\begin{cases} u, \, \dfrac{\partial u}{\partial x_i}, \, \dfrac{\partial^2 u}{\partial x_i \partial x_j}, \, \dfrac{\partial u}{\partial t} \in L^2(Q), \; 1 \leqslant i, j \leqslant N, \\[3mm] u \geqslant \theta \; , \; \dfrac{\partial u}{\partial t} - \Delta u - f \geqslant 0 \text{ and } (u - \theta)\left(\dfrac{\partial u}{\partial t} - \Delta u - f\right) = 0 \text{ in } Q \\[3mm] u = 0 \qquad \text{on } \Sigma \text{ and } u(0, x) = 0 \end{cases}$$

(see Lions [2, p. 291]).

6.5. Let K be a closed convex set of a Hilbert space H and let $T: H \mapsto H$ be such that
    (i) T is Lipschitz continuous, i.e., there exists a $k > 0$ such that

$$\|Tx - Ty\| \leqslant k \|x - y\| \qquad (\forall) \, x, y \in K,$$

(ii) T is strongly monotone, i.e., there exists a $c > 0$ such that

$$(Tx - Ty, x - y) \geqslant c \ \|x - y\|^2 \qquad (\forall) \ x, y \in K.$$

Prove that the sequence $\{x_n\}$ given by the iterative procedure

$$x_{n+1} = P_K(I - rT) x_n \ \text{ with } x_0 \in K \text{ and } r \in \left(0, \frac{2c}{k^2}\right) \tag{3}$$

converges to the unique solution of the variational inequality

$$(Tx, y - x) \geqslant 0 \qquad (\forall) \ y \in K. \tag{4}$$

SOLUTION We can easily see that $c \leqslant k$ and

$$\|(I - rT) x - (I - rT) y\|^2 = (1 - 2rc + k^2 r^2) \|x - y\|^2.$$

Hence $(I - rT)$ is a strict contraction on K provided that $1 - 2rc + k^2 r^2 < 1$. For $r \in \left(0, \dfrac{2c}{k^2}\right)$ the mapping $(I - rT)$ has a unique fixed point $x \in K$, which can be obtained by the iterative scheme (3) and thus x is the solution of the variational inequality (4).

6.6. An element $x \in K$ is said to be a *local solution* of the variational inequality

$$x \in K, \ (Tx - f, y - x) \geqslant 0 \qquad (\forall) \ y \in K, \tag{5}$$

if there exists a ball $\bar{B}(x, r) = \{y \in X| \ \|y - x\| \leqslant r\}$ such that

$$x \in B(0, r) \text{ and } (Tx - f, y - x) \geqslant 0 \text{ for all } y \in K \cap \bar{B}(x, r). \tag{6}$$

Prove that any local solution is also a (global) solution.

SOLUTION Let $x \in K$ be a local solution of (5). Then for any $y \in K$ we may choose an $\varepsilon > 0$ such that $x + \varepsilon (x - y) \in K \cap \bar{B}(x, r)$. Take in (6) $x + \varepsilon(x - y)$ instead of y. We obtain (5).

6.7. Let X be a finite-dimensional Banach space and let $T: K \mapsto X$ be a continuous map. Suppose that either K is bounded or T is coercive in the following sense: There exist an open bounded convex set $B \subset X$ and an element $y_0 \in K \cap B$ such that

$$(Ty, y - y_0) > 0 \text{ for all } y \in K \cap \partial B.$$

Then there exists at least one solution of (5).

HINT Identify X with a finite-dimensional Hilbert space and find the solution $x \in K$ of inequality (1) as a fixed point for the map $P_K(I - T')$ where $T'x = Tx - f$. If K is bounded, it is compact and the existence of fixed points follows by Brouwer's theorem. If K is not bounded we apply the same argument to the variational inequality (6) with $B = B(x, r)$, $r > \|x_0\|$ and consider 6.6.

6.8. Let K be a closed convex set in a reflexive Banach space X and let $\varphi\colon X \mapsto \overline{\mathbb{R}}$ be a proper convex l.s.c. function on K.

Suppose that either K is bounded or φ is coercive in the following sense: There exist $x_0 \in K$ and $r > 0$ such that $\varphi(x_0) < \infty$ and

$$\varphi(y) > \varphi(y_0) \text{ for all } y \in K \text{ with } \|y\| = r.$$

Then there exists at least one solution of problem

$$\inf \{ \varphi(x) \mid x \in K \}. \tag{7}$$

Moreover, if φ is strictly convex, this solution is unique.

HINT Suppose that K is bounded. Then the level set

$$\varphi^{\leqslant}(\varphi(y)) = \{ z \in X \mid \varphi(z) \leqslant \varphi(y) \}$$

is weakly compact. Thus $\cap \{ \varphi^{\leqslant}(\varphi(y)) \mid y \in K \} \neq \emptyset$ and by Theorem 1.2, problem (7) has at least one solution.

If φ is coercive in the above sense, the level sets $\varphi^{\leqslant}(\varphi(y))$ are bounded. In fact, the existence of a $z_1 \in \varphi^{\leqslant}(\varphi(y_0))$ with $\|z_1\| > r$ implies the existence of a $z_2 \in \varphi^{\leqslant}(\varphi(y_0))$ with $\|z_2\| = r$, i.e., $\varphi(z_2) \leqslant \varphi(y_0)$ which contradicts the coercivity of φ.

6.9. Let $\varphi\colon X \mapsto \overline{\mathbb{R}}$ be a proper subdifferentiable function with $D\varphi\colon X \mapsto X^*$ satisfying the following coercivity condition: There exist $y_0 \in K$ and $r > 0$ with $\|y_0\| < r$ such that

$$(D\varphi(y), y - y_0) > 0 \text{ for all } y \in K \text{ with } \|y\| = r.$$

Prove that problem (7) has at least one solution.

SOLUTION Let $z \in K$ be such that $\|z\| > r$. Then $y = \dfrac{t}{1+t} y_0 + \dfrac{1}{1+t} z$ lies in K and $\|y\| = r$ for some $t > 0$. Since $z - y = t(y - y_0)$, by the subgradient inequality, we have

$$\varphi(z) = \varphi(y) + (D\varphi(y), z - y) = \varphi(y) + t(D\varphi(y), y - y_0).$$

Hence the coercivity of $D\varphi$ implies the coercivity of φ in the sense mentioned in 6.8 and thus the assertion follows easily.

6.10. Let K be a closed convex set in a reflexive Banach space X with $0 \in K$ and let T be a bounded demicontinuous mapping of K into X^* which is of type (S_+). Then for each $f \in X^*$ there exists a unique solution of variational inequality (5) (see Browder [8]).

6.11. The following is a variational problem with inhomogeneous boundary conditions corresponding to problem (A, V) in Section 2.1: Given $u_0 \in W^{m,p}(\Omega)$ and $f \in V^*$, find $u \in W^{m,p}(\Omega)$ such that

$$u - u_0 \in V \text{ and } a(u, v) = (f, v) \text{ for all } v \in V$$

(see Nečas [3]).

6.12. Consider a strongly nonlinear variational problem the highest order part of which is a linear uniformly elliptic differential expression of second order with the coefficients in $L^\infty(\Omega)$. Assume that the strongly nonlinear term satisfies hypotheses $(III_1)-(III_3)$ in Section 3.2 and that there exists a continuous function $\varphi : \Omega \times \mathbb{R} \mapsto \mathbb{R}_+$ such that

$$\left| \frac{g(x, t)}{t} \right| > \varphi(x, t) \text{ with } \lim_{t \to \infty} \varphi(x, t) = \infty \text{ a.e. on } \Omega.$$

Then with the notation of Section 3.2, we have

$$a(w, w) + b(w, w) - (f, w) > 0 \qquad (\forall) \ w \in W \text{ with } \|w\|_V = r, \ r > 0.$$

Proof: Assume that, on the contrary, there exists a sequence $\{w_n\}$ in W such that $\|w_n\|_V \to \infty$ and $a(w_n, w_n) + b(w_n, w_n) - (f, w_n) \leq 0$ $(\forall)$ $n \in \mathbb{N}$. Let $v_n = \|w_n\|_V^{-1} w_n$ and write $T_1 = T_0 + T'$, where T_0 is the operator induced by the principal part of $a(u, v)$. Dividing by $\|w_n\|_V^2$ we get

$$(T_0 v_n, v_n) + \int_\Omega \varphi(v, \|w_n\|_V v_n) v_n^2 \, dx \leq - (T'v_n - \frac{1}{\|w_n\|_V} f, v_n). \tag{8}$$

Since V, $(H_0^1 \subseteq V \subseteq H^1)$, is reflexive and completely continuously imbedded in L^2, we may assume that $v_n \rightharpoonup v$ in V, $v_n \to v$ in $L^2(\Omega)$ and $v_n(x) \to v(x)$ a.e. on Ω. Let $\Omega_0 = \{x \in \Omega \mid v(x) \neq 0, v_n(x) \to v(x)\}$. By (8) it follows that $\limsup \int_\Omega \varphi(., w_n) v_n^2 \, dx \leq - (T'v, v) < \infty$. Since $\varphi(x)$, $\|w_n\|_V v_n(x)) v_n^2(x) \to \infty$ for $x \in \Omega_0$, it follows that $\mu(\Omega_0) = 0$ by Fatou's lemma. Consequently $v(x) = 0$ a.e. on Ω. As $v_n \to v = 0$ in $L^2(\Omega)$ and $\|v_n\|_V = 1$, we can deduce that $\int_\Omega |\nabla v_n|^2 \, dx \to 1$. On the other hand, (8) also implies that

$$c \int_\Omega |\nabla v_n|^2 \, dx \leq - (T'v_n, v_n) \to 0,$$

that is, a contradiction.

6.13. Set $V = W^{1, p}(\Omega)$, $W = W^{m, p}(\Omega)$ and assume that W is compactly imbedded in $C(\bar\Omega)$. Consider a function $g : \Omega \times \mathbb{R} \mapsto \mathbb{R}$ satisfying (III_1) and the following modified hypotheses:

(III_2') For any $\tau > 0$, $\displaystyle\int_\Omega \sup_{|t| \leq \tau} |g(x, t)| \, dx < \infty$;

(III_3') There exists a $t_0 > 0$ such that $g(x, t) \, t \geq 0$, $(\forall)$ $|t| > t_0$, for almost all $x \in \Omega$.

With the notation of Section 3.2, let $\{u_n\}$ be a sequence in W such that $u_n \rightharpoonup u$ in V and $b(u_n, u_n) \leq c = $ const., for all $n \in \mathbb{N}$. Prove that $u \in V_1$, $b(u, u) \leq \liminf b(u_n, u_n)$ and $T_2 u_n \rightharpoonup T_2 u$ in W^*.

Proof: Write $g = g_0 + h$, where g_0 and h both satisfy the Caratheodory conditions such that $g_0(x, t) \, t \geq 0$ for a.a. $x \in \Omega$ and all $t \in \mathbb{R}$, $h(x, t) = 0$ for $|t| > t_0$ and a.a. $x \in \Omega$, and

$$\int_\Omega \sup_{|t| < \tau} |h(x, t)| \, dx < \infty \text{ for any } \tau > 0.$$

By the Sobolev theorem, V is compactly imbedded in $L^1(\Omega)$ and we may assume that $u_n \to u$ in $L^1(\Omega)$ as well as point wise a.e. in Ω. For all $n \in \mathbb{N}$ we have

$$\int_{\Omega} g_0(., u_n)u_n \; dx \leqslant \int_{\Omega} g(., u_n)u_n \; dx + \int_{\Omega} |h(., u_n)| \; |u_n| dx \leqslant c_1 = \text{const.}$$

Thus by Fatou's lemma,

$$\int_{\Omega} g_0(,, u)u \; dx \leqslant \lim\inf \int_{\Omega} g_0(., u_n)u_n \; dx \leqslant c_1.$$

Further, the sequence $\{v_n\}$, $v_n = h(., u_n) u_n$ lies in $L^1(\Omega)$ and converges pointwise almost everywhere in Ω. Moreover $|v_n(x)| \leqslant K \sup\limits_{|t| < t_0} |h(x, t)|$, a. a. $x \in \Omega$, for a suitable constant $K > 0$. By Lebesgue dominated convergence theorem we get

$$\int_{\Omega} h(., u_n)u_n \; dx \to \int_{\Omega} h(., u)u \; dx \quad \text{and}$$

$$\int_{\Omega} h(., u_n)w \; dx \to \int_{\Omega} h(., u)w \; dx \qquad (\forall) \; w \in W. \tag{9}$$

Thus

$$\int_{\Omega} g(., u)u \; dx = b(u, u) \leqslant \lim\inf b(u_n, u_n) \leqslant c$$

and therefore $g(., u)u \in L^1(\Omega)$.

On the other hand, let us write $|g_0(x, t)| = \dfrac{1}{|t|} g_0(x, t) t$, for any $t \neq 0$, and estimate

$$|g_0(x, t)| \leqslant g_0(x, t)t + \sup_{|t| \leqslant 1} |g_0(x, t)|.$$

Consequently $g_0(.,u) \in L^1(\Omega)$ and we deduce that $g(., u) \in L^1(\Omega)$. Hence u lies in V_1.

As in Lemma 3.2, we can prove that

$$\int_{\Omega} g_0(., u_n)w \; dx \to \int_{\Omega} g_0(., u)w \; dx \qquad (\forall) \; w \in W. \tag{10}$$

Adding (9) and (10), we also have $b(u_n, w) \to b(u, w)$ for all $w \in W$, that is, $T_2 u_n \to T_2 u$ in W^*. Thus all assertions are proved.

6.14. Consider a strongly nonlinear equation whose highest order part is a nonlinear divergence of second order and the strongly nonlinear term as in 6.13. Assume that the "coefficient functions" in the top order part satisfy hypotheses (II) in Section 2.3 for $m = 1$. Then with the notation of Section 3.2, the mapping $T = i^*T_1 + T_2 : V_1 \mapsto W$ is quasi-bounded of type (M) with respect to (W, V).

Proof: Consider a sequence $\{u_n\} \subset W$ and $u \in V$, $g \in V^*$ such that $u_n \to u$ in V, $Tu_n \to i^*g$ in W^* and $\lim\sup (Tu_n, u_n) \leqslant (g, u)$. Since $\lim\sup (T_2 u_n, u_n) \leqslant (g, u) + \lim\sup |(T_1 u_n, u_n)| \leqslant c_2 = \text{const.}$, by 6.13, we have $u \in V_1 = D(T)$ and $T_2 u_n \to T_2 u$ in W^*.

As T_1 is a bounded mapping and the sequence $\{T_2 u_n\}$ is bounded, we deduce that T is quasi-bounded. The reflexivity of V implies, at least for a subsequence, that $T_1 v_n \to \chi$ in V^* and $i^*\chi + T_2 u = i^*g$. In order to show that T is of type (M) with respect to (W, V), it is sufficient to prove that $\chi = T_1 u$.

As we have seen in Section 2.3, T_1 is demicontinuous of type (S_+) and hence of type (M). Thus the equality $\chi = T_1 u$ follows if we prove that $\limsup (T_1 u_n, u_n - u) \leqslant 0$. For this we apply the same argument as in the proof of Lemma 3.2.

6.15. Let $Au = -\sum_{i=1}^{M} D_i A_i(., u, \nabla u)$ with "coefficient functions" satisfying hypotheses (II) in Section 2.3. Let $g: \Omega \times \mathbb{R} \mapsto \mathbb{R}$ satisfy the Caratheodory conditions and be such that for given $f \in L^{p'}(\Omega)$ there exist constants $R_1 < 0 < R_2$ for which

$$\int_\Omega \sup_{R_1 < t < R_2} |g(x, t)| \, dx < \infty \quad \text{and} \quad \left. \begin{array}{c} g(x, R_1) \leqslant f(x) \\ g(x, R_2) \geqslant f(x) \end{array} \right\} \quad \text{for a.a. } x \in \Omega.$$

Prove that the variational problem

$$\text{(PN)} \quad \left\{ \begin{array}{l} Au(x) + g(x, u) = f(x), \quad x \in \Omega, \\ \dfrac{\partial u}{\partial n_A} \equiv \sum_{i=1}^{N} A_i(x, u, \nabla u) \cos (n, x_i) = 0, \quad x \in \partial\Omega, \end{array} \right.$$

admits a variational solution $u \in V$ with $R_1 < u(x) < R_2$ a.a. $x \in \Omega$.

SOLUTION We introduce the modified function

$$g_1(x, t) = \left\{ \begin{array}{ll} g(x, R_1) & \text{if} \quad t \leqslant R_1, \\ g(x, t) & \text{if} \quad R_1 < t < R_2, \\ g(x, R_2) & \text{if} \quad t > R_2, \end{array} \right.$$

and consider the function $\beta: \mathbb{R} \mapsto \mathbb{R}$ given by

$$\beta(t) = \left\{ \begin{array}{ll} -(R_1 - t)^{p-1} & \text{if} \quad t \leqslant R_1, \\ 0 & \text{if} \quad R_1 < t < R_2, \\ (t - R_2)^{p-1} & \text{if} \quad t > R_2. \end{array} \right.$$

Under these assumptions, we can easily check whether the function

$$g_2(x, t) = g_1(x, t) - f(x) + \beta(t)$$

verifies the hypotheses $(\text{III}_2') - (\text{III}_3')$ in 6.13 with $t_0 = \max \{|R_1|, |R_2|\}$. Since $|g_2(x, t)| \to \infty$ as $|t| \to \infty$ for a.a. $x \in \Omega$, Proposition 3.2 guarantees the existence of a function $u \in V$ such that

$$g_2(., u) \in L^1(\Omega), \quad g_2(., u)u \in L^1(\Omega) \quad \text{and} \quad a(u, w) + \int_\Omega g_2(., u) w \, dx = 0 \quad (\forall) \ w \in W.$$

It remains to show that this u is a variational solution of problem (PN).

For this purpose, let $\gamma: \mathbb{R} \mapsto \mathbb{R}$ be a uniformly bounded, Lipschitz continuous and monotone increasing function such that

$$\gamma(t) = 0 \text{ if and only if } R_1 < t < R_2.$$

Under these conditions $\gamma(u) \in V \cap L^\infty(\Omega)$. For any constant $k > 0$ there exists a sequence $\{\varphi_j\} \subset C^\infty(\overline{\Omega})$ such that $|\varphi_j(x)| \leqslant k$ and $\varphi_j \rightharpoonup \gamma(u)$ in V, (see proposition 3.1). Setting $w = \varphi_j$ in (9) of Section 3.1, we get

$$a(u, \varphi_j) + \int_\Omega g_2(.,u) \, \varphi_j \, dx = 0 \quad (\forall) \, j \in \mathbb{N}.$$

Passing to the limit as $j \to \infty$, by Lebesgue's theorem, we obtain

$$a(u, \gamma(u)) + \int_\Omega g_2(., u) \, \gamma(u) \, dx = 0. \tag{11}$$

The monotonicity of γ implies that $\gamma(u)u \geqslant 0$. Using (II_2) we can deduce that $a(u, \gamma(u)) \geqslant 0$. We assert that

$$\int_\Omega (g_1(., u) - f) \, \gamma(u) \, dx = 0. \tag{12}$$

Indeed, if $x \in \Omega$ is such that $R_1 \leqslant u(x) \leqslant R_2$, then $\gamma(u(x)) = 0$. For $x \in \Omega$ such that $u(x) < R_1$, we have $g_1(x, u(x)) = g(x, R_1) \leqslant f(x)$ and $\gamma(u(x)) < 0$. Consequently the integrand in (12) is non-negative there. The same holds on the set where $u(x) > R_2$. Inequality (12) follows now easily. From (11)–(12) we can deduce that $\int_\Omega \beta(u)\gamma(u) \, dx \leqslant 0$. Then $R_1 \leqslant u(x) \leqslant R_2$ because $\beta(t) \, \gamma(t) \geqslant 0$ whenever $t < R_1$ and $t > R_2$. In this interval we have $g_2(., u) = g_1(., u) - f$ and thus u is a variational solution of problem (11).

6.16. Consider the variational inequality

$$u \in K \cap V_1, \quad a(u, w - u) + b(u, w - u) \geqslant (f, w - u) \quad (\forall) \, w \in K \cap L^\infty(\Omega) \tag{13}$$

associated to the strongly nonlinear equation $Au + g(u) = f$ in Ω, where A is, as in Section 2.3, a linear uniformly elliptic operator of second order, $K = \{u \in V = H^1(\Omega) \mid u \geqslant 0 \text{ on } \partial\Omega\}$, $f \in L^2(\Omega)$ and $g : \mathbb{R} \mapsto \mathbb{R}$ is a bounded continuous functions satisfying the following modified sign condition: There exists a constant $C \geqslant 0$ such that $g(t)$ sign $t \geqslant -C$ for all $t \in \mathbb{R}$. Prove that there exists $u \in K \cap V_1$ which is a solution for (13), provided that

$$g_+ = \lim_{t \to \infty} g(t) > \frac{1}{\mu(\Omega)} \int_\Omega f \, dx. \tag{14}$$

HINT Changing f and g by adding a constant we can take $g_+ > 0$. The above sign condition guarantees a decomposition $g = g_0 + h$, where g_0 and h are continuous functions with the properties that $g_0(t)t \geqslant 0$ and $|h(t)| \leqslant$ const. for all $t \in \mathbb{R}$. It is easy to see that the associated operator $T = i^*T_1 + T_2 : V_1 \mapsto W$ satisfies conditions $(A_1)-(A_2)$ in Lemma 2.3. Show (by the method of contradiction) that condition (14) implies that T is coercive with respect to f and apply Theorem 2.3 (see Hess [16]).

Bibliographical comments

In this chapter we applied the surjectivity results for various classes of monotone mappings to show the existence of solutions for nonlinear equations of divergence type. Many mathematical models of problems in physics, especially those from mechanics of continua, can be described either as variational inequalities or as variational problems.

Variational inequalities derive from minimum problems. Our discussion on the equivalent formulations of these problems follows the lectures of Mosco [1]. The first results concerning variational inequalities associated to coercive bilinear forms were proved by Stampacchia [1] and Hartman-Stampacchia [1]. The study of variational inequalities with monotone mappings was initiated by Browder [4], [8] and the ones with pseudo-monotone operators by Brézis [1], [4]. For the existence of solutions of elliptic variational inequalities, we have used the penalty method (see e.g. Lions [2, p. 371]). We mention the survey works due to Stampacchia [2], Lions [3] and Duvaut – Lions [1]. An interesting extension of variational inequalities are the so-called quasi-variational inequalities introduced by Bensoussan – Lions [1] and Lions [4], with respect to some stochastic impulse control problems. The progress in this area is outlined in the report of Mosco [2] and Baiocchi – Capelo [1].

Variational boundary-value problems for generalized divergence equations are at the core of the chapter. Contributions to the study of these equations have been made by Višik [1], Browder [9], Dubinskii [1], Kachurovskii [2], Strauss [1], Kemnochi [1]–[2] and Kato [3]. The existence results in Sections 2.2–2.3 follow Browder's work [9]. We also mention Lions [1], Ladyzhenskaya – Uraltzeva [1], Gajewski – Gröger – Zacharias [1], Gilbarg – Trudinger [1] and Rektorys [1].

In connection with a different dependence of the divergence operators on its highest order part and its lower order terms, the concept of monotonicity was subsequently weakened, and various classes of nonlinear operators, which we gather under the name "mappings of monotone type", were introduced. After some pioneering work of Browder [13] and Ton [1], Hess established in [8]–[10] a proper abstract setting for perturbed divergence equations and introduced the mappings of monotone type with respect to two Banach spaces. This method, described in Section 3.1, is applied in 3.2 to a strongly nonlinear problem (see Hess [12], [18]). A similar approach is given by Schmitt [1] and Tiba [1].

The existence of bounded solutions for variational inequalities was investigated by Ton [2], Cǎc [1] and Hess [16].

Proposition 4.1, due to Strauss [2], combined with the truncation method has allowed the use of homotopic arguments in the study of these problems (see Hess [12]). For the method of upper and lower solutions in Section 4.2, we have chosen the approach made by Hess [20]. By a refinement of this method, Hess [21] and Duel–Hess [1] have studied the boundary-value problems associated with elliptic differential expressions of the more general form. For different approaches, see also Puel [1] and Bose – Brill [1].

The necessary and sufficient conditions for the existence of solutions for nonlinear perturbations of linear elliptic problems at resonance were given by Landesman – Lazer [1] and Williams [1]. The simple method described in Section 4.3 is due to Hess [15]. An improvement of this method and some up-to-date references can be found in Fučik's lectures [3].

For the unbounded domains the compactness properties of the Sobolev imbeddings fail to hold, in general. To overcome these obstacles, two methods can be used. In Section 5.1, we discussed the method of approximation domains due to Hess [17], [19] and Pascali [6] by using mappings of type (M). The second method described in Section 5.2 is based on the elliptic super-regularization method of Browder and Ton [1]. This method was first used in the case of bounded domains by Ton [1] and Hess [2]. By a refinement of this method combined with a suitable truncation of elements in the Sobolev space involved, Edmunds — Webb [1], Edmunds — Moscatelli — Webb [1] and Webb [1] have obtained existence results for strongly nonlinear Dirichlet problems in unbounded domains. Let us also mention some different approaches in this area, due to Landes [1], Lehtonen [1], Mustonen [1] and Simader [1].

Suggestions for further study

In general, the mappings studied in this book act in reflexive Banach spaces. Monotonicity concepts in non-reflexive spaces were discussed among others, by Bénilan—Brézis—Crandall [1] and Da Prato [3]. In particular, Gossez [2] and Vainberg [2] used Orlicz spaces in the study of elliptic differential and integral nonlinear equations. Other related topics were investigated by Martin Jr. [1] and Deimling [2].

We have seen the important role played by the coercivity conditions in the existence theory of operatorial equations. A larger class of the so-called semi-coercive nonlinear problems have recently been studied by Schatzman [1], Hess [14] and Edmunds-Moscatelli [1].

The theory of semigroups of nonlinear contractions and its applications to discussions of the existence and regularity of solution for differential equations can be extended well beyond the results presented in the book. We refer the reader to Barbu [2], Brézis [5], Iannelli [1] and Pavel [1]. Also, in relation to this topic, nonlinear integral Volterra equations have been intensively studied of late (see, e.g. Barbu [3], C. Corduneanu [1], Luca [1] and Crandall-Londen-Nohel [1]).

The differential operators are in some sense related to the structure of the underlying functional spaces. This fact is at the base of the investigations of nonlinear operators carried out by Amann [6], Calvert-Picard [1] and Konishi [1] in ordered Banach spaces.

Let us also mention the recent research of evolution equations involving time-dependent constraints (see e.g. Attouch-Bénilan-Damlamian-Picard [1]).

The results discussed in the book can be applied to the study of some mathematical models in mechanics of continua and plasma physics. In this area, we mention the works of Beju [1], D. Ciorănescu [1], Dincă [1], Gröger [1], Mossimo [1], Mazilu-Sburlan [1], Oden-Reddy [1] and Hlaváček-Nečas [1]. Concerning the Navier-Stokes equations let us point out the stochastic approach of Foiaş [1] and the numerical treatment of Temam [1].

Finally, numerical studies on solutions of operatorial equations can be found in the papers by Cesary [1] and Petryšyn [2] (for the approximation schemes) and in the book by Oden-Reddy (for the finite-element method). The numerical analysis concerning the variational inequalities was developed by Glowinski-Lions-Trémolières [1].

References

A. F. ABEASIS, J. P. DIAS and A. LOPES-PINTO
  [1] *Sur les valeurs propres du sous-différentiel d'une fonction convexe avec un noyau borné*, C.R.A.S., **278**, (1974), 1197–1199.

R. A. ADAMS
  [1] *Sobolev Spaces*, Academic Press, 1975.

S. AGMON
  [1] *Lectures on Elliptic Boundary Value Problems*, Van Nostrand, 1965.

S. AIZICOVICI
  [1] *Spaţii de distribuţii*, "*Analiză neliniară şi aplicaţii în mecanică*", 11–40, Editura Academiei, Bucureşti, 1977.

H. AMANN
  [1] *Über die Existenz und Eindeutigkeit einer Lösung der Hammersteinschen Gleichungen in Banachräumen*, J. Math. Mech., **19**, (1969), 143–154.
  [2] *Ein Existenz- und Eindeutigkeitssatz für die Hammersteinsche Gleichung in Banachräumen*, Math. Z., **111**, (1969), 175–190.
  [3] *Über die näherungsweise Lösung nichtlinearer Integralgleichungen*, Numer. Math., **19**, (1972), 19–45.
  [4] *Existence theorems for equations of Hammerstein type*, Appl. Anal., **1**, (1972), 385–397.
  [5] *Lectures on some fixed points theorems*, Monografias de Matematica 17, Rio de Janeiro, 1974.
  [6] *Nonlinear operators in ordered Banach spaces and some applications to nonlinear boundary-value problems*, Lect. Notes Math., **543**, 1–55, Springer-Verlag, 1976.

E. ASPLUND
  [1] *Positivity of duality mappings*, Bull. A.M.S., **73**, (1967), 200–203.
  [2] *Averaged norms*, Israel J. Math., **5**, (1967), 227–233.
  [3] *Topics in the theory of convex functions*, "*Theory and Applications of Monotone Operators*", 1–33, Ed. Oderisi, Gubbio, 1969.

H. ATTOUCHE, PH. BÉNILAN, A. DAMLAMIAN and C. P. PICARD
  [1] *Equations d'évolution avec conditions unilatérales*, C.R.A.S., **279**, (1974), 607–609.

J. P. AUBIN
  [1] *Approximation of Elliptic Boundary-Value Problems*, Wiley & Sons, 1972.

A. AVANTAGGIATI
  [1] *Spazi di Sobolev con peso ed alcune applicationi*, Boll. UMI 13–A, (1976), 1–52.

C. AVRAMESCU
  [1] *Sur l'existence des solutions des équations intégrales dans certains espaces fonctionnels*, Ann. Univ. Sci. Budapest, Eötvös, **13**, (1970), 19–34.

M. BACKWINKEL-SCHILLINGS
  [1] *Existence theorems for generalized Hammerstein equations*, J. Funct. Anal., **23**, (1976), 177–194.

References

J. B. Baillon and G. Haddad
    [1] *Quelques propriétés des opérateurs angle-bornés et n-cycliquement monotones*, Israel J. Math., **26**, (1977), 137−150.

C. Baiocchi and A. Capelo
    [1] *Disequazioni Variazionali e Quasivariazionali. Applicazioni a Problemi di Frontiera Libera*, Vol. 2, *Problemi Quasivariazionali*, Pitagora, Bologna, 1978.

V. Barbu
    [1] *Continuous perturbations of m-accretive operators in Banach spaces*, Bull. U.M.I., **6**, (1972), 270−278.
    [2] *Nonlinear Semigroups and Differential Equations in Banach spaces*, Ed. Academiei − Noordhoff Intern. Publ., 1976.
    [3] *On a nonlinear Volterra integral equation on of Hilbert space*, SIAM J. Math. Anal., **8**, (1977), 345−355.

V. Barbu and T. Precupanu
    [1] *Convexity and Optimisation in Banach Spaces*, Ed. Academiei − Sijthoff & Noordhoff Intern. Publ., 1978.

H. Beirao-Da-Veiga and J. P. Dias
    [1] *Sur la surjectivité de certains opérateurs non linéaires liés aux inéquations variationnelles*, Boll. U.M.I., **10**, (1974), 52−59.

I. Beju
    [1] *Theorems on existence, uniqueness and stability of the solution of the place boundary-value problem, in statics, for hyperelastic materials*, Arch. Rational Mech. Annal., **42**, (1971), 1−23.

Ph. Bénilan, H. Brézis and M. G. Crandall
    [1] *A semilinear equation in $L^1(R^N)$*, Ann. Scuola Norm. Sup. Pisa, **2**, (1975), 523−555.

A. Bensoussan and J. L. Lions
    [1] *Contrôle impulsionnel et inéquations quasi-variationnelles d'évolution*, C.R.A.S., **276**, (1973), 1333−1338.

M. Berger and M. Berger
    [1] *Perspectives in Nonlinearity. An Introduction to Nonlinear Analysis*, W. A. Benjamin, New York, 1968.

M. S. Berger and M. Schecter
    [1] *Embedding theorems and quasilinear elliptic boundary-value problems for unbounded domains*, Trans. A.M.S., **172**, (1973), 261−278.

M. Biroli
    [1] *Gli operatori monotoni: teoria ed applicazioni*, Rend. Sem. Mat. Fis. Milano, **17**, (1972), 143−228.

D. K. Bose and H. Brill
    [1] *Nonlinear elliptic variational inequalities*, Math. Nachr. (to appear).

N. Bourbaki
    [1] *Espace vectoriels topologiques*, Chap. 3−5, Herman, 1964.

H. Brézis
    [1] *Équations et inéquations non linéaires dans les espaces vectoriels en dualité*, Ann. Inst. Fourier, **18**, *1*, (1968), 115−175.
    [2] *Inéquations variationnelles associées à des opérateurs d'évolution*, "Theory and Applications of Monotone Operators", 249−258, Ed. Oderisi, Gubbio, 1969.
    [3] *Monotonicity methods in Hilbert spaces and some applications to nonlinear differential equations*, "Contributions to Nonlinear Functional Analysis", 101−156, Academic Press, 1971.
    [4] *Problèmes unilatéraux*, J. Math. Pures Appl., **51**, (1972), 1−164.
    [5] *Opérateurs Maximaux Monotones et Semi-groupes de Contractions dans les Espaces de Hilbert*, Math. Studies, **5**, North-Holland, 1973.

[6] *Nonlinear perturbations of monotone operators*, Techn. Report, **25**, (1972), Univ. of Kansas.

[7] *Quelques propriétés des opérateurs monotones et des semi-groupes non linéaires*, Lect. Notes Math., **543**, 56−82, Springer-Verlag, 1976.

H. BRÉZIS and F. E. BROWDER

[1] *Some new results about Hammerstein equations*, Bull. A.M.S., **80**, (1974), 568−572.

[2] *Existence theorems for nonlinear integral equations of Hammerstein type*, Bull. A.M.S., **81**, (1975), 73−78.

[3] *Nonlinear integral equations and systems of Hammerstein type*, Advances in Math., **18**, (1975), 115−144.

[4] *Singular Hammerstein equations and maximal monotone operators*, Bull. A.M.S., **82**, (1976), 623−625.

[5] *Linear maximal monotone operators and singular nonlinear integral equations of Hammerstein type*, 31−42, E. Rothe Festschrift, Academic Press, 1978.

H. BRÉZIS, M. CRANDALL and A. PAZY

[1] *Perturbations of nonlinear maximal monotone sets in Banach space*, Comm. Pure Appl. Math., **23**, (1970), 123−144.

H. BRÉZIS and A. HARAUX

[1] *Image d'une somme d'opérateurs monotones et applications*, Israel J. Math., **23**, (1976), 165−186.

H. BRÉZIS and L. NIRENBERG

[1] *Image d'une somme d'opérateurs non linéaires et applications*, C.R.A.S., **284**, (1977), 1365−1368.

[2] *Characterizations of the ranges of some non-linear operators and applications to boundary value problems*, Ann. Scuola Norm. Sup. Pisa, **5** (1978), 225−326.

H. BRÉZIS, L. NIRENBERG and G. STAMPACCHIA

[1] *A remark on Ky Fan's minimax principle*, Boll, U.M.I., **6**, (1972), 293−300.

L. E. J. BROUWER

[1] *Über Abbildungen von Mannigfaltigkeiten*, Math. Ann., **71**, (112), 97−115.

F. E. BROWDER

[1] *Nonlinear elliptic boundary-value problems*, Bull. A.M.S., **69**, (1963), 862−874.

[2] *Continuity properties of monotone nonlinear operators in Banach spaces*, Bull. A.M.S., **70**, (1964), 551−553.

[3] *Multi-valued monotone nonlinear mappings*, Trans. A.M.S., **118**, (1965), 338−551.

[4] *Existence and uniqueness theorems for solutions of nonlinear boundary-value problems*, Proc. Sympos. Appl. Math., **17**, 24−49, A.M.S. Providence, 1965.

[5] *Problèmes Non-linéaires*, Presse de l'Univ. de Montréal, **15**, 1966.

[6] *Nonlinear maximal monotone mappings in Banach spaces*, Math. Ann., **175**, (1968), 81−113.

[7] *The fixed point theory of multi-valued mappings in topological vector spaces*, Math. Ann., **177**, (1968), 283−301.

[8] *Nonlinear variational inequalities and maximal monotone mappings in Banach spaces*, Math. Ann., **183**, (1969), 213−231.

[9] *Existence theorems for nonlinear partial differential equations*. Proc. Sympos. Pure Math., **16**, 1−62, A.M.S. Providence, 1970.

[10] *Nonlinear Operators and Nonlinear Equations in Banach spaces*, Proc. Sympos. Pure Math., **18**, part 2, (1968), A.M.S. Providence, 1966.

[11] *Pseudo-monotone operators and the direct method of the calculus of variations*, Arch. Rational Mech. Anal., **38**, (1970), 268−277.

[12] *Nonlinear functional analysis and nonlinear integral equations of Hammerstein and Urysohn type*, "Contributions to Nonlinear Functional Analysis", 425−500, Academic Press, 1971.

[13] *Existence theory for boundary-value problems for quasilinear elliptic systems with strongly nonlinear lower order terms*, Proc. Sympos. Pure Math., **23**, (1971), 269−286, A.M.S. Providence, 1973.

References

[14] *Singular nonlinear integral equations of Hammerstein type*, Lect. Notes Math., **446**, 75—95, Springer-Verlag, 1975.
[15] *Continuation of solutions of equations under homotopies of single-valued and multivalued mappings*, "*Fixed Point Theory and its Applications*", 13—22, Academic Press, 1976.
[16] *On the constructive solution of nonlinear functional equations*, J. Funct. Anal., **25**, (1977), 345—355.

E. E. BROWDER and D. G. DE FIGUEIREDO
   [1] *J-monotone nonlinear operators in Banach spaces*, Indag. Math., **28**, (1966), 412—420.

F. E. BROWDER, D. G. DE FIGUEIREDO and C. P. GUPTA
   [1] *Maximal monotone operators and nonlinear integral equations of Hammerstein type*, Bull. A.M.S., **76**, (1970), 700—705.

F. E. BROWDER and C. P. GUPTA
   [1] *Monotone operators and nonlinear integral equations of Hammerstein type*, Bull. A.M.S., **75**, (1969), 1347—1353.

E. E. BROWDER and P. HESS
   [1] *Nonlinear mappings of monotone type in Banach spaces*, J. Funct. Anal., **11**, (1972), 251—294.

F. E. BROWDER and B. A. TON
   [1] *Nonlinear functional equations in Banach spaces and elliptic super-regularization*, Math. Z., **105**, (1968), 1—16.

N. P. CAC
   [1] *On strongly nonlinear variational inequality*, J. Math. Pures Appl., **54**, (1975), 1—10.

B. CALVERT
   [1] *Maximal accretive is not m-accretive*, Boll. U.M.I., **6**, (1970), 1042—1044.
   [2] *Perturbation of Nemytskii operators of m-T-accretive operators in L^q*, Rev. Roum. Math., **22**, (1977), 883—906.

B. CALVERT and C. P. GUPTA
   [1[*Nonlinear elliptic boundary-value problems in L^p-spaces and sums of ranges of accretive operators*, J. Nonlinear Anal., **2**, (1978), 1—26.

B. CALVERT and C. PICARD
   [1] *Opérateurs accrétifs et Φ-accrétifs dans un espace de Banach*, Hiroshima Math. J., **8**, (1978), 11—30.

B. CALVERT and J. R. L WEBB
   [1] *An existence theorem for quasimonotone operators*, Atti Accad. Naz. Lincei, **50**, (1972), 362—368.

C. CASTAING and M. VALADIER
   [1] *Convex Analysis and Measurable Multifunctions*, Lect. Notes Math., **580**, Springer-Verlag, 1977.

L. CEASARI
   [1] *Functional analysis, nonlinear differential equations and the alternative method*, "*Nonlinear Functional Analysis and Differential Equations*", Lect. Notes in Pure and Appl. Math., **19**, 1—197, M. Dekker Inc., 1976.

D. CIORĂNESCU
   [1] *Sur quelques équations aux dérivées partielles posées par la mécanique des milieux continus. Thèse*, Univ. de Paris, VI, 1977.

I. CIORĂNESCU
   [1] *Aplicaţia de dualitate în analiza neliniară*, Editura Academiei, Bucureşti, 1974.

A. CORDUNEANU
   [1] *Some remarks on the sum of two m-dissipative mappings*, Rev. Roum. Math., **20**, (1975), 411—415.

326

C. CORDUNEANU
   [1] *Integral equations and stability of feedback systems*, Academic Press, 1973.

M. G. CRANDALL, S. O. LONDEN and J. A. NOHEL
   [1] *An abstract nonlinear Volterra integrodifferential equation*, M.R.C. Techn. Summary Report, Wisconsin, 1976.

R. CRISTESCU
   [1] *Topological Vector Spaces*, Ed. Academiei – Noordhoff Intern. Publ., 1977.

J. DANEŞ and J. KOLOMÝ
   [1] *Fixed point, surjectivity and invariance of domain theorems for weakly continuous mappings*, Boll. U.M.I., **13** – B, (1976), 369–394.

G. DA PRATO
   [1] *Weak solution for linear abstract differential equations*, Advances in Math., **5**, (1970).
   [2] *Somme d'applications non linéaires*. Symposia Math., **7**, 233–268, Academic Press, 1971.
   [3] *Applications Croissantes et Équations d'Évolutions dans les espaces de Banach*, Inst. Math., **2**, Academic Press, 1976.

K. DEIMLING
   [1] *Nichtlineare Gleichungen und Abbildungsgrade*, Hochschultext, Springer-Verlag, 1974.
   [2] *Ordinary Differential Equations in Banach Spaces*, Lect. Notes Math., **596**, Springer-Verlag, 1977.

J. DEUEL and P. HESS
   [1] *A criterion for the existence of solutions of nonlinear elliptic boundary-value problems*, Proc. Royal Soc. Edinburgh (A), **74**, *3*, (1974–75), 49–54.

J. P. DIAS
   [1] *Un théorème de Strum-Liouville pour une classe d'opérateurs non linéaires maximaux monotones*, J. Math. Anal. Appl., **47**, (1974), 400–405.
   [2] *Variational inequalities and eigenvalue problems for nonlinear maximal monotone operators in a Hilbert space*, Amer. J. Math., **97**, (1975), 905–914.

J. DIESTEL
   [1] *Geometry of Banach Spaces-Selected Topics*, Lect. Notes Math., **485**, Springer-Verlag, 1975.

G. DINCĂ
   [1] *Operatori monotoni în teoria plasticității*, Editura Academiei, București, 1972.

N. DINCULEANU
   [1] *Vector Measure*, VEB Deutscher-Verlag, 1966.

C. L. DOLPH and G. J. MINTY
   [1] *On nonlinear integral equations of the Hammerstein type*, "Nonlinear Integral Equations", 99–154, The Univ. Wisconsin Press, Madison, 1964.

YU. A. DUBINSKII
   [1] *Quasilinear elliptic and parabolic equations of arbitrary order*, Russian Math. Surveys, **23**, *1*, (1968), 45–91.

N. DUNFORD and J. T. SCHWARTZ
   [1] *Linear Operators*, Part I, Interscience Publ., New York, 1958.

G. DUVAUT and J. L. LIONS
   [1] *Les Inéquations en Mécanique et en Physique*, Dunod, Paris, 1972.

D. E. EDMUNDS and W. D. EVANS
   [1] *Elliptic and degenerate-elliptic operators in unbounded domains*, Ann. Scuola Norm. Sup. Pisa, **27**, (1973), 591–640.

References

D. E. EDMUNDS and V. B. MOSCATELLI
[1] *Semi-coercive nonlinear problems*, Boll. U.M.I., **11**, (1975), 144−153.

D. E. EDMUNDS, V. B. MOSCATELLI and J. R. L. WEBB
[1] *Strongly nonlinear elliptic operators in unbounded domains*, Publ. Math. Bordeaux, **4**, (1974), 6−32.

D. E. EDMUNDS and J. R. L. WEBB
[1] *Quasilinear elliptic problems in unbounded domains*, Proc. Royal Soc. London, (A), **334**, (1973), 397−410.

I. EKELAND and R. TEMAM
[1] *Analyse Convexe et Problèmes Variationnels*, Dunod, Paris, 1974.

G. FICHERA
[1] *Linear elliptic differential systems and eigenvalue problems*, Lect. Notes Math., **5**, Springer-Verlag, 1965.

D. G. DE FIGUEIREDO
[1] *Topics in Nonlinear Analysis*, Lect. Notes, **48**, Univ. Maryland, 1967.
[2] *An existence theorem for pseudo-monotone operator equations*, J. Math. Anal. Appl., **34**, (1971), 151−156.

D. G. DE FIGUEIREDO and C. P. GUPTA
[1] *Nonlinear integral equations of Hammerstein type involving unbounded monotone linear mappings*, J. Math. Anal. Appl., **39**, (1972), 37−48.
[2] *Nonlinear integral equations of Hammerstein type with indefinite linear kernel in a Hilbert space*, Indag. Math., **34**, (1972), 335−358.
[3] *On the variational method for the existence of solutions of nonlinear equations of Hammerstein type*, Proc. Amer. Math. Soc., **40**, (1973), 470−476.

P. M. FITZPATRICK
[1] *Surjectivity results for nonlinear mappings from a Banach space to its dual*, Math. Ann., **204**, (1973), 177−188.
[2] *Continuity of nonlinear monotone operators*, Proc. A. M. S., **62**, (1977), 111−116.

C. FOIAŞ
[1] *Statistical study of Navier-Stokes equations*, Rend. Sem. Mat. Univ. Padova, **48**, (1973), 219−348; **49**, (1973), 9−123.

A. FRIEDMAN
[1] *Partial Differential Equations*, Holt, Rinehart and Winston, 1969.

K. O. FRIEDRICHS
[1] *On the boundary-value problems of the theory of elasticity and Korn's inequality*, Ann. Math., **48**, (1948), 184−202.

S. FUČIK
[1] *Remarks on monotone operators*, Comm. Math. Univ. Carolinae, **11**, (1970), 435−448.
[2] *Fredholm alternative for nonlinear operators in Banach spaces and its applications to differential and integral equations*, Casopis Pešt. Mat., **96**, (1971), 371−390.
[3] *Ranges of Nonlinear Operators*, I−V, Charles Univ. Prague, 1977.

S. FUČIK and J. NEČAS
[1] *Spectral theory of nonlinear operators*, Proc. Equadiff, **3**, 163−174, Brno, 1972.

S. FUČIK, J. NEČAS, J. SOUČEK and S. SOUČEK
[1] *Spectral Analysis of Nonlinear Operators*, Lect. Notes Math., **346**, Springer-Verlag, 1973.

R. E. GAINES and J. MAWHIN
[1] *Coincidence Degree and Nonlinear Differential Equations*, Lect. Notes Math., **568**, Springer-Verlag, 1977.

H. GAJEWSKI, K. GRÖGER and K. ZACHARIAS
[1] *Nichtlineare Operatorgleichungen und Operatordifferentialgleichungen*, Akademie-Verlag, Berlin, 1974.

D. GILBARG and N. S. TRUDINGER
[1] *Elliptic Partial Differential Equations of Second Order*, Springer-Verlag, 1977.

R. GLOWINSKI, J. L. LIONS and R. TRÉMOLIÈRES
[1] *Analyse Numérique des Inéquations Variationnelles*, vols. 1−2, Dunod-Bordas, 1976.

M. GOLOMB
[1] *Zur Theorie der nichtlinearen Integragleichungen, Integralgleichungssysteme und Funktionalgleichungen*, Math. Z., **39**, (1935), 45−75.

J. P. GOSSEZ
[1] *Opérateurs monotones non linéaires dans les espaces de Banach non réflexifs*, J. Math. Anal. Appl., **34**, (1971), 371−395.
[2] *Nonlinear elliptic boundary-value problems for equations with rapidly or slowly increasing coefficients*, Trans. A.M.S., **190**, (1974), 163−205.

K. GRÖGER
[1] *Evolution equations in the theory of plasticity*, Proc. Summer School "Theory of Nonlinear Operators", Berlin, 1977.

C. P. GUPTA
[1] *On nonlinear integral equations of Hammerstein type with unbounded linear mappings*. Lect. Notes Math., **384**, 184−238, Springer-Verlag, 1974.
[2] *On compact perturbation of certain nonlinear equations in Banach spaces*, J. Math. Appl., **45**, (1974), 497−505.
[3] *A new existence theorem for nonlinear integral equations of Hammerstein type involving unbounded linear mappings*, "*Analyse Fonctionnelle et Applications*", 133−144, Hermann, Paris, 1975.
[4] *On an operator equation involving mappings of monotone type*, Proc. A.M.S., **53**, (1975), 143−148.
[5] *Nonlinear equations of Urysohn's type in a Banach space*, Comm. Math. Univ. Carolinae, **16**, (1975), 377−386.
[6] *On a class of nonlinear integral equations of Uryshon's type*, J. Math. Anal. Appl., **58**, (1977), 344−360.
[7] *Sum of ranges of operators and applications*, "*Nonlinear Systems and Applications*", 547−559, Academic Press, 1977.

C. P. GUPTA and P. HESS
[1] *Existence theorems for nonlinear noncoercive operator equations and nonlinear elliptic boundary-value problems*, J. Diff. Equat., **22**, (1976), 305−313.

A. HAMMERSTEIN
[1] *Nichtlineare Integralgleichungen nebst Anwendungen*, Acta Math., **54**, (1930), 117−176.

P. HARTMAN and G. STAMPACCHIA
[1] *On some nonlinear elliptic differential functional equations*, Acta Math., **115**, (1966), 271−310.

E. HEINZ
[1] *An elementary analytic theory of the degree of mapping in n-dimensional space*, J. Math. Mech., **8**, (1959), 231−247.

P. HESS
[1] *A variational approach to a class of nonlinear eigenvalue problems*, Proc. A.M.S., **29**, (1971), 272−276.
[2] *On a method of singular perturbations type for proving the solvability of nonlinear functonal equations in Banach spaces*, Math. Z., **122**, (1971), 355−362.
[3] *On nonlinear equations of Hammerstein type in Banch spaces*, Proc. A.M.S., **30**, (1971), 308 − 312.

329

References

[4] *On the Fredholm alternative for nonlinear functional equations in Banach spaces*, Proc. A.M.S., **23**, (1972), 55−61.

[5] *On nonlinear mappings of monotone type homotopic to odd operators*, J. Funct. Anal., **11**, (1972), 138−167.

[6] *A remark on a class of linear monotone operators*, Math. Z., **125**, (1972), 104−106.

[7] *Théorème d'existence pour des perturbations d'opérateurs maximaux monotones*, C.R.A.S., **275**, (1972), 1171−1173.

[8] *On nonlinear mappingg of monotone type with respect to two Banach spaces*. J. Math. Pures Appl., **52**, (1973), 13−26.

[9] *Variational inequality for strongly nonlinear elliptic operators*, J. Math. Pures Appl., **53**, (1973), 285−298.

[10] *A strongly nonlinear elliptic boundary-value problem*, J. Math. Anal. Appl., **43**, (1973), 241−249.

[11] *On a unilateral problem associated with elliptic operators*, Proc. A.M.S., **39**, (1973), 94−100.

[12] *On some nonlinear elliptic boundary-value problems*. Lect. Notes Math., **399**, 235−247, Springer-Verlag, 1974.

[13] *A homotopy argument for mappings of monotone type in Banach spaces*, Math. Ann., **207**, (1974), 63−65.

[14] *On semi-coercive nonlinear problems*, Indiana Univ. Math. J., **23**, (1974), 645 − 654.

[15] *On a theorem by Landesman and Lazer*, Indiana Univ. Math. J., **23**, (1974), 827−829.

[16] *On a class of strongly nonlinear elliptic variational inequalities*, Math. Ann., **211**, (1974), 289−297.

[17] *Problèmes aux limites non linéaires dans des domaines non bornés*, C.R.A.S., **281**, (1975), 555−557.

[18] *On strongly nonlinear elliptic problems*, "*Functional Analysis*" Lect. Notes in Pure and Appl. Math., **18**, 91−109, M. Dekker Inc., 1977.

[19] *Nonlinear elliptic problems in unbounded domains*, "*Theory of Nonlinear Operators, Constructive Aspects*", 105−110, Akademie Verlag, Berlin, 1978.

[20] *On the solvability of nonlinear elliptic boundary-value problems*, Indiana Univ. Math. J., **25**, (1976), 461−466.

[21] *On a second order nonlinear elliptic boundary-value problem*, (in print).

E. HEWITT and K. STROMBERG
[1] *Real and Abstract Analysis*, Springer-Verlag, 1965.

I. HLAVČEK and J. NEČAS
[1] *Introduction to the mathematical theory of elastic-inelastic materials*, SNTL, Prag, 1978.

R. B. HOLMES
[1] *Geometric Functional Analysis and Its Applications*, GTM 24, Springer-Verlag, 1975.

L. HÖRMANDER
[1] *Linear Paralel Differential Operators*, Springer-Verlag, 1964.

M. IANNELLI
[1] *Opérateurs dérivables et semi-groupes non-linéaires non-contractifs*, J. Math. Anal. Appl., **46**, (1974), 700−724.

A. D. IOFFE and V. M. TIHOMIROV
[1] *Theory of Extremal Problems*, North-Holland, 1978.

V. ISTRĂȚESCU
[1] *Introducere în teoria punctelor fixe*, Editura Academiei, București, 1973.

M. JOHSI
[1] *Existence theorem for a genealized Hammerstein type equation*, Comm. Math. Univ. Carolinae, **15**, (1974), 283−291.

R. I. KACHUROVSKII
[1] *On monotone operators and convex functionals*, Uspehi Mat. Nauk, **15**, *4*, (1960), 213−215 (in Russian).

[2] *Nonlinear monotone operators in Banach spaces*, Russian Math. Surveys, **23**, 2, (1968), 117−165.

L.V. KANTOROVICH and G. P. AKILOV
[1] *Functional Analysis in Normed Spaces*, Pergamon Press, 1964.

T. KATO
[1] *Demicontinuity, hemicontinuity and monotonicity*, Bull. A.M.S., **70**, (1964), 548−550; **73**, (1967), 886−889.
[2] *Perturbations Theory for Linear Operators*, Springer-Verlag, 1975.
[3] *Quasilinear equations of evolution, with applications to partial differential equations*, Lect. Notes. Math., **448**, 25−70, Springer-Verlag, 1975.

P. S. KENDEROV
[1] *The set-valued monotone mappings are almost everywhere single-valued*, Studia Math., **56**, (1976), 199−203.

N. KEMNOCHI
[1] *Nonlinear operators of monotone type in reflexive Banach spaces and nonlinear perturbations*, Hiroshima Math. J., **4**, (1974), 229−263.
[2] *Pseudo-monotone operators and nonlinear elliptic boundary-value problems*, J. Math. Soc. Japan, **27**, (1975), 121−149.

R. KLUGE
[1] *Approximation methods for nonlinear problems with constraints in form of variational inequalities*, Banach Center Publ., **1**, 131−138, Warszawa, 1976.

I. I. KOLODNER
[1] *Equations of Hammerstein type in Hilbert spaces*, J. Math. Mech., **13**, (1964), 701−750.

J. KOLOMY
[1] *Application of some existence theorems for the solutions of Hammerstein integral equations*, Com. Math. Univ. Carolinae, **7**, (1966), 461−478.
[2] *The solvability of nonlinear integral equations*, Comm. Math. Univ. Carolinae, **8**, (1976), 273−279.

Y. KONISHI
[1] *Some examples of nonlinear semi-groups in Banach lattices*, J. Fac. Sci. Univ. Tokyo, **18**, (1972), 537−543.

M. E. KOSITSKII
[1] *Nonlinear equations of Hammerstein type with a monotone operator*, Soviet Math. Dokl., **11**, (1970), 25−28.

M. A. KRASNOSELSKII
[1] *Topological Methods in the Theory of Nonlinear Integral Equations*, Macmillan, New York, 1964.

M. A. KRASNOSELSKII, P. P. ZABREYKO, E. I. PUSTYLNIK and P. E. SOBOLEVSKI
[1] *Integral Operators in Spaces of Summable Functions*, Noordhoff Intern. Publ., Leyden, 1976.

D. A. KUFNER, O. JOHN and S. FUČIK
[1] *Functional Spaces*, Academia, Prague 1977.

G. E. LADAS and V. LAKSHMIKANTHAM
[1] *Differential Equations in Abstract Spaces*, Academic Press, 1972.

R. LANDES
[1] *Quasilineare elliptische Differentialoperatoren mit starken Wachtum in den Termen höchster Ordnung*, Math. Z., 157, (1977), 23−36.

E. M. LANDESMAN and A. C. LAZER
[1] *Nonlinear perturbations of linear elliptic boundary-value problems at resonance*, J. Math. Mech., **19**, (1969/70), 609−623.

References

S. Lang
  [1] *Introduction to Differentiable Manifolds*, Wiley & Sons, 1962.

O. A. Ladyzhenskaya and N. N. Uraltzeva
  [1] *Linear and quasilinear elliptic equations*, Academic Press, 1968.

A. Langenbach
  [1] *Monotone Potentialoperatoren in Theorie und Anwendung*, Akademie-Verlag, Berlin, 1977.

A. Lehtonen
  [1] *On boundary value problems for quasilinear elliptic systems*, Report, Univ. of Jyväskyla, 1977.

J. Leray and J. L. Lions
  [1] *Quelques résultats de Višik sur les problèmes elliptiques non linéaires par les méthodes de Minty-Browder*, Bull. Soc. Math. France, **93**, (1965), 97−107.

J. Leray and J. Schauder
  [1] *Topologie et équations fonctionnelles*, Ann. Sci. Ecole Norm. Sup., **51**, (1934), 45−78.

J. L. Lions
  [1] *Problèmes aux Limites dans les Équations aux Dérivées Partielles*, Presse de l'Univ. Montréal, *1*, 1967.
  [2] *Quelques Méthodes de Résolution des Problèmes aux Limites Non Linéaires*, Dunod, Paris, 1969.
  [3] *Partial differential inequalities*, Russian Math. Survey, **27**, 2, (1972), 91−160.
  [4] *Sur Quelques Questions d'Analyse, de Mécanique et de Contrôle Optimal*, Presse de l'Univ. Montréal, 1976.

J. L. Lions and E. Magenes
  [1] *Problèmes aux Limites Non Homogènes et Applications*, Vol. 1, Dunod, Paris, 1968.

N. Luca
  [1] *The behaviour of solutions of a class of nonlinear integral equations of the Volterra type*, Atti Accad. Naz. Lincei, **62**, (1977), 9−16.

M. Marcus and V. J. Mizel
  [1] *Nemitsky operators on Sobolev spaces*, Arch. Rational Mech. Anal., **51**, (1973), 347−370.
  [2] *Continuity of certain Nemitsky operators on Sobolev spaces and the chain rule*, J. d'Anal. Math., **28**, (1975), 303−334.

G. Marinescu
  [1] *Tratat de analiză funcţională*, vol. 2, Editura Academiei, Bucureşti, 1972.

R. H. Martin Jr.
  [1] *Nonlinear Operators and Differential Equations in Banach Spaces*, Wiley & Sons, 1976.

P. Mazilu and S. Sburlan
  [1] *Metode funcţionale in rezolvarea ecuaţiilor teoriei elasticităţii*, Editura Academiei, Bucureşti, 1973.

N. Meyers and J. Serrin
  [1] *H = W*, Proc. Nat. Acad. Sci. USA, **51**, (1964), 1055−1056.

G. J. Minty
  [1] *Monotone (nonlinear) operators in Hilbert spaces*, Duke Math. J., **29**, (1962), 341−346.
  [2] *On some aspects of theory of monotone operators*, "Theory and Applications of Monotone Operators", 67−82, Oderesi, Gubbio, 1969.

J. J. Moreau
  [1] *Proximité et dualité dans un espace hilbertien*, Bull. Soc. Math. France, **93**, (1965), 273−299.
  [2] *Fonctionnelles Convexes*, Collège de France, 1966−1967.

332

G. MOROŞANU
Asymptotic behaviour of resolvent for a monotone mapping in a Hilbert space, Atti Accad. Naz. Lincei, **61**, (1976), 565−570.

C. B. MORREY Jr
[1] *Integrals in the Calculus of Variations*, Springer-Verlag, 1966.

U. MOSCO
[1] *An introduction to the approximate solution of variational inequalities*, "Constructive Aspects of Functional Analysis", 497−684, Cremonese, Roma, 1973.
[2] *Implicit variational problems and quasi-variational inequalities*, Lect. Notes Math., **543**, 82−156, Springer-Verlag, 1976.

J. MOSSIMO
[1] *Étude de quelques problèmes non linéaires d'un type nouveau apparaissant en physique des plasmas*, Thèse, Paris, XI, 1977.

V. MUSTONEN
[1] *An elliptic boundary value problem for strongly non-linear equations in unbounded domains*, Proc. Roy. Soc Edinburgh, **77** A (1977), 217−230.

J. NAUMANN
[1] *On the existence of solutions to hyperbolic variational inequalities*, Boll. U.M.I., **13**−A, (1976, 312−321.

J. NEČAS
[1] *Les Méthodes Directes en Théorie des Équations Elliptiques*, Prague, 1967.
[2] *Sur l'alternative de Fredholm pour les opérateurs non linéaires avec applications aux problèmes aux limites*, Ann. Scuola Norm. Sup. Pisa, **23**, (1969), 331−345.
[3] *Les équations elliptiques linéaires*, Czech. Math. J., **19**, (1969), 252−274.
[4] *Remark on the Fredholm alternative for nonlinear operators with applications to nonlinear equations of generalized Hammertein type*, Comm. Math. Univ. Carolinae, **13**, (1972), 109−120.
[5] *Fredholm alternative for nonlinear operators and applications to partial differential equations and integral equations*, Čas. Pešt. Math., **97**, (1972), 65−71.
[6] *Fredholm theory of boundary-value problems for nonlinear ordinary differential operators*, "Theory of Nonlinear Operators", 85−120, Prague, 1973.

M. NICOLESCU
[1] *Analiză Matematică*, Editura tehnică, Bucureşti, vol. 2, 1958; vol. 3, 1960.

L. NIRENBERG
[1] *Topics in Nonlinear Functional Analysis*, Courant Inst., 1974.

J. T. ODEN and J. N. REDDY
[1] *Variational Methods in Theoretical Mechanics*, Springer-Verlag, 1976.
[2] *An Introduction to the Mathematical Theory of Finite Elements*, Wiley & Sons, 1976.

D. PASCALI
[1] *Operatori Neliniari*, Ed. Academiei Bucureşti, 1974.
[2] *Hammerstein equations in general Banach spaces*, Seminari di Analisi, 1974/75, Istituto Matematico, Roma.
[3] *On a Fredholm alternative for nonlinear operators of type (S)*, "Proc. Conf. Diff. Equat. and their Appl.", 25−30, Editura Academiei, Bucureşti, 1977.
[4] *Metode de monotonie pe spaţii Hilbert*, "Analiză neliniară şi aplicaţii în mecanică", 41−114, Editura Academiei, Bucureşti, 1977.
[5] *On variational methods for Hammerstein equations*, Lect. Notes Math., Springer-Verlag, 1978 (to appear).
[6] *On nonlinear divergence equations*, Proc. Summer School "Theory of Nonlinear Operators", 205−213, Akademie-Verlag, Berlin, 1978.

References

D. PASCALI and S. SBURLAN
[1] *Strongly non-linear perturbation of divergence equations on unbounded domains,* Rev. Roum. Math. Pures et Appl., **21**, (1976), 726–732.
[2] *Operatori neliniari impari,* St. cerc. mat. **30**, (1978), 413–424.

N. PAVEL
[1] *Ecuaţii diferenţiale asociate unor operatori neliniari pe spaţii Banach.* Editura Academiei, 1977.

A. PAZY
[1] *Semi-groups of nonlinear contractions in Hilbert spaces,* "Problems in Nonlinear Analysis", 343–430, Cremonese, Roma, 1971.

D. PETROVANU
[1] *Equations Hammerstein intégrales et discrètes,* Ann. Mat. Pura Appl., **74**, (1966), 227–254.

W. PETRY
[1] *Generalized Hammerstein equations and quasi-linear differential equations with non-linear boundary conditions,* Math. Nachr., **57**, (1973), 141–161.
[2] *Generalized Hammerstein equations and integral equations of Hammerstein type,* Math. Nachr., **59**, (1974), 51–62.

W. V. PETRYSHYN
[1] *A characterization of strict convexity of Banach spaces and other uses of duality mappings,* J. Funct. Anal., **6**, (1970), 282–291.
[2] *On the approximation-solvability of equations involving A-proper and pseudo-A-proper,* Bull. A.M.S., **81**, (1975), 223–312.
[3] *On the relationship of A-properness to mappings of monotone type with applications to elliptic equations,* "Fixed Point Theory and its Applications", 149–174, Academic Press, 1976.

W. V. PETRYSHYN and P. M. FITZPATRICK
[1] *New existence theorems for nonlinear equations of Hammerstein type,* Trans. A.M.S., **160**, (1971), 39–63.

S. I. POHOŽAEV
[1] *Solvability of nonlinear equations with odd operators,* Anal. and Appl., **1**, (1967), 227–233.

G. PRODI and A. AMBROSETTI
[1] *Analisi Non Lineare,* I Quaderno, Scuola Norm. Sup. Pisa, 1973.

J. P. PUEL
[1] *Existence, comportement à l'infini et stabilité dans certains problèmes quasilinéaires elliptiques et paraboliques d'ordre 2,* Ann. Scuola Norm. Sup. Pisa, **3**, (1976), 89–119.

P. H. RABINOWITZ
[1] *Some aspects of nonlinear eigenvalue problems,* Rocky Mountain J. Math., **3**, (1973), 161–202.
[2] *Théorie du degré topologique et applications à des problèmes aux limites non linéaires,* Univ. Paris, VI, 1975.

L. B. RALL
[1] *Variational methods for nonlinear integral equations,* "Nonlinear Integral Equations", 155–190, Univ. Wiscosin Press, Madison, 1964.

K. REKTORYS
[1] *Variational Methods in Mathematics, Sciences and Engineering,* D. Reidel Publ. Company, 1977.

P. RICCIARDI and L. TUBARO
[1] *Local existence for differential Equations in Banach space,* Boll. U.M.I., **8**, (1973), 306–316.

References

R. T. ROCKAFELLAR
[1] *Characterization of the subdifferentials of convex functions*, Pacific J. Math., **17**, (1966), 497—510.

[2] *Local boundedness of nonlinear monotone operators*, Michigan Math. J., **16**, (1969), 397—407.

[3] *Convex Analysis*, Princeton Univ., 1970.

[4] *Maximality of sums of nonlinear monotone operators*, Trans. A.M.S., **149**, (1970), 75—88.

[5] *Integral functionals, normal integrands and measurable selections*, Lect. Notes Math., **543**, 157—207, Springer-Verlag, 1976.

I. A. RUS
[1] *Principii şi Aplicaţii ale Teoriei Punctului Fix*, Editura Dacia, Cluj-Napoca, 1978.

S. SBURLAN
[1] *On a particular class of optimal problems with application in the projection method*, Opns. Res. Vrfn., **19**, (1973), 102—108.

[2] *Some remarks on the existence theorems for functional equations with odd operators*, Rev. Roum. Math., **21**, (1976), 1107—1116.

[3] *Constraint strongly monotone operators*, Lect. Notes Math., Springer-Verlag, 1978 (to appear).

[4] *Duality principles in nonlinear programming and continuous mechanics*, Proc. 4-th Conf. Probl. Theory, 586—596, Ed. Academiei, Bucureşti, 1973.

H. H. SCHAEFER
[1] *Topological Vector Spaces*, GMT 3, Springer-Verlag, 1971.

M. SCHATZMAN
[1] *Problèmes aux limites non-linéaires non-coercifs*, Ann. Scuola Norm. Sup. Pisa, **27**, (1973), 641—686.

J. SCHAUDER
[1] *Die Fixpunktsatz in Funktionalräume*, Studia Math., **2**, (1930), 171—180.

M. SCHECHTER
[1] *Principles of Functional Analysis*, Academic Press, 1971.

[2] *Spectra of Partial Differential Operators*, North-Holland, 1971.

K. SCHMITT
[1] *Boundary value problems for quasilinear second order elliptic equations*, Nonlinear Anal., **2**, (1978), 263—309.

J. T. SCHWARTZ
[1] *Nonlinear Functional Analysis*, Gordon and Breach, 1969.

J. SERRIN
[1] *The problem of Dirichlet for quasi-linear elliptic differential equations with many independent variables*, Phylos. Trans. Royal Soc. London (A), **264**, (1969), 413—496.

M. SHINBROT
[1] *Lecture on Fluid Mechanics*, Gordon and Breach, 1973.

R. E. SHOWALTER
[1] *Hilbert Space Methods for Partial Differential Equations*, Pitman, 1977.

C. G. SIMADER
[1] *Another approach to the Dirichlet problem for very strongly nonlinear elliptic equations*, Lect. Notes Math., **564**, 425—437, Springer-Verlag, 1976.

[2] *Über Schwach Lösungen des Dirichletproblems für streng nichtlineare elliptische Differentialgleichungen*, Math. Z., **150**, (1976), 1—26.

References

E. SINESTRARI
[1] *Accretive differential operators*, Boll. U.M.I., **13**−B, (1976), 19−31.

I. V. SKRIPNIK
[1] *Nonlinear Elliptic Equations of Higher Order*, Kiev, 1973 *(Russian)*.

V. I. SIRNOV
[1] *Course of Higher Mathematics*, vol. 5, Pergamon Press, 1964.

S. L. SOBOLEV
[1] *Applications of Functional Analysis in Mathematical Physics*, Leningrad, 1950; A.M.S. Trans., 7, (1963).

M. SPIVAK
[1] *Calculus on Manifolds*, W.A. Benjamin, New York, 1965.

G. STAMPACCHIA
[1] *Formes bilinéaires coercitives sur les ensembles convexes*, C.R.A.S., **258**, (1964), 4413−4416.
[2] *Variational inequalities*, "Theory and Applications of Monotone Operators", 101− 192, Ed. Oderisi, Gubbio, 1969.

W. STRAUSS
[1] *The energy method for nonlinear partial differential equations*, Notas de Matem., **47**, (1969), Rio de Janeiro.
[2] *On weak solutions of semi-linear hyperbolic equations*, An. Acad. brasil. Ciênc., **42**, (1970), 645−651.

E. STROESCU
[1] *Perturbation of mon-linear partial differential variational inequalities (I)*, Proc. Royal Soc. Edinburg 76 A (1976), 1−12.

R. TEMAM
[1] *Navier-Stokes Equations (Theory and Numerical Analysis)*, Studies in Math. and its Appl. 2, Nörth-Holland, 1977.

D. TIBA
[1] *Nonlinear boundary-value problems for second order differential equations*, Funkcialaj Ekvacioj, (1977).

B. A. TON
[1] *Pseudo-monotone operators in Banach spaces and nonlinear elliptic equations*, Math. Z., **121**, (1971), 243−252.
[2] *On strongly nonlinear elliptic variational inequalities*, Pacific J. math., **48**, (1973), 279−291.

S. L. TROJANSKI
[1] *On locally uniformly convex and differentiable norms in certain non-separable Banach spaces*, Studia Math., 37, (1971), 173−180.

M. M. VAINBERG
[1] *Variational Methods for the Study of Nonlinear Operators*, Moscow, 1956; Holden-Day, San Francisco, 1964.
[2] *Variational Method and Method of Monotone Operators in the Theory of Nonlinear Equations*, Moscow, 1972; Wiley & Sons, 1974.

M. M. VAINBERG and I. M. LAVRENTIEV
[1] *Equations with monotonic and potential operators in Banach spaces*, Soviet Math. Dokl., **10**, (1969), 907−910.

M. I. VIŠIK
[1] *Quasilinear strongly elliptic systems of differential equations in divergence form*, Trans. Moscow Math. Soc., (1963), 140−208 (in Russian).

J. R. L. Webb

    [1] *On the Dirichlet problem for strongly non-linear elliptic operators in unbounded domains*, J. London Math. Soc., (2), **10**, (1975), 163−170.

S. Williams

    [1] *A sharp sufficient condition for solution of a nonlinear elliptic boundary-value problem*, J. Diff. Equat., **8**, (1970), 580−586.

K. Yosida

    [1] *Functional Analysis*, Springer-Verlag, 1966.

E. H. Zarantenello

    [1] *Projections on convex sets in Hilbert spaces and spectral theory*, "*Contributions to Nonlinear Analysis*", 237−424, Academic Press, 1971.

    [2] *Dense single-valuedness of monotone operators*, Israel J. Math., **15**, (1973),158−166.

E. Zeidler

    [1] *Vorlesungen über nichtlineare Funktionalanalysis*; I. *Fixpunktsätze* (1976); II. *Monotone Operatoren* (1977), Teubner-Texte zur Math. Leipzig.

E. Zini

    [1] *À propos de quelques opérateurs non linéaires*, Math. Z., **141**, (1975), 111−138.

    [2] *Une remarque à propos de certaines équations de Hammerstein.*, C.R.A.S., **282**, (1976), 1101−1103.

Subject index